Welding for Collision Repair

Larry Jeffus

Delmar Publishers

an International Thomson Publishing company I(T)P®

Albany • Bonn • Boston • Cincinnati • Detroit • London • Madrid
Melbourne • Mexico City • New York • Pacific Grove • Paris • San Francisco
Singapore • Tokyo • Toronto • Washington

NOTICE TO THE READER

Cover Design: Paul Roseneck

Delmar Staff
Publisher: Alar Elken
Acquisitions Editor: Vernon Anthony
Editorial Assistant: Betsy Hough
Fast Cycle Production Editor: Dianne Jensis
Marketing Manager: Mona Caron

COPYRIGHT © 1999
By Delmar Publishers

an International Thomson Publishing company I(T)P®

The ITP logo is a trademark under license.
Printed in the United States of America

Online Services

Delmar Online
To access a wide variety of Delmar products and services on the World Wide Web, point your browser to:
  http://www.delmar.com
  or email: info@delmar.com

A service of I(T)P®

For more information, contact:

Delmar Publishers
3 Columbia Circle, Box 15015
Albany, New York 12212-5015

International Thomson Publishing Europe
Berkshire House
168-173 High Holborn
London, WC1V7AA
United Kingdom

Nelson ITP, Australia
102 Dodds Street
South Melbourne,
Victoria, 3205 Australia

Nelson Canada
1120 Birchmont Road
Scarborough, Ontario
M1K 5G4, Canada

International Thomson Publishing France
Tour Maine-Montparnasse
33 Avenue du Maine
75755 Paris Cedex 15, France

International Thomson Editores
Seneca 53
Colonia Polanco
11560 Mexico D. F. Mexico

International Thomson Publishing GmbH
Königswinterer Strasse 418
53227 Bonn
Germany

International Thomson Publishing Asia
60 Albert Street
#15-01 Albert Complex
Singapore 189969

International Thomson Publishing Japan
Hirakawa-cho Kyowa Building, 3F
2-2-1 Hirakawa-cho, Chiyoda-ku,
Tokyo 102, Japan

ITE Spain/Paraninfo
Calle Magallanes, 25
28015-Madrid, Spain

    4 5 6 7 8 9 10 XXX 04

Library of Congress Cataloging-in-Publication Data

Jeffus, Larry F.
    Welding for collision repair / Larry Jeffus.
        p. cm.
    Includes index.
    ISBN 0–7668–0966–8
    1. Welding. 2. Motor vehicles—Welding. 3. Automobiles-
-Collision damage—Repairing. I. Title.
TS227.J4178   1998
629.26′7—dc21
                                                98–31153
                                              CIP

Contents

Chapter 1

Chapter 2

Chapter 3

Chapter 4

Chapter 5

Chapter 6

Chapter 7

Preface

Many things affect how collision repairs are made. For example insurance company policies can significantly affect auto body repairs. Federal and state rules, regulations and laws about the restoration of vehicles have led to improvements to ensure that all repairs made to motor vehicles are done in a way that does not adversely affect the future performance and safe operation of the repaired vehicle. Changes in vehicle design and construction, of course, have a major effect on the repair industry. The introduction of high-strength, low-alloy (HSLA) and high-strength steel (HSS) into vehicles has enabled manufacturers both to strengthen components and to make them thinner. The thinner the component, the lighter the vehicle. While these changes have been beneficial to the design and construction of vehicles, their universal application has resulted in more components that are impractical or impossible to repair and therefore must be replaced.

Modern material technology allows manufacturers to custom-design components so that they serve both aesthetic and structural purposes. The exterior surfaces, fenders, doorskins, and roof panels on most of today's cars function as part of the structural component. This is very similar to the function of the skin of an aircraft, which gives it both shape and strength.

All of these factors have combined to produce an attractive, functional vehicle that is more difficult to repair using outdated techniques. Cars manufactured before the introduction of these design concepts and materials could easily be repaired using low-tech welding processes such as gas welding, brazing, and shielded metal arc (stick) welding. Many of these vehicles' surface panels could be readily repaired and made as good as new with welding.

VEHICLE REPAIRS

Because of changes in vehicle construction, insurance regulations, and increased operating costs, collision repair facilities have changed. Some of those changes include such things as dealer-specific shops in which virtually all of the work performed is on a single manufacturer's vehicles. Some large shops specialize in new or nearly new vehicles. Such large shops can employ helpers to assist the auto body technician with such matters as vehicle teardown and component reassembly following repairs. Smaller shops and those located outside major metropolitan areas, on the other hand, repair any vehicle brought in. There are also shops that specialize in the restoration of classic or antique vehicles, and in these shops welding repairs are more widespread and diverse, since many replacement components are no longer available or have become prohibitively expensive.

This text will cover all aspects of welding, including those skills required in high-volume production repair shops and those needed by the classic car repair technician. When appropriate, reference will be made to a particular process as "commonly used on most vehicles" or "not commonly used today." In becoming a well-rounded and talented auto body repair technician, you should master all of the welding techniques.

JOINING PROCESSES

The term *welding* encompasses all metal-joining processes that use some source of heat such as a torch or arc. The term, however, specifically refers to a process in which the base metal itself is melted and allowed to flow together, forming a single piece as the molten weld pool solidifies. During welding, additional filler metal may or may not be added to assist in forming a weld bead. Welding is most commonly accomplished using one of the following processes:

- Oxyacetylene (gas) welding
- Shielded metal arc (stick) welding
- MIG (gas metal arc) welding
- Resistance spot welding
- Brazing
- Soldering

The term *welding* is also sometimes used to refer to cutting processes such as:

- Oxyacetylene cutting
- Plasma cutting

Acknowledgments

I would like to extend my sincere thanks to some very helpful people without whose assistance this book would not have been possible: Carol Jeffus for proofreading, Tina Ivey and Kathy Cott for their typing, and Larry Maupin for his photography skills. Also, I would like to thank The Body Shop of Garland for their allowing us to photograph many jobs in progress.

DEDICATION

This book is dedicated to three very special people, my wife Carol and my daughters Wendy and Amy.

Chapter

1

Welding Safety

Objectives

After reading this chapter, you should be able to:
- Describe the type of protection that should be worn for welding.
- Explain the various methods of ventilation.
- Describe how to maintain a safe work area.
- Explain how to use hand tools and electrical equipment safely.
- Describe the proper method of handling, storing, and setting up cylinders.
- Explain how to prevent and put out fires in the shop.
- Discuss the proper way to ventilate a welding area.
- Explain how to avoid electric shock.
- Describe how to avoid possible health hazards from welding.

All welding processes have potential safety problems. Welding motor vehicles can present additional hazards and concerns. Strict safety measures and precautions must be maintained in order to prevent injuries and property damage during welding.

There are federal, state, and local laws, codes, standards, and regulations related to the welding industry that must be followed. Such rules cover a variety of areas related to welding, such as health, safety, and environmental areas. Many health rules address the fumes created by welding. Safety issues address such areas as potential fire hazards and electrical hazards, as well as the safe operation of the repaired vehicle. Environmental issues include the impact of welding on air quality, water quality, and hazardous waste. It is the responsibility of the welding shop operator to make certain that all such concerns are addressed appropriately. In addition to the laws enacted by governments and the rules and regulations established by government agencies, there are many requirements established by insurance companies that cover the welding business. Many insurance companies will help the shop to

establish appropriate operating procedures that will ensure that all work performed meets current safety standards.

Many trade and professional associations offer recommendations for the various welding processes. The Inter-Industry Conference on Auto Collision Repair (I-CAR), Automotive Service Excellence (ASE), the American Welding Society (AWS), the American National Standards Institute (ANSI), the American Society for Testing and Materials (ASTM), National Fire Protection Association (NFPA), the Compressed Gas Association (CGA), and the National Safety Council (NSC) have established many voluntary safety standards for the welding trade. Although these organizations' standards are not laws, they are used by most government regulators to establish legal requirements.

Although there are serious threats to safety in the welding shop, welding can be a safe occupation if the proper precautionary measures are taken. Under the proper conditions, welding is no more hazardous or injurious to health than any other metalworking occupation.

1.1   BURNS

Burns are among the most common and painful injuries that occur in a collision repair facility. Burns can be caused by ultraviolet light rays as well as by contact with hot material. The chance of infection is high with burns because they kill tissue. It is important that all burns receive proper medical treatment to reduce the chance of infection. Burns are divided into three classifications, depending upon the degree of severity. These include first-degree, second-degree, and third-degree burns.

FIRST-DEGREE BURNS

First-degree burns have occurred when the surface of the skin is reddish in color, tender, and painful and no skin is broken. The first step in treating a first-degree burn is to immediately put the burned area under cold water (not iced) or apply cold water compresses (clean towel, washcloth, or handkerchief soaked in cold water) until the pain decreases. Then cover the area with sterile bandages or a clean cloth. Do not apply butter or grease. Do not apply any other home remedies or medications without a doctor's recommendation (Figure 1-1).

SECOND-DEGREE BURNS

Second-degree burns have occurred when the surface of the skin is severely damaged, resulting in the formation of blisters and possible breaks in the skin. Again, the most important first step in treating a second-degree burn is to put the area under cold water (not iced) or apply cold water compresses until the pain decreases. Gently pat the area dry with a clean towel and cover the area with a sterile bandage or clean cloth to prevent infection. Seek medical attention. If the burns are around the mouth or nose or involve singed nasal hair, breathing problems may develop. Do not apply ointments, sprays, antiseptics, or home remedies. Reduce the skin temperature as quickly as possible to reduce tissue damage (Figure 1-2).

THIRD-DEGREE BURNS

Third-degree burns have occurred when the surface of the skin and possibly the tissue below the skin appear white or charred. Initially little pain may be present because nerve endings have been destroyed. Do not remove any clothes that are stuck to the burn. Do

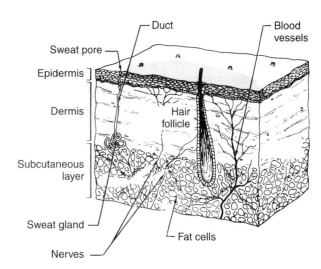

FIGURE 1–1 *First-degree burn: only the skin surface (epidermis) is affected.*

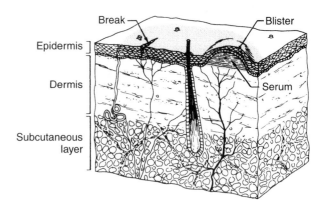

FIGURE 1–2 *Second-degree burn: the epidermal layer is damaged, forming blisters or shallow breaks.*

not put ice water or ice on the burns; this could intensify the shock reaction. Do not apply ointments, sprays, antiseptics, or home remedies to burns. If the victim is on fire, smother the flames with a blanket, rug, or jacket. Breathing difficulties are common with burns around the face, neck, and mouth. Be sure that the victim is breathing. Place a cold cloth or cool (not iced) water on burns of the face, hands, or feet to cool the burned areas. Cover the burned area with thick, sterile, non-fluffy dressings. Call for an ambulance immediately. People with even small third-degree burns need to consult a doctor (Figure 1-3).

BURNS CAUSED BY LIGHT

Some types of light can cause burns. The three types of light include ultraviolet, infrared, and visible. Ultraviolet and infrared are not visible to the unaided human

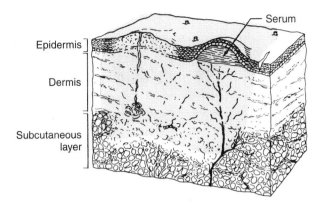

FIGURE 1–3 Third-degree burn: the epidermis, dermis, and subcutaneous layers of tissue are destroyed.

FIGURE 1–4 Portable welding curtains. (Courtesy of Frommelt Safety Products)

eye. They are the types of light that can cause burns. During welding, one or more of the three types of light may be present. Arc welding produces all three types of light, but gas welding produces visible and infrared light only.

The light from the welding process can be reflected from walls, ceilings, floors, or any other large surface. This reflected light is as dangerous as the direct welding light. To reduce the danger from reflected light, the welding area, if possible, should be painted flat black. Flat black will reduce the reflected light by absorbing more of it than any other color. When the welding is to be done on a job site, in a large shop or other area that cannot be painted, weld curtains can be placed to absorb the welding light (Figure 1-4). These special portable welding curtains may be either transparent or opaque. Transparent welding curtains are made of a special high-temperature, flame-resistant plastic that will prevent the harmful light from passing through.

CAUTION: Welding curtains must always be used to protect other workers in an area that might be exposed to the welding light.

Ultraviolet Light

Ultraviolet light waves are the most dangerous. They can cause first-degree and second-degree burns to a technician's eyes or to any exposed skin. Because the technician cannot see or feel ultraviolet light while being exposed to it, he or she must stay protected when in the area of any of the arc welding processes. The closer a technician is to the arc and the higher the current, the quicker a burn may occur.

The ultraviolet light is so intense during some welding processes that a technician's eyes can receive a flash burn within seconds, and the skin can be burned within minutes. Ultraviolet light can pass through loosely woven clothing, thin clothing, light-colored clothing, and damaged or poorly maintained arc welding helmets.

Infrared Light

Infrared light is the light wave that is felt as heat. Although infrared light can cause burns, a person will immediately feel this type of light. Therefore, burns can easily be avoided.

Visible Light

Visible light is the light that we see. It is produced in varying quantities and colors during welding. Too much visible light may cause temporary night blindness (poor eyesight under low light levels). Too little visible light may cause eye strain, but visible light is not hazardous.

Whether burns are caused by ultraviolet light or hot material, they can be avoided if proper clothing and other protection are worn.

TABLE 1-1: HUNTSMAN SELECTOR CHART (COURTESY OF KEDMAN CO., HUNTSMAN PRODUCT DIVISION.)

Selection Chart for Eye and Face Protectors for Use in Industry and Schools

1 Goggles, flexible fitting, regular ventilation

2 Goggles, flexible fitting, hooded ventilation

3 Goggles, cushioned fitting, rigid body

4 Spectacles

5 Spectacles, eyecup type eyeshields

6 Spectacles, semi-flat-fold sideshields

7 Welding goggles, eyecup type, tinted lenses

7A Chipping goggles, eyecup type, tinted lenses

8 Welding goggles, coverspec type, tinted lenses

8A Chipping goggles, coverspec type, clear safety lenses

9 Welding goggles, coverspec type, tinted plate lens

10 Face shield, plastic or mesh window (see caution note)

11 Welding helmet

Non-sideshield spectacles are available for limited hazard use requiring only frontal protection

Applications

Operation	Hazards	Protectors
Acetylene (burning) Acetylene (cutting) Acetylene (welding)	Sparks, harmful rays, molten metal, flying particles	7,8,9
Chemical handling	Splash, acid burns, fumes	2 (for severe exposure add 10)
Chipping	Flying particles	1,2,4,5,6,7A,8A
Electric (Arc) welding	Sparks, intense rays, molten metal	11 (in combination with 4,5,6 in tinted lenses advisable)
Furnace operations	Glare, heat, molten metal	7,8,9 (for severe exposure add 10)
Grinding-light	Flying particles	1,3,5,6 (for severe exposure add 10)
Grinding-heavy	Flying particles	1,3,7A,8A (for severe exposure add 10)
Laboratory	Chemical splash, glass breakage	2 (10 when in combination with 5,6)
Machining	Flying particles	1,3,5,6 (for severe exposure add 10)
Molten metals	Heat, glare, sparks, splash	7,8 (10 in combination with 5,6 in tinted lenses)
Spot welding	Flying particles, sparks	1,3,4,5,6 (tinted lenses advisable; for severe exposure add 10)

CAUTION:
Face shields alone do not provide adequate protection. Plastic lenses are advised for protection against molten metal splash. Contact lenses, of themselves, do not provide eye protection in the industrial sense and shall not be worn in a hazardous environment without appropriate covering safety eyewear.

1.2 EYE AND EAR PROTECTION

FACE AND EYE PROTECTION

Eye protection must be worn in the shop at all times. Eye protection can be safety glasses (Figure 1-5), goggles, or a full face shield. For better protection when working in brightly lit areas or outdoors, some auto body technicians wear flash glasses, which are special, lightly tinted safety glasses. These safety glasses provide protection from both flying debris and reflected light.

Suitable eye protection is important, because eye damage caused by excessive exposure to arc light can occur without warning. Its effects are not noticed until damage is done, like a sunburn that is felt the day after sunbathing. Therefore, you must take appropriate precautions in selecting filters or goggles that are suitable for the process being used (Table 1-1). Selecting the correct shade of lens is important, because both extremes of too light or too dark of a lens can cause eye strain. Beginning auto body technicians often select too dark a lens, assuming it will give them better protection; however, this results in eye strain, just as if they were trying to read in a poorly lit room. In reality, any approved arc welding lenses will filter out the harmful ultraviolet light. Select a lens that lets you see comfortably.

Ultraviolet light can burn the eye in two ways. This light can injure either the white of the eye or the retina, which is the back of the eye. Burns on the retina are not painful but may cause some loss of eyesight (Figure 1-6). The whites of the eyes are very sensitive, and burns are very painful. The eyes are easily infected because, as with any burn, many cells

are killed. These dead cells in the moist environment of the eyes will promote the growth of bacteria that cause infection. When the eye is burned, it feels as though there is something in the eye. Without a professional examination, however, it is impossible to tell if there is something in the eye. Because there may be something in the eye and because of the high risk of infection, home remedies or other medicines should never be used for eye burns. Any time you receive an eye injury, you should see a doctor.

Even with quality welding helmets, like those shown in Figure 1-7, the auto body technician must check for potential problems that may occur from accidents or daily use. Small, undetectable leaks of ultraviolet light in an arc welding helmet can cause a technician's eyes to itch or feel sore after a day of welding. To prevent these leaks, make sure the lens gasket is installed correctly (Figure 1-8). The outer clear lens may be either glass or plastic, but the inside clear lens must be plastic. As shown in Figure 1-9, the lens can be checked for cracks by twisting it between your fingers. Worn or cracked spots on a helmet must be repaired. Tape can be used for temporary repair until the helmet can be replaced or permanently repaired.

Safety glasses with side shields are adequate for general use, but if heavy grinding, chipping, or overhead work is being done, goggles or a full-face shield should be worn in addition to safety glasses (Figure 1-10). Safety glasses are best for general protection because they can be worn under an arc welding helmet.

EAR PROTECTION

The welding environment can be very noisy. The sound level is at times high enough to cause pain and some loss of hearing if the auto body technician's ears are unprotected. Hot sparks can also drop into an open ear, causing severe burns.

Ear protection is available in several forms. One form of protection is earmuffs that cover the outer ear completely (Figure 1-11). Another form of protection is

FIGURE 1–5 _Safety glasses with side shields._

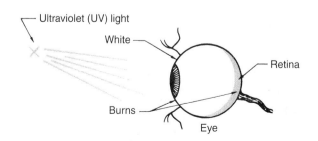

FIGURE 1–6 _The eye can be burned on the white or on the retina by ultraviolet light._

FIGURE 1-7 Typical arc welding helmets used to provide eye and face protection during welding. (Photos courtesy of [top] Thermacote Welco, [bottom] Hornell Speedglas, Inc. [R.H. Blake Inc.])

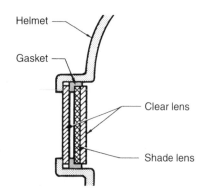

FIGURE 1-8 The correct placement of the gasket around the shade lens is important, because it can stop ultraviolet light from bouncing around the lens assembly.

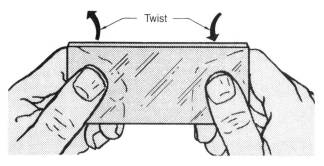

FIGURE 1-9 To check the shade lens for possible cracks, gently twist it.

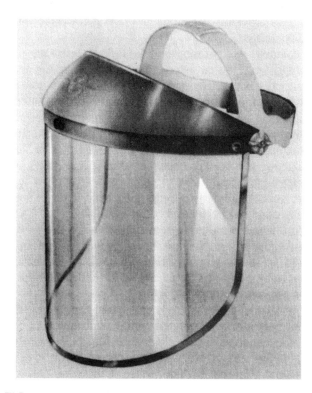

FIGURE 1-10 Full-face shield. (Courtesy of Jackson Products)

FIGURE 1-11 Earmuffs provide complete ear protection and can be worn under a welding helmet. (Courtesy of Mine Safety Appliances Company)

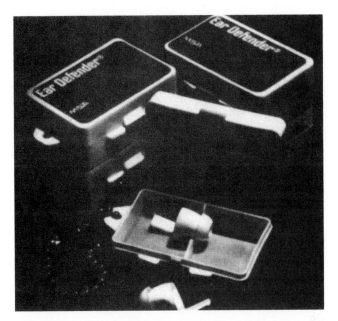

FIGURE 1-12 _Earplugs used as protection from noise only. (Courtesy of Mine Safety Appliances Company)_

plugs that fit into the ear canal (Figure 1-12). Both of these protect a person's hearing, but only the earmuffs protect the outer ear from burns.

CAUTION: Damage to your hearing caused by high sound levels may not be detected until later in life, and the resulting loss in hearing is non-recoverable. It will not get better with time. Each time you are exposed to high levels of sound, your hearing will become worse.

1.3 RESPIRATORY PROTECTION

All welding and cutting processes produce undesirable by-products such as fumes and gases. Production of these by-products cannot be avoided. They are created when the temperature of metals and fluxes is raised above the temperatures at which they boil or decompose. Some by-products do escape into the atmosphere, producing the haze that occurs in improperly ventilated welding shops. Some fluxes used in brazing or soldering produce fumes that may irritate the auto body technician's nose, throat, and lungs.

CAUTION: Welding or cutting must never be performed on drums, barrels, tanks, or other containers until they have been emptied and cleaned thoroughly,

---

eliminating all flammable materials and all substances (such as detergents, solvents, greases, tars, or acids) that might produce flammable, toxic, or explosive vapors when heated.

Some materials that can cause respiratory problems are used in paints, coating, or plating on metals. Any metal that has any grease, oil, or chemicals on its surface must be thoroughly cleaned. This cleaning may be done by grinding, sandblasting, or using an approved solvent. Metals that are plated may not be able to be cleaned before welding or cutting begins.

Most paints containing lead have been removed from the market. Some old machinery and farm equipment may still have lead-based paints on them. Solder often contains lead alloys. The welding and cutting of lead-bearing solders or metals whose surfaces have been painted with lead-based paint can generate lead oxide fumes. The inhalation and ingestion of lead oxide fumes and other lead compounds will cause lead poisoning. Symptoms include a metallic taste in the mouth, loss of appetite, nausea, abdominal cramps, and insomnia. In time, anemia and a general weakness, chiefly in the muscles of the wrists, develop.

Both cadmium and zinc are plating materials used to prevent rusting. Cadmium is often used on bolts, nuts, hinges, and other hardware items, and it gives the surface a yellowish-gold appearance. Acute exposure to high concentrations of cadmium fumes can produce severe lung irritation. Long-term exposure to low levels of cadmium in air can result in emphysema (a disease affecting the ability of the lung to absorb oxygen) and can damage the kidneys.

Zinc, often in the form of galvanizing, may be found on body panels, sheet metal, bolts, nuts, and some other types of hardware. Zinc plating that is thin may appear as a shiny metallic patchwork of a crystal pattern. Thicker hot-dipped zinc appears rough and may appear dull. Zinc is also used in large quantities in the manufacture of brass. It is found in brazing rods. The inhalation of zinc oxide fumes can occur when welding or cutting on these materials. Exposure to these fumes is known to cause metal fume fever. Symptoms of metal fume fever are very similar to those of common influenza.

Some concern has also been expressed about the possibility of lung cancer being caused by some of the chromium compounds that are produced when one is welding stainless steels.

Despite these fumes and other potential hazards in welding shops, people who weld have been found to be as healthy as workers employed in other industrial occupations. Rather than take chances, you should recognize that fumes of any type, regardless of their source, should not be inhaled. The best way to avoid problems is to provide adequate ventilation. If this is not possible, breathing protection should be used. Protective devices for use in poorly ventilated or confined areas are shown in Figures 1-13 and 1-14.

Phosgene is formed when ultraviolet radiation decomposes chlorinated hydrocarbon. Fumes from chlorinated hydrocarbons can come from solvents, such as those used for degreasing metals, and from refrigerants from some air-conditioning systems. They decompose in the arc to produce a potentially dangerous chlorine acid compound. This compound reacts with the moisture in the lungs to produce hydrogen chloride, an acid that destroys lung tissue. For this reason, any use of chlorinated solvents should be well away from welding operations in which ultraviolet radiation or intense heat is generated. Any welding or cutting around leaking refrigerants must be done only after the refrigerant has been completely removed in accordance with EPA regulations.

 ## MATERIALS SPECIFICATIONS DATA SHEETS (MSDSs)

All manufacturers of potentially hazardous materials must provide to the users of their products detailed information about possible hazards resulting from the use of their products. These materials specifications data sheets are often called MSDSs. They must be provided to anyone using the products or anyone working in the area where the products are in use. Often companies will post these sheets on a bulletin board or put them in a convenient place near the work area.

 ## VENTILATION

The welding area should be well ventilated. Excessive fumes, ozone, or smoke may collect in the welding area. Ventilation should be provided for their removal. Natural ventilation is best, but forced ventilation may be required. Areas that have 10,000 cubic feet (283 cubic meters) or more per welder or that have ceilings 16 feet (4.9 meters) high or higher may not require forced ventilation unless fumes or smoke begin to collect (Figure 1-15).

Small shops or shops with large numbers of auto body technicians welding require forced ventilation. Forced ventilation can be general or localized using

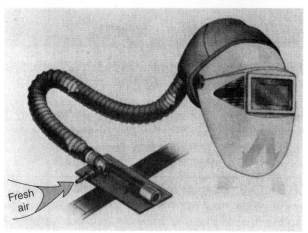

FIGURE 1–13   *Filtered fresh air is forced into the welder's breathing area. The air can come from a belt-mounted respirator or through a hose for a remote location. (Courtesy Hornell Speedglas, Inc. [R.H. Blake Inc.])*

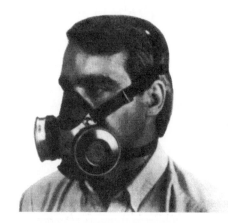

FIGURE 1–14   *Typical respirator for contaminated environments. The filters can be selected for specific types of containment. (Courtesy of Mine Safety Appliances Company)*

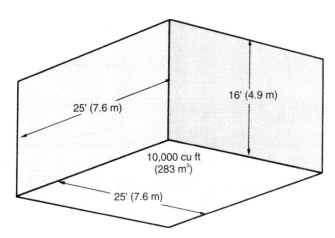

FIGURE 1-15 A room with a ceiling sixteen feet (4.9 m) high may not require forced ventilation for one welder.

FIGURE 1-16 A flexible exhaust pickup.

fixed or flexible exhaust pickups (Figure 1-16). General room ventilation must be at a rate of 2,000 cu ft (564 m³) or more per person welding. Localized exhaust pickups must have a suction strong enough to provide 100 linear feet (30.5 m) per minute velocity of welding fumes away from the auto body technician. Local, state, or federal regulations may require that welding fumes be treated to remove hazardous components before they are released into the atmosphere.

Any system of ventilation should draw the fumes or smoke away before they rise past the level of the auto body technician's face. Forced ventilation is always required when welding on metals that contain zinc, lead, beryllium, cadmium, mercury, copper, austenitic manganese, or other materials that give off dangerous fumes.

 ## ELECTRICAL SAFETY

Injuries and even death can be caused by electric shock unless proper precautions are taken. Most welding and cutting operations involve electrical equipment in addition to the arc welding power supplies. Grinders, lights, and drills are examples. Most electrical equipment in a welding shop is powered by alternating-current (AC) sources having input voltages ranging from 115 volts to 220 volts. However, fatalities have occurred with equipment operating at less than 80 volts. Most electric shock in the welding industry occurs not from contact with welding guns but from accidental contact with bare or poorly insulated conductors. Electrical resistance is lowered in the presence of water or moisture, so auto body

WARNING	
READ AND UNDERSTAND THIS WARNING	PROTECT YOURSELF AND OTHERS
ELECTRIC SHOCK can kill. • Do not permit electrically live parts or electrodes to contact skin . . . or your clothing or gloves if they are wet. • Insulate yourself from work and ground. FUMES AND GASES can be dangerous to your health. • Keep fumes and gases from your breathing zone and general area. • Keep your head out of fumes. • Use enough ventilation or exhaust at the arc or both. ARC RAYS can injure eyes and burn skin. • Wear correct eye, ear, and body protection.	ENGINE EXHAUST GASES can kill. • Use in open, well ventilated areas or vent the engine exhaust to the outside. HIGH VOLTAGE can kill. • Do not touch electrically live parts. • Stop engine before servicing. MOVING PARTS can cause serious injury. • Keep clear of moving parts. • Do not operate with protective covers, panels, or guards removed. • Only qualified personnel should install, use, or service this equipment. • Read operating manual for details.

READ AND UNDERSTAND THE MANUFACTURER'S INSTRUCTIONS AND YOUR EMPLOYER'S SAFETY PRACTICES.

See American National Standard Z49.1, "Safety in Welding and Cutting," published by the American Welding Society, 550 Le Jeune Rd., Miami, Florida 33126; OSHA Safety and Health Standards, 29 CFR 1910, available from U.S. Government Printing Office, Washington, D.C. 20402.

DO NOT REMOVE THIS WARNING   E202

FIGURE 1-17 Note the warning information for electrical shock and high voltage contained on this typical label, which is attached to welding equipment by the manufacturer. (Courtesy of the Lincoln Electric Company)

technicians must take special precautions when working under damp or wet conditions, such as those caused by leaking radiator fluids. This warning also applies to perspiration. Figure 1-17 shows a typical warning label attached to welding equipment.

The vehicle being welded and the frame or chassis of all electrically powered machines must be connected to a good electrical ground. The work lead from the welding power supply is not an electrical ground and is not sufficient. A separate lead is required to ground the workpiece and power source.

Electrical connections must be tight. Terminals for welding leads and power cables must be shielded from accidental contact by people or by metal objects. Cables must be used within their current carrying and duty cycle capacities; otherwise, they will overheat and break down the insulation rapidly. Cable connectors for lengthening leads must be insulated. Cables must be checked periodically to be sure that they have not become frayed. If they have, they must be replaced immediately.

CAUTION: Welding cables must never be spliced within 10 feet (3 m) of the electrode holder.

Never allow the metal parts of electrodes or electrode holders to touch your skin or wet coverings on your bodies. Dry gloves in good condition must always be worn. Rubber-soled shoes are advisable. Precautions against accidental contact with bare conducting surfaces must be taken when the auto body technician is required to work in cramped kneeling, sitting, or lying positions. Insulated mats or dry wooden boards are desirable protection from the damp shop floor.

Welding circuits must be turned off when the workstation is left unattended. Circuits must be turned off and locked or tagged to prevent electrocution when someone is working on the welder, welding leads, electrode holder, torches, wire feeder, guns, or other parts. Since the electrode holder is energized when you change coated electrodes, you must wear dry gloves.

ELECTRICAL SAFETY SYSTEMS

For protection from electric shock, the standard portable tool is built with either of two equally safe systems: external grounding or double insulation.

A tool with external grounding has a wire that runs from the housing through the power cord to a third prong on the power plug. When this third prong is connected to a grounded, three-hole electrical outlet, the grounding wire will carry any current that leaks past the electrical insulation of the tool away from the user and into the ground. In most electrical systems, the three-prong plug fits into a three-prong, grounded receptacle. If the tool is operated at less than 150 volts, it has a plug like that shown in Figure 1-18A. If it is for use at 150 to 250 volts, it has a plug like that shown in Figure 1-18B. In either type, the green (or green and yellow) conductor in the tool cord is the grounding wire. Never connect the grounding wire to a power terminal.

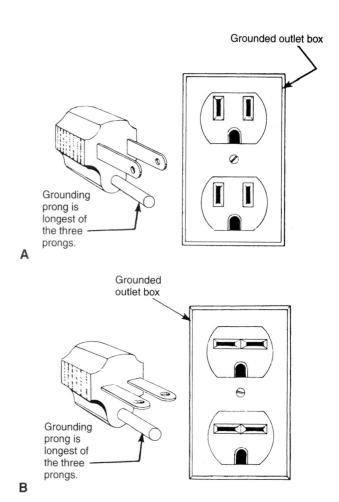

FIGURE 1-18   (A) A three-prong grounding plug for use with up to 150-volt tools and (B) a grounding plug for use with 150- to 250-volt tools.

Adapters (Figure 1-19) are available for connecting three-prong grounding-type plugs to two-prong receptacles. The rigid ear lug extending from the adapter is the grounding means and must be connected to a permanent ground—for example, to a properly grounded outlet box. If there is uncertainty about whether or not the receptacle in question is properly grounded, have it checked by a certified electrician.

A double-insulated tool has an extra layer of electrical insulation that eliminates the need for a three-prong plug and grounded outlet. Double-insulated tools do not require grounding and thus have a two-prong plug. In addition, double-insulated tools are always labeled as such on their nameplate or case (Figure 1-20).

VOLTAGE WARNINGS

Before connecting a tool to a power supply, be sure the voltage supplied is the same as that specified on the nameplate of the tool. A power source with a

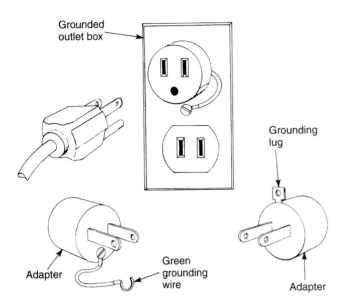

FIGURE 1–19 Three-prong plug adapters.

FIGURE 1–20 Typical portable power tool nameplate.

voltage greater than that specified for the tool can lead to serious injury to the user as well as damage to the tool. Using a power source with a voltage lower than the nameplate rating is harmful to the motor.

Tool nameplates also bear a figure with the abbreviation _amps_ (for amperes, a measure of electric current). This refers to the current-drawing requirement of the tool. The higher the input current, the more powerful the motor.

EXTENSION CORDS

If there is some distance from the power source to the work area or if the portable tool is equipped with a stub power cord, an extension cord must be used. When using extension cords on portable power tools, the size of the conductors must be large enough to prevent an excessive drop in voltage. A voltage drop is the lowering of the voltage at the power tool from the voltage at the supply. This occurs because of resistance to electrical flow in the wire. A voltage drop

causes loss of power, overheating, and possible motor damage. Table 1-2 shows the correct size extension cord to use based on cord length and nameplate amperage rating. If in doubt, use the next larger size. The smaller the gauge number of an extension cord, the larger the cord.

Two-wire extension cords with two-prong plugs are not acceptable in most collision repair facilities. Only three-wire, grounded extension cords connected to properly grounded three-wire receptacles should be used in auto body shops. Current specifications require outdoor receptacles to be protected with ground-fault circuit interpreter (GFCI) devices. These safety devices are often referred to as a GFI.

When using extension cords, keep in mind the following safety tips:

- Always connect the cord of a portable electric power tool into the extension cord before the extension cord is connected to the outlet. Always unplug the extension cord from the receptacle before the cord of the portable power tool is unplugged from the extension cord.
- Extension cords should be long enough to make connections without being pulled taut, creating unnecessary strain and wear.
- Be sure that the extension cord does not come in contact with sharp objects or hot surfaces. The cords should not be allowed to kink, nor should they be dipped in or splattered with oil, grease, or chemicals.
- Before using a cord, inspect it for loose or exposed wires and damaged insulation. If a cord is damaged, it must be replaced. This also applies to the tool's power cord.
- Extension cords should be checked frequently while in use to detect unusual heating. Any cable that feels more than slightly warm to a bare hand placed outside the insulation should be checked immediately for overloading.
- See that the extension cord is positioned so that no one trips or stumbles over it.
- To prevent the accidental separation of a tool cord from an extension cord during operation, make a knot as shown in Figure 1-21A or use a cord connector as shown in Figure 1-21B.
- Use an extension cord that is long enough for the job but not excessively long.

SAFETY RULES FOR PORTABLE ELECTRIC TOOLS

In all tool operation, safety is simply the removal of any element of chance. A few safety precautions that should be observed are listed below. These are general rules that apply to all power tools. They should be

TABLE 1–2: RECOMMENDED EXTENSION CORD SIZES FOR USE WITH PORTABLE ELECTRIC TOOLS															
Nameplate Ampere Rating															
Cord Length	0 to 5	6	8	8	9	10	11	12	13	14	15	17	17	18	19	20
25′	18	18	18	18	18	18	16	16	16	14	14	14	14	14	12	12
50′	18	18	18	18	18	18	16	16	16	14	14	14	14	14	12	12
75′	18	18	18	18	18	18	16	16	16	14	14	14	14	14	12	12
100′	18	18	18	16	16	16	16	16	14	14	14	14	14	14	12	12
125′	18	18	16	16	16	14	14	14	14	14	14	12	12	12	12	12
150′	18	16	16	16	14	14	14	14	14	12	12	12	12	12	12	12

Note: Wire sizes shown are AWG (American Wire Gauge) based on a line voltage of 120.

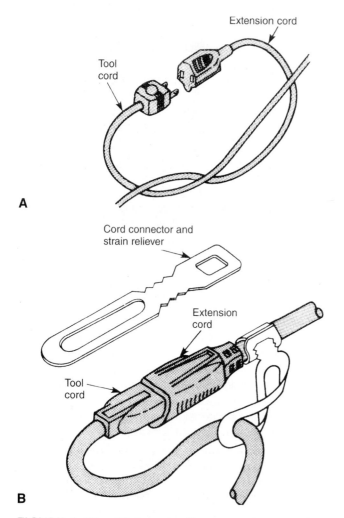

FIGURE 1–21 (A) A knot will prevent the extension cord from accidentally pulling apart from the tool cord during operation. (B) A cord connector will serve the same purpose.

strictly obeyed to avoid injury to the operator and damage to the power tool.

- Know the tool. Learn the tool's applications and limitations as well as its specific potential hazards by reading the manufacturer's literature.

- Ground the portable power tool unless it is double-insulated. If the tool is equipped with a three-prong plug, it must be plugged into a three-hole electrical receptacle. If an adapter is used to accommodate a two-pronged receptacle, the adapter wire must be attached to a known ground. Never remove the third prong.
- Do not expose the power tool to rain. Do not use a power tool in wet locations.
- Keep the work area well lighted. Avoid chemical or corrosive environments.
- Because of electric tools sparking, portable electric tools should never be started or run where there is any possibility of a fire or explosion due to the presence of propane, natural gas, gasoline, paint thinner, acetylene, or other flammable vapors.
- Do not force a tool. It will do the job better and more safely if operated at the rate for which it was designed.
- Use the right tool for the job. Never use a tool for any purpose except that for which it was designed.
- Wear eye protectors. Safety glasses or goggles will protect the eyes while you operate power tools.
- Wear a face or dust mask if the operation creates dust.
- Take care of the power cord. Never carry a tool by its cord or yank it to disconnect it from the receptacle.
- Secure your work. Use clamps to hold the work, because it is safer than using your hands and it frees both hands to operate the tool.
- Do not overreach when operating a power tool. Keep proper footing and balance at all times.
- Maintain power tools. Follow the manufacturer's instructions for lubricating and changing accessories. Replace all worn, broken, or lost parts immediately.
- Disconnect the tools from the power source when not in use.

- Remove adjusting keys and wrenches before operation. Form the habit of checking to see that any keys or wrenches are removed from the tool before turning it on.
- Avoid accidental starting. Do not carry a plugged-in tool with your finger on the switch. Be sure the switch is off when plugging in the tool.
- Be sure accessories and cutting bits are attached securely to the tool.
- Do not use tools with cracked or damaged housings.
- When operating a portable power tool, give it your full and undivided attention; avoid dangerous distractions.

 ## HAND TOOLS

Hand tools are used by the auto body technician to do necessary assembly and disassembly of parts for repair and welding.

The adjustable wrench is a common tool used by the auto body technician. When in use, it should be adjusted tightly on the nut and pushed so that most of the force is on the fixed jaw (Figure 1-22). When a wrench is being used on a tight bolt or nut, the wrench should be pushed with the palm of an open hand or pulled to prevent injuring the hand. If a nut or bolt is too tight to be loosened with a wrench, obtain a longer wrench. A cheater bar should not be used. The fewer points a box end wrench or socket has, the stronger it is and the less likely it is to slip or damage the nut or bolt (Figure 1-23).

Striking a hammer directly against a hard surface such as another hammer face or anvil may result in chips flying off and causing injury.

The mushroomed heads of chisels, punches, and faces of hammers should be ground off (Figure 1-24 and see Figure 1-26). Chisels and punches that are going to be hit harder than with a slight tap should be held in a chisel holder or with pliers to eliminate the danger of injuring your hand. A handle should be placed on the tang of a file in order to avoid injuring your hand (Figure 1-25). A file can be kept free of chips by rubbing a piece of soapstone on it before it is used.

It is important to remember to use the correct tool for the job. Do not try to force a tool to do a job it was not designed to do.

HAND TOOL SAFETY

Hand tools used in the collision repair facility should be treated properly and not abused. Many accidents

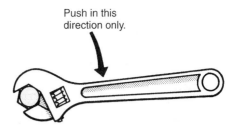

FIGURE 1-22 The adjustable wrench is stronger when used in the direction indicated.

FIGURE 1-23 The fewer the points, the less likely the wrench is to slip.

FIGURE 1-24 Any mushroomed heads must be ground off.

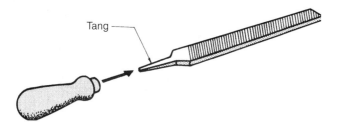

FIGURE 1–25 *To protect yourself from the sharp tang of a file, always use a handle with a file.*

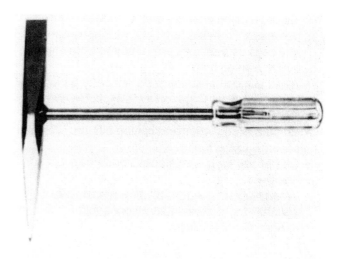

FIGURE 1–26 *Welder's chipping hammer. Grind the head periodically. (Courtesy of MAC Tools, Inc.)*

can be avoided by using the right tool for the job. For instance, use a tool that is the correct size for the work instead of one that is too large or too small.

Keep hand tools clean to protect them against the damage caused by corrosion. Wipe off any accumulated dirt and grease. Dip the tools occasionally in cleaning fluids or solvents and wipe them clean. Lubricate adjustable and other moving parts to prevent wear and misalignment.

Make sure that hand tool cutting edges are sharp. Sharp tools make work easier, improve the accuracy of the work, save time, and are safer than dull tools. When sharpening, redressing, or repairing tools, shape, grind, hone, file, fit, and set them properly using other tools suited to each purpose. For sharpening tools, either an oilstone or a grindstone is preferable. If grinding on an abrasive wheel is required, grind only a small amount at a time with the tool rest set not more than 1/16 inch from the wheel. Hold the tool lightly against the wheel to prevent overheating, and frequently dip the part being ground in water to keep it cool. This will protect the hardness of the metal and help to retain the sharpness of the cutting edge. Tools struck by hammers, such as chisels or punches, should have their heads ground periodically to prevent mushrooming (Figures 1-24 and 1-26). Be sure to wear safety goggles when sharpening or redressing tools.

Keep handles secure and safe. Do not rely on friction tape to secure split handles or to prevent handles from splitting. Check wedges and handles frequently. Be sure heads are wedged tightly on handles. Keep handles smooth and free of rough or jagged surfaces. Protect their tips before driving them into tools or use a proper mallet to avoid splitting or mushrooming them. Replace handles that are split, chipped, or that cannot be refitted securely.

When swinging any tool, be absolutely certain that no one is within range or can come within range of the swing or be struck by flying material. Always allow plenty of room for arm and body movements and for handling the work. When carrying tools, protect the cutting edges and carry the tools in such a way that you will not endanger yourself or others. Carry

pointed or sharp-edged tools in pouches or holsters. Never create sparks in the presence of flammable materials or explosive vapors.

HAMMER AND MALLET SAFETY

The following safety precautions generally apply to all hammers and mallets:

- Check to see that the handle is tight before using any hammer or mallet. Never use a hammer or mallet with a loose or damaged handle.
- Always use a hammer or mallet of suitable size and weight for the job.
- Discard or repair any tool if the face shows excessive wear, dents, chips, mushrooming, or improper redressing.
- Rest the face of the hammer on the work before striking to get the feel or aim; then, grasp the handle firmly with the hand near the end of the handle. Get the fingers out of the way before striking with force.
- A hammer blow should always be struck squarely, with the hammer face parallel to the surface being struck. Always avoid glancing blows and over-and-under strikes.
- For striking another tool (cold chisel, punch, wedge, and so on), the face of the hammer should be proportionately larger than the head of the tool. For example, a 1/2-inch cold chisel requires at least a 1-inch hammer face.
- Never use one hammer to strike another hammer.
- Do not use the end of the handle of any tool for tamping or prying; it might split.

1.8   DRILLS

Secure the workpiece as necessary and fasten it in a vise or clamp. Holding a small item in your hand can cause injury if it is suddenly seized by the bit and whirled from your grip. This is most likely to happen just as the bit breaks through the hole at the backside of the work. All sheet metal tends to cause the bit to grab as it goes through. This can be controlled by reducing the pressure on the drill just as the bit starts to go through the workpiece.

Carefully center the drill bit in the jaws of the chuck and securely tighten it. Avoid inserting the bit off-center, because it will wobble and probably break when it is used. Drill bits that are 1/4 inch (6 mm) may be hand tightened in the drill chuck to prevent them from snapping if they accidently grab. Hand tightening the small bits allows them to spin in the chuck if necessary, thus reducing bit breakage. This technique does not always work, because some chucks cannot hold the bits securely enough to prevent them from spinning during normal use. In these cases the chuck must be tightened securely with a chuck key.

When possible, center-punch the workpiece before drilling to prevent the drill bit from walking across the surface as the drilling begins. After centering the drill bit tip on the exact point at which the hole is to be drilled, start the motor by pulling the trigger switch. Never apply a spinning drill bit to the work. With a variable-speed drill, run it at a very low speed until the cut has begun. Then gradually increase to the optimum drill speed.

Except when it is desirable to drill a hole at an angle, hold the drill perpendicular to the face of the work. Align the drill bit and the axis of the drill in the direction the hole is to go and apply pressure only along this line, with no sideways or bending pressure. Changing the direction of pressure will distort the dimensions of the hole and might snap a small drill bit.

Use just enough steady and even pressure to keep the drill cutting. Guide the drill by leading it slightly, if needed, but do not force it. Too much pressure can cause the bit to break or overheat. Too little pressure will keep the bit from cutting and dull its edges due to the friction created by sliding over the surface.

If the drill becomes jammed in the hole, release the trigger immediately, remove the drill bit from the work, and determine the cause of the stalling or jamming. Do not squeeze the trigger on or off in an attempt to free a stalled or jammed drill. When using a reversing-type model, the direction of the rotation may be reversed to help free a jammed bit. Be sure the direction of the rotation is reset before trying to continue the drilling.

Reduce the pressure on the drill just before the bit cuts through the work to avoid stalling in metal. When the bit has completely penetrated the work and is spinning freely, withdraw it from the work while the motor is still running, and then turn off the drill.

1.9   POWER TOOLS

All power tools must be properly grounded to prevent accidental electrical shock. If even a slight tingle is felt while using a power tool, stop and have the tool checked by an electrical technician. Power tools should never be used with force or allowed to overheat from excessive or incorrect use. If an extension cord is used, it should have a large enough current rating to carry the load. An extension cord that is too small will cause the tool to overheat.

Safety glasses must be worn at all times when using any power tools.

GRINDERS

Grinding using a portable grinder or a pedestal grinder is required on most collision repair jobs. Often it is necessary to grind welds, remove rust, or smooth a surface. Grinding stones have the maximum revolutions per minute (r/min) listed on the paper blotter (Figure 1-27). They must never be used on a machine

FIGURE 1-27 *Always check to be sure that the grinding stone and the grinder are compatible before installing a stone.*

with a higher rated r/min. If grinding stones are turned too fast, they can explode.

GRINDING STONES

Before a grinding stone is put on the machine, it should be tested for cracks. This is done by tapping the stone in four places and listening for a sharp ring that indicates it is good (Figure 1-28). A dull sound indicates that the grinding stone is cracked and should not be used. Once a stone has been installed and has been used, it may need to be trued and balanced by using a special tool designed for that purpose (Figure 1-29). Truing keeps the stone face flat and sharp for better results.

Each grinding stone is made for grinding specific types of metal. Most stones are for ferrous metals, meaning iron, cast iron, steel, and stainless steel, among others. Some stones are made for nonferrous metals such as aluminum, copper, and brass. If a ferrous stone is used to grind nonferrous metal, the stone will become glazed (the surface clogs with metal) and may explode due to frictional heat building up on the surface. If a nonferrous stone is used to grind ferrous metal, the stone will be quickly worn away.

When the stone wears down, keep the tool rest adjusted to within 1/16 inch (2 mm) (Figure 1-30), so that the metal being ground cannot be pulled between the tool rest and the stone surface. Stones should not be used when they are worn down to the size of the paper blotter. If small parts become hot from grinding, pliers can be used to hold them. Gloves should never be worn when grinding. If a glove gets caught in a stone, the whole hand may be drawn in.

The sparks from grinding should be directed down and away from other workers and vehicles, especially glass. If it's not possible to direct sparks to where they cannot cause damage, the area the sparks are hitting must be covered with an approved blanket.

EQUIPMENT MAINTENANCE

A routine schedule of equipment maintenance will aid in detecting potential problems such as leaking shielding gas, loose wires, poor grounds, frayed insulation, or split hoses. Small problems, if fixed in time, can prevent the loss of valuable time due to equipment breakdown or injury.

Any maintenance beyond routine external maintenance should be referred to a trained service

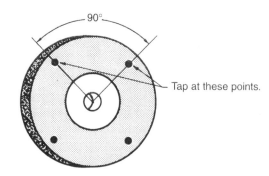

FIGURE 1–28 Tap the stone to check for cracks.

FIGURE 1–29 Use a grinding stone redressing tool as needed to keep the stone in balance.

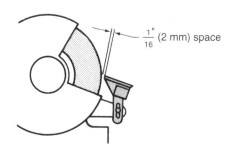

FIGURE 1–30 Keep the tool rest adjusted.

technician. In most areas, it is against the law for anyone but a licensed electrician to work on equipment such as arc welders and anyone but a factory-trained repair technician to work on regulators. Electrical shock and exploding regulators can cause serious injury or death.

Hoses must be used only for the gas or liquid for which they were designed. Green hoses are to be used only for oxygen, and red hoses are to be used only for acetylene or other fuel gases. Orange or black hoses are used for compressed air. Using unnecessarily long lengths of hoses should be avoided. Never use oil, grease, lead, or other pipe-fitting compounds for any joints. Hoses should also be kept out of the direct line of sparks. Any leaking or bad joints in gas hoses must be repaired.

1.11 — WORK AREA

The work area should be kept picked up and swept clean. Collections of body panels, used parts, wire, hoses, and power cables are difficult to work around and easy to trip over. Hooks can be made to hold hoses and cables, and scrap parts should be thrown into scrap bins.

Portable screens should be used whenever arc welding is to be done in an area where others are working (Figure 1-31).

If a piece of hot metal is going to be left unattended, write the word _hot_ on it before leaving. This procedure can also be used to warn people of hot tables, vises, and tools.

1.12 — MATERIAL HANDLING

Proper lifting, moving, and handling of large or heavy parts is important to the safety of the auto body technicians. Improper work habits can cause serious personal injury.

When you are lifting a heavy object, the weight of the object should be distributed evenly between both hands, and your legs should be used to lift, not your back. Do not try to lift a large or bulky object without help if the object is heavier than you can lift with one hand.

1.13 — WORK CLOTHING

GENERAL WORK CLOTHING

Because of the amount and temperature of hot sparks, metal, and slag produced during welding, cutting, or brazing, and the fact that special protective clothing cannot be worn at all times, it is important to choose general work clothing that will minimize the possibility of getting burned.

Wool clothing (100 percent wool) is the best choice but difficult to find. All-cotton (100 percent cotton) clothing is a good second choice, and it is the most popular material used. Synthetic materials, including nylon, rayon, and polyester, should be avoided because they are easily burned, produce a hot, sticky ash (because it sticks, burns can be more severe), and some produce poisonous gases. The clothing must also stop ultraviolet light from passing through it. This is accomplished if the material chosen is a dark color, thick, and tightly woven.

The following are some guidelines for selecting work clothing:

- Shirts must be long sleeved to protect the arms, have a high-buttoned collar to protect the neck (Figure 1-32), be long enough to tuck into the pants to protect the waist, and have flaps on the pockets to keep sparks out or have no pockets.
- Pants must have legs long enough to cover the tops of the boots and must be without cuffs that would catch sparks.

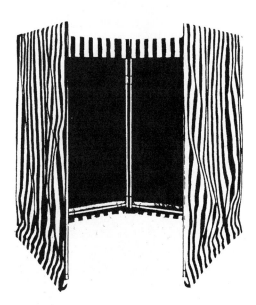

FIGURE 1–31 Portable safety screen.

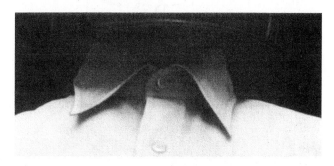

FIGURE 1–32 _The top button of the shirt worn by the welder should always be buttoned to prevent severe burns to that person's neck._

- Boots must have high tops to keep out sparks, have steel toes to prevent crushed toes (Figure 1-33), and have smooth tops to prevent sparks from being trapped in seams.
- Caps should be thick enough to prevent sparks from burning the top of an auto body technician's head.

All clothing must be free of frayed edges or holes. The clothing must be relatively tight-fitting in order to prevent excessive folds or wrinkles that might trap sparks.

CAUTION: There is no safe place to carry butane lighters and matches while welding or cutting. They may catch fire or explode if they are subjected to welding heat or sparks. Butane lighters may explode with the force of 1/4 stick of dynamite. Matches can erupt into a ball of fire. Both butane lighters and matches must always be removed from the auto body technician's pockets and placed a safe distance away before any work is started.

SPECIAL PROTECTIVE CLOTHING

General work clothing is worn by each person in the shop. In addition to this clothing, extra protection is needed for each person who is in direct contact with hot materials. Leather is often the best material to use because it is lightweight, flexible, resists burning, and is readily available. Synthetic insulating materials are also available. Ready-to-wear leather protection includes capes, jackets, aprons, sleeves, gloves, caps, pants, knee pads, and spats, among other items.

Hand Protection

All-leather, gauntlet-type gloves should be worn when doing any welding (Figure 1-34). Gauntlet gloves that

have a cloth liner for insulation are best for hot work. Noninsulated gloves will give greater flexibility. Some leather gloves are available with a canvas gauntlet top; these should be used for light work only. All-cotton gloves are sometimes used when doing very light welding.

Body Protection

Full leather jackets and capes will protect an auto body technician's shoulders, arms, and chest (Figure 1-35). A jacket, unlike the cape, protects the

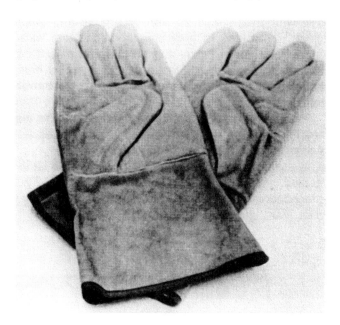

FIGURE 1-34   *All leather, gauntlet-type welding gloves.*

FIGURE 1-33   *Safety boots with steel toes are required by many welding shops.*

Steel

FIGURE 1-35   *Full leather jacket. (Courtesy of Elliott Glove Co., Inc.)*

technician's back and complete chest. A cape is open and much cooler but offers less protection. The cape can be used with a bib apron to provide some additional protection while leaving the back cooler. Either the full jacket or the cape with a bib apron should be worn for any welding under a vehicle.

Waist and Lap Protection

Bib aprons or full aprons will protect a technician's lap. You will especially need to protect your laps if you squat or sit while working and if you bend over or lean against the vehicle.

1.14   HANDLING AND STORING CYLINDERS

Oxygen and fuel gas cylinders or other flammable materials must be stored separately. The storage areas must be separated by 20 ft (6.1 m) or by a wall 5 ft high (1.5 m) with at least a 1/2-hour burn rating (Figure 1-36). The purpose of the distance or wall is to keep the heat of a small fire from causing the oxygen cylinder safety valve to release. If the safety valve releases the oxygen, a small fire would become a raging inferno.

Inert gas cylinders may be stored separately or with either fuel cylinders or oxygen cylinders.

Empty cylinders must be stored separately from full cylinders, although they may be stored in the same room or area. All cylinders must be stored vertically and have the protective caps screwed on firmly.

SECURING GAS CYLINDERS

Cylinders must be secured with a chain or other device so that they cannot be knocked over accidentally. Even though more stable, cylinders attached to a manifold or stored in a special room used only for cylinder storage should be chained.

STORAGE AREAS

Cylinder storage areas must be located away from halls, stairwells, and exits so that in case of an emergency they will not block an escape route. Storage areas should also be located away from heat, radiators, furnaces, and welding sparks. The location of storage areas should be such that unauthorized people cannot tamper with the cylinders. A warning sign that reads "Danger—No Smoking, Matches, or Open Lights," or similar wording, should be posted in the storage area (Figure 1-37).

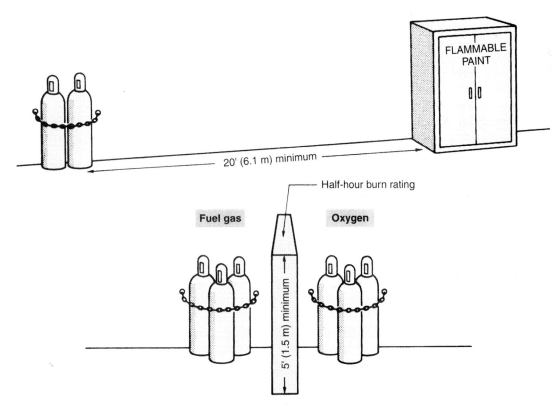

FIGURE 1–36   _Stored fuel gas cylinders must be separated from any flammable material by at least 20 feet (6.1 m) or a wall 5 feet (1.5 mm) high._

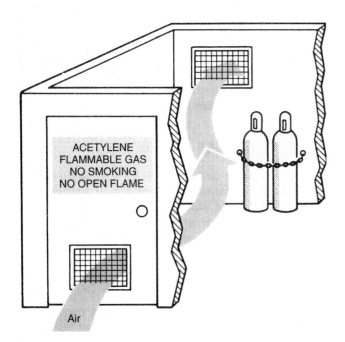

FIGURE 1-38 *Move a leaking fuel gas cylinder out of the building or any work area. Slowly release the pressure after posting a warning of the danger.*

FIGURE 1-37 *A separate room used to store acetylene must have good ventilation and should have a warning sign posted on the door.*

CYLINDERS WITH VALVE PROTECTION CAPS

Cylinders equipped with a valve protection cap must have the cap in place unless the cylinder is in use. The protection cap prevents the valve from being broken off if the cylinder is knocked over. If the valve of a full high-pressure cylinder of argon, oxygen, carbon dioxide (CO_2), and mixed gases is broken off, the cylinder can fly around the shop like a missile if it has not been secured properly. Never lift a cylinder by the safety cap or the valve. The valve can easily break off or be damaged.

When you are moving cylinders, the valve protection cap should be replaced, especially if the cylinders are mounted on a truck or trailer for out-of-shop work. The cylinders must never be dropped or handled roughly.

GENERAL PRECAUTIONS

Use warm water, not boiling, to loosen cylinders that are frozen to the ground. Any cylinder that leaks, has a bad valve, or has gas-damaged threads must be identified and reported to the supplier. A piece of soapstone is used to write the problem on the cylinder. If the leak cannot be stopped by closing the cylinder valve, the cylinder should be moved to a vacant lot or open area. The pressure should then be slowly released after posting a warning sign (Figure 1-38).

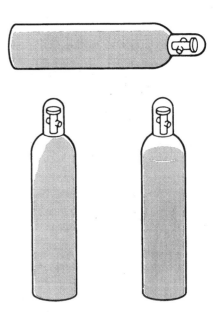

FIGURE 1-39 *The acetone in an acetylene cylinder must have time to settle before the cylinder can be used safely.*

Acetylene cylinders that have been lying on their sides must stand upright for fifteen minutes or more before they are used. The acetylene is absorbed in acetone, and the acetone is absorbed in a filler. The filler does not allow the liquid to settle back away from the valve very quickly (Figure 1-39). If the cylinder has been in a horizontal position, using it too soon after it is placed in a vertical position may draw acetone out of the cylinder. Acetone lowers the flame temperature and can damage regulator or torch valve settings.

FIRE PROTECTION

Fire is a constant danger during welding or cutting. The potential for fires cannot always be eliminated, but it should be minimized. Highly combustible materials should be 35 ft (10.7 m) or more away from any welding. When it is necessary to weld within 35 feet of combustible materials, when sparks can reach materials farther than 35 feet away, or when anything more than a minor fire might start, a fire watch is needed.

FIRE WATCH

A fire watch can be provided by any person who knows how to sound the alarm and use a fire extinguisher. The fire extinguisher must be of the type required to put out a fire of combustible materials used in the welding area. In the event of a fire, combustible materials that cannot be removed from the welding area should be covered with a noncombustible insulating blanket.

FIRE EXTINGUISHERS

The four types of fire extinguishers are type A, type B, type C, and type D. Each type is designed to put out fires on certain types of materials. Some fire extinguishers can be used on more than one type of fire. However, using the wrong type of fire extinguisher can add to the danger by causing the fire to spread, causing electrical shock, or causing an explosion.

Type A Extinguishers

Type A extinguishers are used for combustible solids (articles that burn) such as paper, wood, and cloth. The symbol for a type A extinguisher is a green triangle with the letter A in the center (Figure 1-40).

Type B Extinguishers

Type B extinguishers are used for combustible liquids, such as oil, gas, and paint thinner. The symbol for a type B extinguisher is a red square with the letter B in the center (Figure 1-41).

Type C Extinguishers

Type C extinguishers are used for electrical fires. For example, they are used on fires involving motors, fuse boxes, and welding machines. The symbol for a type C extinguisher is a blue circle with the letter C in the center (Figure 1-42).

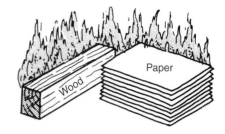
FIGURE 1–40   Type A fire extinguisher symbol.

FIGURE 1–41   Type B fire extinguisher symbol.

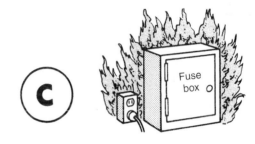
FIGURE 1–42   Type C fire extinguisher symbol.

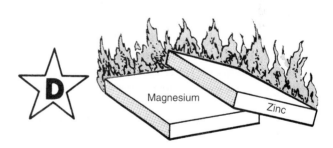
FIGURE 1–43   Type D fire extinguisher symbol.

Type D Extinguishers

Type D extinguishers are used on fires involving combustible metals, such as zinc, magnesium, and titanium. The symbol for a type D extinguisher is a yellow star with the letter D in the center (Figure 1-43).

LOCATION OF FIRE EXTINGUISHERS

Fire extinguishers should be of a type that can be used on the types of combustible materials located nearby (Figure 1-44). The extinguishers should

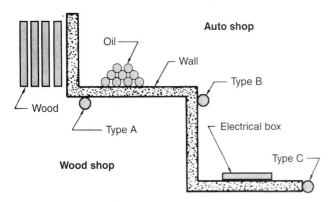

FIGURE 1–44 *The type of fire extinguisher provided should be appropriate for the materials being used in the surrounding area.*

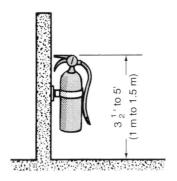

FIGURE 1–45 *Mount the fire extinguisher so that it can be lifted easily in an emergency.*

be placed so that they can be easily removed without reaching over combustible material. They should also be placed at a level low enough to be easily lifted off the mounting (Figure 1-45). The locations of fire extinguishers should be marked with red paint and signs high enough so that their location can be seen from a distance over people and equipment. The extinguishers should also be marked near the floor so that they can be found even if a room is full of smoke (Figure 1-46).

USE OF FIRE EXTINGUISHERS

A fire extinguisher works by breaking the fire triangle of heat, fuel, and oxygen. Most extinguishers both cool the fire and remove the oxygen. Fire extinguishers use a variety of materials to extinguish the fire. The ones most commonly found in welding shops use foam, carbon dioxide, soda-acid gas cartridges, pump tanks, or dry chemicals.

When using a foam extinguisher: Don't spray the stream directly into the burning liquid. Allow the foam to fall lightly on the fire.

When using a carbon dioxide extinguisher: Direct the discharge as close to the fire as possible, first at the edge of the flames and gradually to the center.

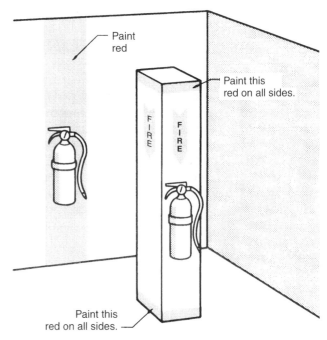

FIGURE 1–46 *The locations of fire extinguishers should be marked so they can be located easily in an emergency.*

FIGURE 1–47 *Point the fire extinguisher at the material burning, not at the flames.*

When using a soda-acid gas cartridge extinguisher: Place your foot on the footrest and direct the stream at the base of the flames (Figure 1-47).

When using a dry chemical extinguisher: Direct the extinguisher at the base of the flames. In the case of type A fires, follow up by directing the dry chemicals at the remaining material that is burning.

1.16 VEHICLE HANDLING IN THE SHOP

When handling a vehicle in the shop, keep the following safety precautions in mind:

- Set the parking brake when working on the vehicle. If the car has an automatic transmission, set it in Park unless instructed otherwise for a specific service operation. If the vehicle has a manual transmission, it should be in Reverse (engine off) or Neutral (engine on) unless instructed otherwise for a specific service operation.
- Use safety stands whenever a procedure requires work under the vehicle (Figure 1-48).
- To prevent serious burns, avoid contact with hot metal parts such as the radiator, exhaust manifold, tailpipe, catalytic converter, and muffler.
- Keep clothing and yourself away from moving parts when the engine is running, especially the radiator fan blades and belts (Figure 1-49). Electric cooling fans can start to operate at any time by an increase in temperature under the hood even though the ignition is in the Off position. Therefore, care should be taken to insure that the electric cooling fan is completely disconnected when working under the hood.
- Be sure that the ignition switch is always in the Off position unless otherwise required by the procedure.
- When moving a vehicle around the shop, be sure to look in all directions and make certain that nothing is in the way.
- Do not smoke while working on vehicles.
- Remove the battery ground lead before you start any welding.

FIGURE 1-48 *Use safety stands whenever working under a vehicle. (Courtesy of Larry Maupin)*

 **HORSEPLAY**

Horseplay, running, and otherwise fooling around should not be part of any shop. Horseplay is also unprofessional and wastes time and can result in your being fired.

SUMMARY

The safety of the auto body technician working in the shop is of utmost importance to the industry. A sizable amount of money is spent for the protection of workers. Some large shops may have a safety person, usually a foreman, in charge of shop safety. This person's job is to make sure that all auto body technicians comply with safety rules during production. The proper clothing, shoes, and eye protection are emphasized in these shops. Any worker who does not follow established safety rules is subject to dismissal.

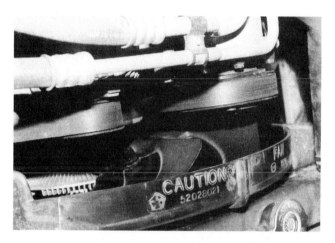

FIGURE 1-49 *Keep fingers and other body parts away from moving parts. Do not turn on the ignition switch (only required by the checking procedure). (Courtesy of Larry Maupin)*

If an accident does occur, it is important that appropriate and immediate first aid steps be taken. All collision repair facilities should have established plans for accidents. You should take time to learn the proper procedures for accident response and reporting before you need to respond in an emergency. After the situation has been properly taken care of, you should fill out an accident report.

Equipment is periodically checked to be sure that it is safe and in proper working condition. Maintenance workers are sometimes employed to see that the equipment is in proper working condition at all times.

Further safety information is available in *Safety for Welders* by Larry F. Jeffus, published by Delmar Publishers Inc.; from the American Welding Society; and from the U.S. Department of Labor (OSHA) Regulations.

REVIEW QUESTIONS

1. What is essential in order to avoid injuries to individuals and damage to property during welding?
2. Whose responsibility is it to make certain that safety concerns are addressed appropriately?
3. What are the three classifications of burns?
4. What color should the welding area be painted to reduce the danger from reflected light?
5. List the types of eye protection.
6. What two ways can ultraviolet light burn the eye?
7. Why should ear protection be used?
8. Some fluxes used in brazing or soldering produce fumes that may irritate what?
9. What are the symptoms of lead poisoning?
10. How are both cadmium and zinc used?
11. When is phosgene formed?
12. What are MSDSs?
13. When is forced ventilation required?
14. When do most electric shock accidents in the welding industry occur?
15. What are some of the safety precautions used for welding cables and grounds?
16. How do you properly ground a three-prong grounding-type plug to a two-prong receptacle?
17. Why is it beneficial to true a grinding stone?
18. How does a routine schedule of equipment maintenance save equipment breakdown time or prevent injury?
19. What are green gas hoses used for?
20. What is the best material to use in protective clothing?

Chapter
2

Restoring a Vehicle's Structural Integrity

Objectives

After reading this chapter, you should be able to:

- List the general guidelines used to determine whether a part should be repaired or replaced.
- Name the two basic procedures used for part replacement.
- Explain how to remove structural panels.
- Describe how to install new panels.
- Identify caution areas to be avoided when making cuts for a section.
- Identify the basic types of sectioning joints.
- Explain what is involved in the preparation for sectioning.
- Explain the sectioning methods used for different structural members.
- Define a full-body section.

Most late-model vehicles are manufactured as unibody structures. Older vehicles have a separate frame; new vehicles, however, have a structure that integrates the body and frame as one unit or body. The load-bearing components of a frame are called structural members. In unibody construction, both the frame and body are structural, load-bearing members.

Modern material technology allows manufacturers to custom-design components so that they serve as both aesthetic and structural components of the vehicle. With unibody construction, some exterior surfaces, including fenders, doorskins, and roof panels, function as part of the structural component. This is very similar to the function of the skin of an aircraft, which gives it shape, form, and strength.

These manufacturing advances have combined to produce attractive and functional vehicles that are more difficult to repair than those built with the old technology. Cars manufactured before the introduction of these design concepts and materials could easily be repaired using low-tech welding processes such as gas welding, brazing and stick, or shielded metal arc welding. Many of these vehicles' surface

panels could be readily repaired and made as good as new with welding.

All structural members of a collision-damaged vehicle must be restored or replaced to bring them to their original structural integrity. Such repairs are required so that the vehicle's passengers will receive the protection that the vehicle was designed to provide. If another accident occurred that resulted in injury or death to anyone in the car, the shop where the repairs were made could be held liable if improper repair techniques were used.

WORKING WITH HIGH-STRENGTH STEEL

For many years, low-carbon or "mild" steel was used in most automotive structural applications. This type of steel has a carbon content of 0.30 percent or less, making it extremely weldable. (The higher the carbon content of the steel, the more difficult it is to weld.)

However, by the mid-1970s carmakers had begun to design smaller, unitized vehicles in an effort to reduce vehicle weight and thereby improve fuel economy. Low-carbon steel was now inadequate for the structural members, which were forced to handle far greater load-carrying and energy-absorption requirements (Figure 2-1). Stronger, lighter, high-strength steels filled these needs while also improving crashworthiness.

One of the things that complicates structural repairs and even some cosmetic repairs to external sheet metal is not knowing exactly where high-strength steels are apt to be found. Lists and diagrams are available from the vehicle manufacturers and aftermarket sources that specify the locations of special steels on various makes and models (Figure 2-2). It can be a time-consuming job to find the information needed. One good way to approach the

FIGURE 2–1 Low-carbon steel was inadequate for the structural members of the smaller, unitized cars built after the mid-1970s. (Courtesy of Larry Maupin)

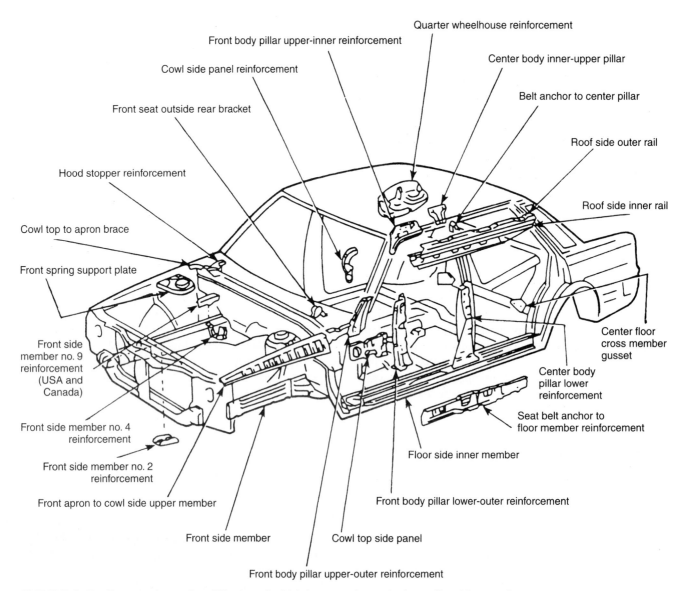

FIGURE 2–2 Because it can be difficult to find high-strength steel, charts like this are often necessary.

problem is to treat all late-model thin-gauge panels and structural members as if they were made from high-strength steel.

MILD VERSUS HIGH-STRENGTH STEEL

It is important to treat high-strength steels differently than ordinary mild steel for two reasons. First and foremost is heat sensitivity. Although some high-strength steels can withstand temperatures of 1,200°F (650°C) for up to 3 minutes without weakening significantly, others have temperature limits as low as 700°F to 900°F (370°C to 480°C). Most, but not all, high-strength steels are heat sensitive and will be weakened if heated excessively. Naturally, this presents a unique challenge to the auto repair industry. Such steels cannot be welded with an oxyacetylene torch. The recommended method of welding in this case is with a MIG welder.

What many body technicians do not realize is that when highly sensitive steel is subjected to temperatures in excess of 1,200°F, (650°C), it is, for all intents and purposes, converted to mild steel. What this means is that a weld or section of a car that was originally designed with a yield strength of 65,000 psi and sections of 20-gauge high-strength steel is now much weaker than anticipated. Upon visual inspection and subjecting the panel to hammering and prying, it may very well appear undamaged and perfectly capable of meeting original specifications. However, the 20-gauge high-strength steel has now been converted to 20-gauge mild steel and is no longer capable of withstanding the high stresses required. This is now a potentially dangerous situation that can cause serious damage to the car and injury to the occupants if the car should be involved in another collision, or even hit a large pothole. Simply put, this section of the car will now yield much earlier than its engineers originally intended in its design.

The second reason for treating high-strength steels differently is because most are brittle. A hard steel will not pull like a mild steel: the harder the steel, the more it tends to tear or crack if straightened. With some of the new high-strength steels, it is safe to use heat to relax stress when pulling and straightening, but with other steels the heat will weaken the metal too much.

TYPES OF HIGH-STRENGTH STEEL

In spite of the fact that most types of steel look alike to the naked eye, there can be considerable differences in chemical makeup and crystalline structure that affect strength and sensitivity to heat. There is a variety of high-strength steels, all of which have

unique properties that relate back to the way in which they can or cannot be repaired.

The following types of steel can be found in this class:

- Chromium alloy
- Carbon molybdenum
- Chromium molybdenum
- Chromium vanadium
- Manganese alloy
- Nickel molybdenum
- Manganese molybdenum
- Nickel chromium
- Nickel chromium molybdenum
- Nickel copper

Other high-strength, low-alloy steels are available under a variety of trade names and are listed in the _Metals Handbook_ published by the American Society for Metals.

Two primary numbering systems have been developed to classify standard grades of steel, including both carbon and alloy steels. These systems classify the types according to their basic chemical composition. One classification system was developed by the Society of Automotive Engineers (SAE); the other is sponsored by the American Iron and Steel Institute (AISI). The so-called Unified Numbering System currently being promoted for all metals may eventually replace these systems.

The numbers used in both systems are now just about the same. However, the AISI system uses a letter before the number to indicate the method used in the manufacture of the sheet. Both systems usually have a four-digit series of numbers. In some cases, a five-digit series is used for certain alloy steels. The entire number is a code for the approximate composition of the steel.

In both classification systems, the first number usually refers to the basis type of steel, as follows:

1XXX Carbon
2XXX Nickel
3XXX Nickel chrome
4XXX Molybdenum
5XXX Chromium
6XXX Chromium vanadium
7XXX Tungsten
8XXX Nickel chromium vanadium
9XXX Silicomanganese

The first two digits together indicate the series within the basis alloy group. There may be several series within a group, depending upon the amount of principal alloying elements. The last two or three digits refer to the approximate permissible range of carbon content.

The letters in the AISI system, if used, indicate the manufacturing process as follows:

- **C.** Basic open-hearth or electric furnace steel and basic oxygen furnace steel
- **E.** Electric furnace alloy steel

Table 2-1 shows the AISI and SAE numerical designations of alloy steels. Note that the elements in the table are expressed in percentages.

TABLE 2–1:	AISI AND ASE ALLOY STEEL DESIGNATIONS*
13XX	Manganese 1.75
23XX**	Nickel 3.50
25XX**	Nickel 5.00
31XX	Nickel 1.25; chromium 0.65
E33XX	Nickel 3.50; chromium 1.55; electric furnace
40XX	Molybdenum 0.25
41XX	Chromium 0.50 or 0.95; molybdenum 0.12 or 0.20
43XX	Nickel 1.80; chromium 0.50 or 0.80; molybdenum 0.25
E43XX	Same as above, produced in basic electric furnace
44XX	Manganese 0.80; molybdenum 0.40
45XX	Nickel 1.85; molybdenum 0.25
47XX	Nickel 1.05; chromium 0.45; molybdenum 0.20 or 0.35
50XX	Chromium 0.28 or 0.40
51XX	Chromium 0.80, 0.88, 0.93, 1.00
E5XXXX	High carbon; high chromium; electric furnace bearing steel
E50100	Carbon 1.00; chromium 0.50
E51100	Carbon 1.00; chromium 1.00
E52100	Carbon 1.00; chromium 1.45
61XX	Chromium 0.60, 0.80, or 0.95; vanadium 0.12, or 0.10, or .015 minimum
7140	Carbon 0.40; chromium 1.60; molybdenum 0.35; aluminum 1.15
81XX	Nickel 0.30; chromium 0.40; molybdenum 0.12
86XX	Nickel 0.55; chromium 0.50; molybdenum 0.20
87XX	Nickel 0.55; chromium 0.50; molybdenum 0.25
88XX	Nickel 0.55; chromium 0.50; molybdenum 0.35
92XX	Manganese 0.85; silicon 2.00; 9262 chromium 0.25 to 0.40
93XX	Nickel 3.25; chromium 1.20; molybdenum 0.12
98XX	Nickel 1.00; chromium 0.80; molybdenum 0.25
14BXX	Boron
50BXX	Chromium 0.50 or 0.28; boron
51BXX	Chromium 0.80; boron
81BXX	Nickel 0.33; chromium 0.45; molybdenum 0.12; boron
86BXX	Nickel 0.55; chromium 0.50; molybdenum 0.20; boron
94BXX	Nickel 0.45; chromium 0.40 molybdenum 0.12; boron

*Consult current AISI and SAE publications for the latest revisions
**Nonstandard steel

STRENGTH

Before you can understand the differences between the types of high-strength steel, *strength* must be defined. There are four basic kinds of strength, all of which relate to the ability of the steel to resist permanent deformation (Figure 2-3).

- **Tensile Strength.** This is the property of a material that resists forces applied to pull it apart. Tension includes both yield stress and ultimate strength. Yield stress is the amount of strain needed to permanently deform a test specimen; ultimate strength is a measure of the load that breaks a specimen. The tensile strength of a metal can be determined by a tensile testing machine.
- **Compressive Strength.** This is the property of a material to resist being crushed.
- **Shear Strength.** This is a measure of how well a material can withstand forces acting to cut or slice it apart.
- **Torsional Strength.** This is the property of a material that withstands a twisting force.

Strength is expressed in pounds per square inch (psi) or kilograms per millimeter squared (kg/mm^2). Heat treatment, cold rolling, and chemical additives are among the manufacturing procedures used to increase the strength of steel. Any further heating beyond the set time and temperature limits can alter the strength significantly and permanently. These limits are determined by the type and extent of the heat and chemical treatment.

HIGH-STRENGTH STEEL CLASSIFICATIONS

There are several types of steel within the classification of high strength, all of which must be treated with extreme care when heating (Figure 2-4). Those most often used in automotive applications are:

- High–tensile strength steel (HSS)
- High-strength, low-alloy steel (HSLA)
- Martensitic steel

Whenever possible, check with the automobile manufacturer regarding the acceptable heat ranges for specific components. Also, be sure to follow the manufacturer's instructions for restoring corrosion protection after any repairs are completed.

HSS Steel

High–tensile strength, or HSS, steel is used in the structural components and some exterior panels of many Japanese imports; its strength is derived from heat treatment. Because HSS has a yield strength of

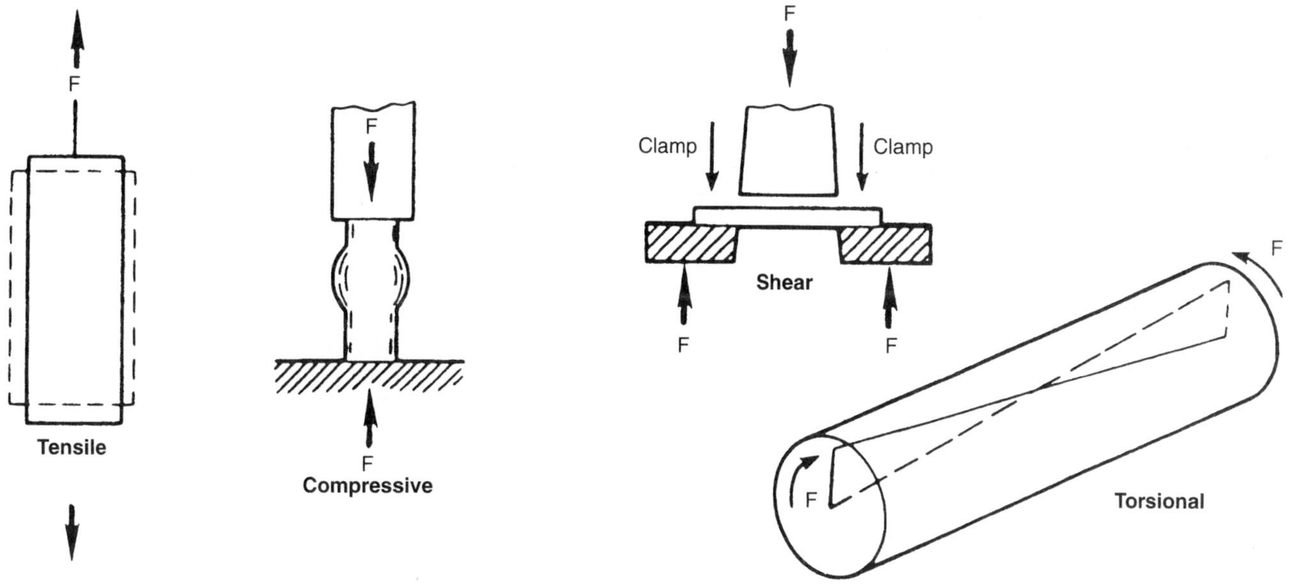

FIGURE 2–3 Types of strength and how they react to force.

more than 34,000 psi and a tensile strength of at least 45,000 psi, conventional heating and welding methods do not weaken it. When high–tensile strength steel is deformed (Figure 2-5), it experiences an increase in stress that exceeds its yield strength. When heat is applied to assist in straightening it, the stresses resulting from the collision are decreased and the strength of the material is restored to precollision levels. HSS steel will tear or fracture if the collision stresses exceed the tensile strength.

Normal MIG welding procedures are used to restore HSS components. Keep in mind that door guard beams cannot be straightened and should instead be replaced; the same is true of some bumper reinforcements. All replacement parts should be MIG-welded using AWS-E-705-6 wire, because it possesses the same strength level as HSS.

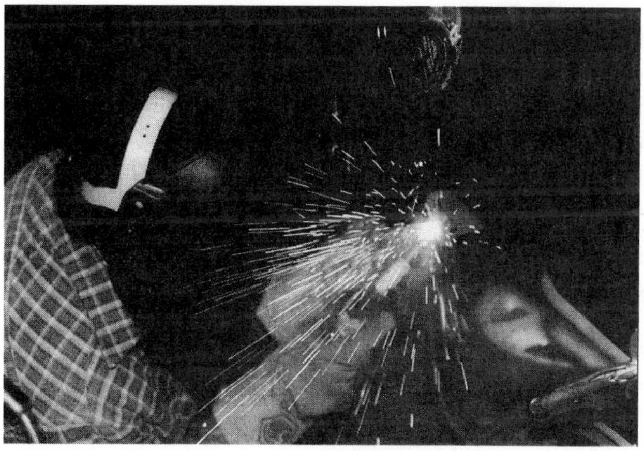

FIGURE 2–4 All heating and welding of high-strength steel must be done with care. (Courtesy of Larry Maupin)

HSLA Steel

High-strength, low-alloy steel, or HSLA, is more often than not found in domestic vehicles. HSLA is used in structural components, including front and rear rails, rocker panels (Figure 2-6), bumper face bars, bumper reinforcements, door hinges, and lock pillars. Special chemical elements added to the steel give HSLA its high strength. After exposure to temperatures of 1,200°F (650°C) or higher for several minutes (the temperature limit can be lower for some parts), the special hardening elements are absorbed by larger, softer elements in the heated area. The result is reduced strength, which in turn substantially reduces the ability of the part to react to normal loads and collision forces. Use AWS-E-705-6 wire when MIG-welding HSLA steel.

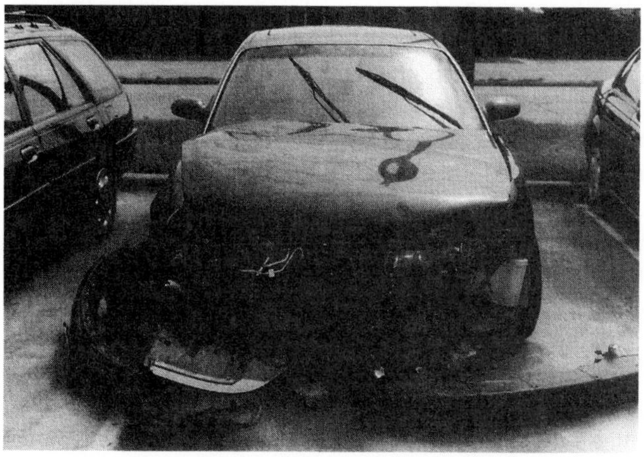

FIGURE 2–5 An example of HSS steel deformed in a collision. (Courtesy of Larry Maupin)

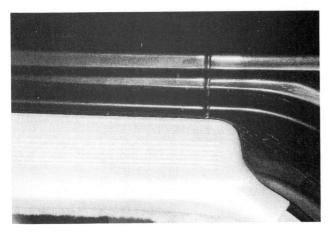

FIGURE 2–6   Rocker panels are frequently made from HSLA steel. (Courtesy of Larry Maupin)

FIGURE 2–7   Door guard beam. (Courtesy of Larry Maupin)

Martensitic Steel

In some domestic models, ultra–high-strength Martensitic steel is used in door guard beams, bumper face bars, and reinforcements. Because it is very resistant to atmospheric corrosion, this type of steel is ideal wherever high strength and wear resistance are required. Martensitic steel is so heat sensitive that heat cannot be applied to it for purposes of repair. Therefore, any damaged parts must be replaced; they cannot be straightened. Replacement parts should be installed by MIG plug-welding with AWS-705-6 wire.

2.2   REPAIR OR REPLACEMENT

In the event of relatively minor damage to high-strength steel, a judgment must be made regarding the question of repair versus replacement. For example, if damage to a door guard beam (Figure 2-7) does not interfere with door alignment or function, it can be ignored. However, if the corrugations in the beam are damaged, the door beam must be replaced. Bumper face bar reinforcements that are covered by separate fascias do not require replacement in the event of minor damage, provided the operation of the bumper is unaffected. Plastic filler and sealer can be applied to shallow dents to restore the original contour.

Changes in insurance companies' policies have significantly affected most auto body repairs. Stricter state and federal rules and regulations and laws affecting the restoration of vehicles have also affected repairs. The stricter policies are the result of the desire to ensure that all repairs made to motor vehicles are done in a way that does not adversely affect the future performance and safe operation of the repaired vehicle. Additionally, changes in vehicle design and construction have occurred largely because of the need to make a safer, more fuel-efficient vehicle. The introduction of high-strength, low-alloy steels (HSLA) and high-strength steel (HSS) into the vehicles has enabled manufacturers to strengthen components while making them thinner. The thinner the components, the lighter the vehicle. While these changes have been beneficial in the design and construction of vehicles, their universal application has resulted in vehicles containing components that are often impractical or impossible to repair and therefore must be replaced.

Today's auto body technician will see more replacement of metal parts, since automobile manufacturers are using more materials such as high-strength steels and thinner-gauge metals that may be impractical to repair. These metals are also more prone to crack during pulling and straightening. High-strength steels are less ductile than mild steels, so they will not straighten as easily.

In addition to changes in vehicle construction, new insurance rules, and increased operating costs, changes in the collision repair industry itself have brought about changes in the way damaged vehicles are repaired. Some of those changes include dealer-specific shops where virtually all of the work performed is on a single manufacturer's vehicles. Large shops specialize in new or nearly new vehicles. Such large shops can employ helpers to assist the auto body technician with such matters as vehicle teardown and component reassembly following repairs.

On the other end of the spectrum are smaller shops and those located outside major metropolitan areas that work on any vehicle brought in for repair. There are also shops that specialize in the restoration of classic or antique vehicles. In these shops, welding repairs on components is more widespread, since many replacement components are no longer available or have become prohibitively expensive.

This text will cover all aspects of welding, from the skills required in high-volume production repair shops as to those needed by the classic car repair technician. When appropriate, reference will be made to a process as "commonly used on most vehicles" or "not commonly used today." As you become a well-rounded and talented auto body repair technician, you should master all of the welding techniques for both newer and older vehicles. For some new large and heavy vehicles such as trucks, buses, and some recreational vehicles, which are constructed of thicker metal members, some low-tech welding processes are still used.

When a component must be replaced, there are two basic procedures that can be used to accomplish a quality restoration. The first procedure is to replace the structural member at the factory seams with a new or used part. The second procedure is called *sectioning,* which is the replacement of a component at practical locations other than the factory seams.

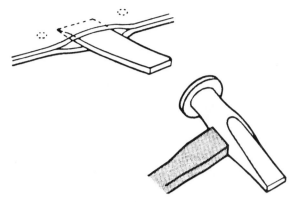

FIGURE 2–8   _Separating a seam with a chisel will make the spot welds show up._

REPLACEMENT AT FACTORY SEAMS

2.3

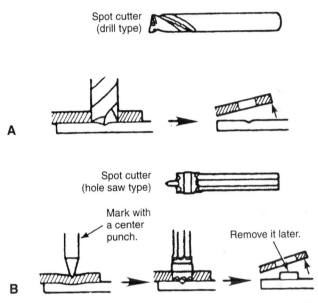

FIGURE 2–9   _Spot weld cutters: (A) drill type and (B) hole saw type._

Structural panels should be replaced at factory seams whenever it is possible and practical. This straightforward method is still the first choice when replacing a structural panel. Body repair manuals are available from many automobile manufacturers. These manuals usually contain the instructions for panel replacement. Use them whenever they are available. This whole operation can be broken down into two segments: removal of the damaged panels and installation of the new.

REMOVING STRUCTURAL PANELS

Structural body panels are joined together in the factory by spot welding. Therefore, removing panels mainly involves the separation of spot-welded seams. Spot welds can be drilled out, blown out with a plasma torch, chiseled out, or ground out with a grinding wheel. The best method for removing a spot-welded panel must be determined by the number and arrangement of the mating panels.

Some spot weld seams may have a number of layers of sheet metal. The removal process is determined by the position of the weld and the arrangement of the panels.

Determining Spot Weld Positions

It is not usually necessary to remove the paint film, undercoat, sealer, or other coatings covering the joint area in order to find the spot welds. If it is necessary to clean the part to locate the spot welds, you can use one of several methods. You can scorch the paint film with an oxyacetylene or propane torch and brush it off with a wire brush. A coarse wire wheel or brush attached to a drill can also be used to grind off the paint.

Scrape off thick portions of undercoating or sealer before scorching the paint. Do not burn through the paint film so that the sheet metal panel begins to turn color. Heat the area just enough to soften the paint and then brush or scrape it off. It is not necessary to remove paint from areas where the spot welds are visible through the paint film.

In areas where the spot weld positions are not visible after the paint is removed, drive a chisel between the panels as shown in Figure 2-8. Doing so will cause the outline of the spot welds to appear.

Separating Spot Welds

After the spot welds have been located, they can be drilled out. Two types of cutting bits can be used: a drill type or a hole saw type Figure 2-9. Table 2-2

TABLE 2–2:   SEPARATION OF SPOT WELDS

Type			Application Method	Characteristics
Spot Cutter	Drill Type	Small	Places where the replacement panel is between other panels and welding cannot be done from the backside. Places where the replacement panel is on top and the weld is small.	The separation can be accomplished without damaging the bottom panel. Since the nugget is not left in the bottom panel, finishing is easy.
		Large	When the replacement panel is on top. When the panel is thick (places where nuggets are large). Places where the weld shape is destroyed.	
	Hole Saw Type		When the replacement panel is on top.	Separation can be accomplished without damaging the bottom panel. Since only the circomference of the nugget is cut, it is necessary to remove the nugget remaining in the bottom panel after the panels are separated.
Drill			When the replacement panel is on bottom. When the replacement panel is between and welding can be done from the backside. (Select a drill diameter that is appropriate for the panel thickness and the weld diameter.)	Lower cost. Recently, a labor-saving spot weld removing tool has been developed that is easy to use and has a built-in attaching clamp.
Self-Starting Multiple-Hole Drill			Same as above.	One drill can be used to cut out multiple-sized spot welds.

shows when each type should be used. Regardless of which is used, be careful not to cut into another panel, be sure to cut out the plugs, and avoid creating an excessively large hole.

Both the spot cutter and hole saw–type tools will allow you to cut through one or more layers without completely penetrating the entire joint assembly (Figure 2-10). Being able to remove only those panel sections that need replacement while leaving the remaining sections intact has several advantages. First, not completely disassembling the seam can result in a faster reassembly. It also insures that the alignment remains intact between those parts not removed. Second, not drilling through the entire joint enables new MIG spot welds to be made without a backing plate.

When using a standard drill bit or a self-starting multiple-hole drill to remove spot welds, you will be drilling completely through the entire spot-welded seam (Figure 2-11). When selecting a drill bit, try to choose one that is only slightly larger than the diameter of the spot welds. The advantage of using a multiple-hole drill is that the farther the drill is pushed through the panel, the larger the hole it cuts (Figure 2-12). Although the multiple-hole drill bits are

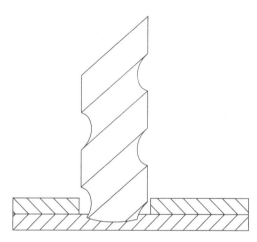

FIGURE 2–10 Drill out one side of a spot weld to separate the panels.

FIGURE 2–11 Drilling out entire weld. (Courtesy of Larry Maupin)

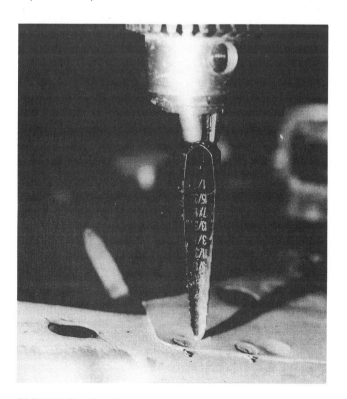

FIGURE 2–12 The deeper the multiple-hole drill is driven, the larger the hole. (Courtesy of Larry Maupin)

more expensive than standard bits, they are commonly used because the entire spot weld can be removed in spite of slight changes in size that might occur.

Portable punches are available and can be used on some spot weld seams to remove the weld (Figure 2-13). These punches can be operated manually, pneumatically, or hydraulically. A major disadvantage with punches is that they cannot be used on all spot weld seams. The punch can only be used on seams in which the flange will allow the anvil portion of the punch to fit securely against the back of the weld (Figure 2-14). When punches can be used, they are considerably faster than drilling.

A plasma arc cutting torch can also be used to remove spot welds. A plasma torch generates a concentrated arc that has a temperature of approximately 43,000°F (24,000°C) (Figure 2-15). In addition to the very high temperature, the plasma arc is concentrated into a small area, which results in the instantaneous vaporization of any metal it comes into contact with. This thermal cutting process is similar to oxyacetylene cutting except that this concentrated arc does not require preheating time before cutting can begin. This results in very little temperature increase in the metal being cut, which results in such a small area being heated that no heat-related distortion occurs (Figure 2-16). Plasma arc cutting, therefore, is a very rapid method of removing spot welds; however, it requires a higher skill level than drilling, punching, or grinding.

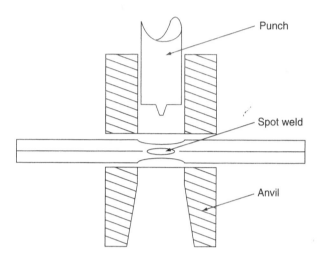

FIGURE 2-14 A portable punch can be used to remove spot welds if the anvil portion fits snugly against the back of the weld.

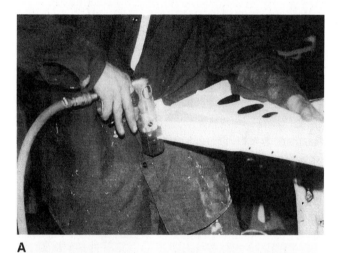

A

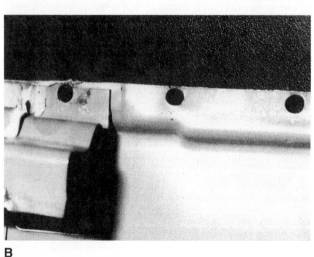

B

FIGURE 2-13 (A) Pneumatic punch; (B) punched holes. (Courtesy of Larry Moupin)

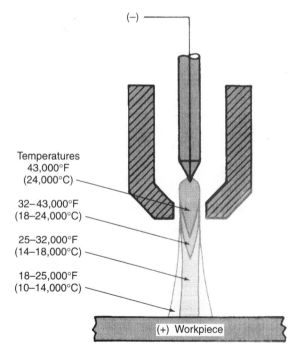

FIGURE 2-15 Approximate temperature of a plasma arc.

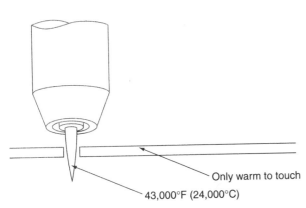

FIGURE 2–16 During a good plasma cut, the plate remains cool even within an inch of the cut.

Only warm to touch

43,000°F (24,000°C)

A grinder can be used to remove spot welds by grinding through to the desired thickness, enabling the layer or layers to be removed (Figure 2-17). Grinding is the slowest method used to remove spot welds and therefore is only employed when the other methods are not practical. Grinding is used when it would be difficult or impossible to align a drill to remove the spot weld. It is also used when only the top layer or layers are to be removed and the depth of cutting provided by a plasma cutting torch would be undesirable. Grinding using a grinding stone or dye grinding tool is most frequently used to remove the last few spot welds on a panel that are not readily accessible with the process you have been using.

Hole saws and weld-removing tools can be used to remove the spot welds or to cut through the surrounding metal to release the top panel or panels without making a through-thickness cut. The primary advantage of using these types of tools is that it is not necessary to disassemble the entire joint.

After the spot welds have been drilled out, blown out, or ground down, drive an air chisel between the panels to separate them. Be careful not to cut or bend the undamaged panel (Figure 2-18). Sometimes a long spot-welded joint can be removed with a pneumatic chisel (Figure 2-19). This makes it possible to then remove the spot welds using a punch or other tool.

Separating Continuous Welds

In some vehicles, panels are joined by continuous welding. Since the welding bead is long, use a grinding wheel or high-speed grinder to separate the panels. As shown in Figure 2-20, cut through the weld without cutting into or through the panels, holding the

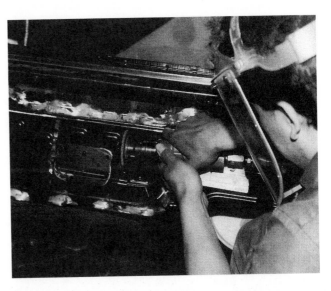

A

B

C

FIGURE 2–17 (A) Grinding a spot weld. (B) A ground spot weld. (C) Removed panel. (Courtesy of Larry Maupin)

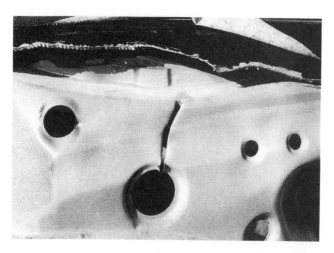

FIGURE 2-18 Underpanel accidently cut with pneumatic chisel while removing spot-welded seam. (Courtesy of Larry Maupin)

FIGURE 2-19 Removing a seam with a pneumatic chisel. (Courtesy of Larry Maupin)

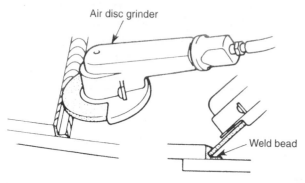

FIGURE 2-20 Using a grinder to remove a continuous weld.

grinding wheel at a 45-degree angle to the lap joint. After grinding through the weld, use a hammer and chisel to separate the panels.

Separating Brazed Areas

Brazing is used at the ends of outer panels or at the joints of the roof and body pillars to improve the finish quality and to seal the body. Small brazed areas are more easily separated than longer brazed joints. For example, the brazed joint found on the front column adjacent to the windshield could be easily separated by remelting the brazed metal column. Longer brazed joints can be separated using a torch; however, the higher heat input on such large surface areas often results in an unacceptable level of distortion. Brazed joints can easily be identified by their brass color.

It is easier to separate brazed areas after all the other welded parts have been separated. In some cases the main portion of the panel can be cut off, leaving only the brazed joint to be removed. First, soften the paint with an oxyacetylene torch and remove it with a wire brush or scraper. Next, heat the brazing metal until it starts to melt and forms a molten pool. Now quickly pull apart the joint. Be careful not to overheat the surrounding sheet metal. If the joint does not separate easily, then drive a chisel or screwdriver between the panels and separate them (Figure 2-21). Keep the panels separated until the brazing metal cools and hardens.

On long joints, a high-speed grinder and grinding wheel can be used to cut through the brazed joint (Figure 2-22). If you are replacing the top panel, do not cut through the panel underneath. The top portion of this panel must remain intact to provide support for the replacement part. After grinding through the braze, separate the lapped panels by using a chisel and a hammer.

INSTALLING NEW PANELS

Before welding replacement panels, you must do thorough preparation and carefully align the new parts.

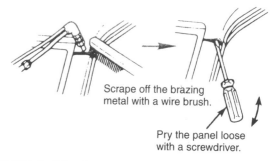

Scrape off the brazing metal with a wire brush.

Pry the panel loose with a screwdriver.

FIGURE 2-21 Separating panels connected by arc brazing.

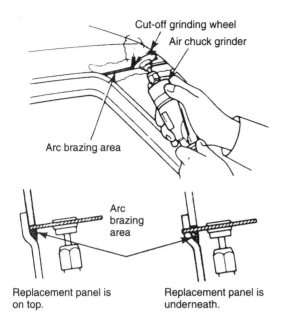

FIGURE 2–22 *Separating brazed joints with a grinder.*

The following procedure is typical of many panel replacement operations. Always refer to the appropriate collision repair manual provided by the manufacturer for the type and placement of welds.

Vehicle Preparation

After removing the damaged panels, prepare the vehicle for the installation of the new panels. To do this, follow these steps:

1. Grind off the weld metal from the spot-welded areas (Figure 2-23). Do not grind the flanges of structural panels. Grinding will remove metal, thinning the section and weakening the joint. Use a wire brush to remove dirt, rust, paint, sealers, zinc coatings, and other material from the joint surfaces.
2. On parts that will be spot welded during installation, remove paint and undercoating from the back sides of the panel-joining surfaces.
3. Smooth the dents and bumps in the mating flanges with a hammer and dolly (Figure 2-24).
4. Apply weld-through primer to areas where the base metal is exposed after the paint and rust have been removed from the joining surfaces (Figure 2-25). It is very important to apply the primer to joining surfaces or to areas where painting cannot be done later.

Replacement Panel Preparation

If the parts are to be electric resistant spot welded, the primer and zinc coating must be removed from the

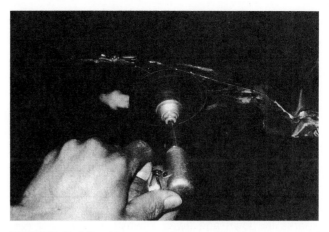

FIGURE 2–23 *After removing panels, clear away old weld metal with a grinder. (Courtesy of Larry Maupin)*

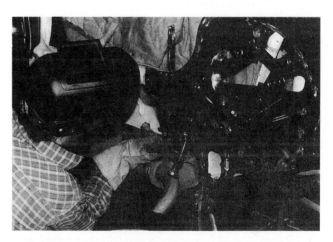

FIGURE 2–24 *Light taps can be used to straighten flange for welding. (Courtesy of Larry Maupin)*

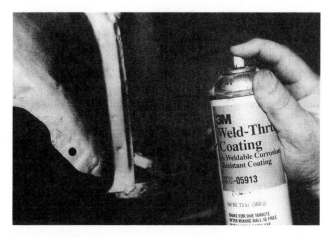

FIGURE 2–25 *Spraying weld-through primer. (Courtesy of Larry Maupin)*

mating flanges to allow the welding current to flow properly during spot welding. If the parts are to be MIG plug welded, then drill holes for plug welds. Plug holes are usually 5/16 inch (8 mm) in diameter. To prepare the new panel for welding, follow these steps:

1. Use a disc sander to remove the paint from both sides of the spot welding area. Do not grind into the steel panel and do not heat the panel so that it turns blue or begins to warp.

CAUTION: Wherever possible, grind so that sparks fly down and away. Always wear proper eye protection when grinding.

---

2. Apply weld-through primer, as an antirust treatment to the welding surfaces where the paint was removed. Apply the weld-through carefully so that it does not ooze out from the joining surfaces. If it does ooze out, it will have a detrimental effect on painting. So remove any excess with a solvent-soaked rag.

3. Make holes for plug welding with a punch or drill. Always refer to the specific colllision repair manual for the required number of plug welding holes. Duplicate the number and location of factory spot welds. Be sure to make plug welding holes the proper diameter. If the size of the holes is too large or too small for the thickness of the panel, either the metal will melt through or the weld will be inadequate (Table 2-3). When drilling high-strength steel, use a variable-speed drill and low RPMs, and strive for uniform cutting. If the metal chips begin to change color to gold, purple, or blue, the RPMs should be reduced. Use high-speed or carbide drill bits for drilling high-strength steel.

Positioning New Panels

Aligning new parts with the existing body is a very important step in repairing vehicles with major damage. Improperly aligned panels can affect both the appearance and the driveability of the repaired vehicle. Basically, there are two methods of positioning body panels. One is to use dimension-measuring instruments to determine the installation position. The other is to determine the position by the relationship between the new part and the surrounding panels. The dimensional accuracy of certain structural panels—such as those that make up the engine compartment, the fender aprons, or the front and rear side members—has a direct effect on wheel alignment and driving characteristics. Therefore, when replacing structural panels in unibody vehicles, use the dimensional measurement positioning method, because it is more

TABLE 2–3:   PLUG HOLE DIAMETER FOR RESPECTIVE WELDED PORTION

Welded Area Panel Thickness	Plug Hole Diameter
Less than 1/32″	3/16″–1/4″
1/32″–3/32″	1/4″–13/32″
More than 3/32″	13/32″ or more

accurate. There is also a definite relationship between the fit of the new and old parts and the finish appearance, so whether structural or cosmetic panels are being replaced, the emphasis is on proper fit. Of course, it is desirable to use both methods together, and thus to assure the accuracy and the finish necessary for a high-quality vehicle repair.

CAUTION: All measurements must be accurate before finish-welding structural panels in position, regardless of the measurement system used for replacing structural panels. Because there are no shims on unibodies, no adjustments can be made to the outer panels. Therefore, each panel must be precisely positioned before welding.

Clamping Tools. Locking jaw (vise) pliers, C-clamps, sheet metal screws, and special clamps are all necessary tools for good welding. Clamping panels together correctly will require close attention to detail (Figure 2-26). As shown in Figure 2-27, a hammer can often be used to fit panels closely together in places that cannot be clamped completely together.

Clamping both sides of a panel is not always possible. In these cases, a simple technique using sheet metal screws can be employed. To hold panels together with sheet metal screws, punch and drill holes through the panel closest to the operator. In the case of plug welding, every other hole is filled with a screw. The empty holes are then plug welded. After the original holes are plug welded, the screws are removed and the remaining holes are plug welded (Figure 2-28).

Fixtures can also be used in some cases to hold panels in proper alignment. Fixtures alone, however, should not be depended upon to maintain tight clamping force at the welded joint. Some additional clamping will be required to make sure that the panels are tight together and not just held in proper alignment.

Positioning by Dimensional Measurement Methods. When you are using a dedicated bench, or a mechanical universal, system of measurement, the vehicle must be properly positioned on the bench before the new panel can be correctly aligned. This will usually have already been done in order to pull

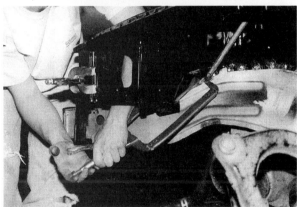

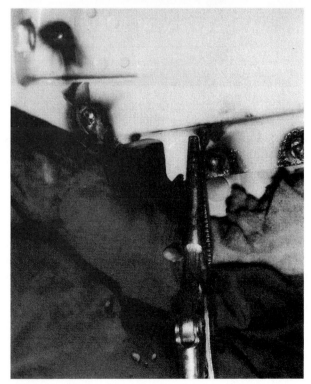

A

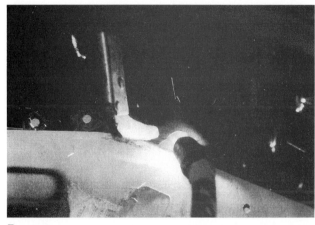

B

FIGURE 2–27 Tapping a spot weld to (A) get a better fit (B) for welding. (Courtesy of Larry Maupin)

FIGURE 2–26 A wide variety of clamping pliers are available. (Courtesy of Larry Maupin)

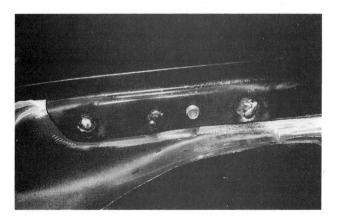

FIGURE 2–28 Self-drilling hex head sheet metal screw holds panel in place for welding. (Courtesy of Larry Maupin)

and straighten damaged panels that do not require replacing. All straightening must be done before replacing any panels. Otherwise, proper alignment of the new panels will be impossible.

So, when using a bench system for panel replacement at a factory joint, place the fixtures or gauges on the bench in their correct locations and tie them to the bench. Then place the new panel in position on the fixtures and see how it lines up with the undamaged panels on the car (Figure 2-29). Make any necessary adjustments, clamp the panel in place, and weld it (Figure 2-30). Measure it again, then plug-weld the panel into place. Grind the welds for appearance, if necessary.

Positioning by Visual Inspection. Nonstructural outer panels can sometimes be visually aligned with adjacent panels without the precise measurements necessary when replacing structural panels. The emphasis here is on appearance. The panel is carefully aligned with adjacent body parts and secured with spot welds.

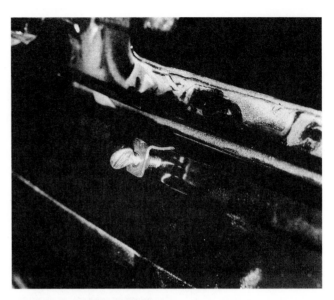

FIGURE 2-29 Thumb screws can be used to temporarily hold panels in place during alignment. (Courtesy of Larry Maupin)

 2.4 ─── **SECTIONING**

Sectioning is a practical and efficient alternative to replacing parts at factory seams. The following procedures are considered proper techniques for unibody structural components. Various manufacturers and independent testing agencies have examined these procedures and found that they provide structurally adequate repairs for the specified areas. However, be sure to follow the carmaker's recommendations on the subject of sectioning.

Major auto manufacturers are continually investigating the sectioning of structural components and are publishing repair procedures as they are developed and verified. Always give careful attention to detailed damage repair techniques where prescribed in technical manuals or notes published by the manufacturers.

This discussion of panel sectioning and replacement covers the repair of the following components: rocker panels, quarter panels, floor pan, front rails, rear rails, trunk floor, B-pillars, and A-pillars (Figure 2-31). These unibody structural components involve two basic types of construction design: closed sections, such as rocker panels, pillars, and body rails (Figure 2-32), and open-surface or single-layer, overlap-joint components, such as floor pans and trunk floors.

The closed-type sections are the most critical, because they provide the principal strength in the unibody structure. They possess much greater strength per pound of material than other types of sections.

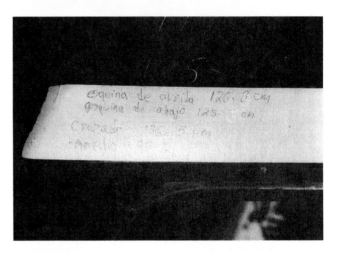

FIGURE 2-30. Writing dimensions on tape can save time for rechecking like this. Top corner 126.8 cm, bottom corner 125.7 cm, across 126.3 cm. (Courtesy of Larry Maupin)

CRUSH ZONES

Certain structural components have crush zones, or buckling points, designed into them for absorbing the impact in a collision. This is particularly true of the front and rear rails, because they take the brunt of the impact in most collisions. Crush zones are in all front and rear rails (Figure 2-33). Identify these crush zones by how they look. Some are in the form of convoluted or crinkled areas, some are in the form of dents or dimples, and others are in the form of holes or slots put in deliberately so the rail will collapse at these points. In virtually all cases, the crush zones are ahead of the front suspension and behind the rear suspension.

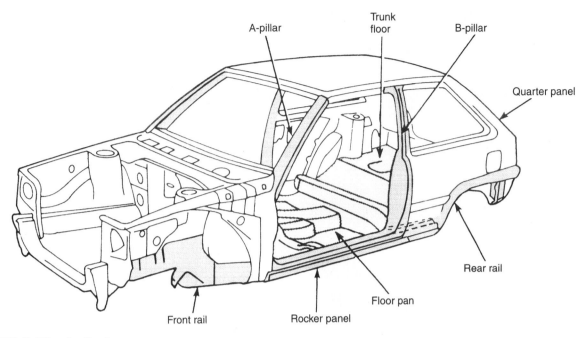

A-pillar

Trunk floor

B-pillar

Quarter panel

Rear rail

Floor pan

Front rail

Rocker panel

FIGURE 2–31 Sectioning areas.

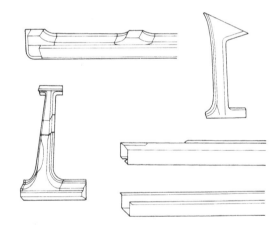

FIGURE 2–32 Closed unibody sections.

Sectioning procedures can change the designed collapsibility if improperly located. So avoid crush zones whenever possible. If a rail has suffered major damage, it will normally already be buckled in the crush zone, so the crush zone will usually be easy to locate. Where only moderate damage has occurred, examine it very carefully. The hit might not have used up the entire crush zone, but if it is damaged, it must be replaced or repaired.

OTHER CAUTION AREAS

There are other areas to stay away from when making cuts. Stay away from holes in the component. Do not cut through any inner reinforcements, meaning double

layers in the metal. Stay away from anchor points, such as suspension anchor points, seat belt anchor points in the floor, and shoulder belt D-ring anchor points. For example, when sectioning a B-pillar, make an offset cut around the D-ring area to avoid disturbing the anchor reinforcement. Other areas to avoid are mounts for the engine or drivetrain and compound shapes.

BASIC TYPES OF SECTIONING JOINTS

Correct structural sectioning procedures and techniques involve three basic types of joints, as well as certain variations and combinations of them. One is a butt joint with an insert. This type is used mainly on closed sections, such as rocker panels, A- and B-pillars, and body rails (Figure 2-34). Inserts help make it easy to fit and align the joints correctly and make the welding process easier.

Another basic type is an offset butt joint without an insert, also known as a staggered butt joint (Figure 2-35). This type is used on A- and B-pillars and front rails. The third type is an overlap joint (Figure 2-36). The lap joint is used on rear rails, floor pans, trunk floors, and B-pillars.

The configuration and makeup of the component being sectioned might call for a combination of joint types. Sectioning a B-pillar, for instance, might require the use of an offset cut with a butt joint in the outside piece and a lap joint in the inside piece.

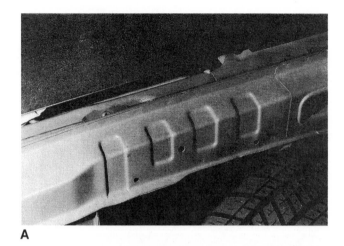

A

B

C

FIGURE 2-33 Crush zones: (A) dimples and slots, (B) holes and dimples, (C) dimples and dents. (Courtesy of Larry Maupin)

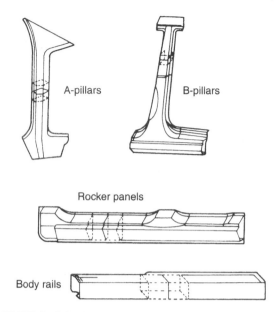

A-pillars   B-pillars

Rocker panels

Body rails

FIGURE 2-34 Butt joints with inserts.

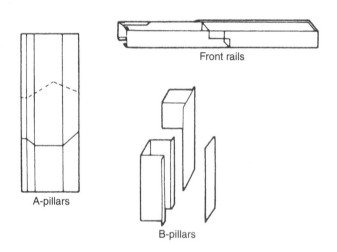

Front rails

A-pillars

B-pillars

FIGURE 2-35 Offset butt joints without inserts.

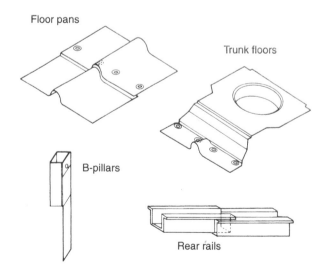

Floor pans

Trunk floors

B-pillars

Rear rails

FIGURE 2-36 Overlap joints.

PREPARING TO SECTION

When the sectioning method is used to replace a damaged part, the replacement part may either be new or recycled. In preparing to section and replace any structural member with a recycled part, certain steps must be taken to ensure the quality of the repair. Specific instructions must be provided to the recycler as to the placement of the section and the method of sectioning to be used. Other important considerations are the welding techniques used and the cleanliness of the joint metal.

USING RECYCLED OR SALVAGED PARTS

When using recycled parts, tell the recycler exactly where to make the cuts. It is preferabe to have the required part removed with a metal saw. If the recycler uses a cutting torch, make sure that at least two inches of extra length are left on the part to insure that the head dispersion from the cut does not invade the joint area. Instruct the recycler to make the cut so that reinforcing pieces welded inside the component are not cut through.

When a recycled or salvaged part is received, examine it for corrosion. If it has a lot of rust on it, do not use it. Before installing a recycled part, check it for possible damage and make sure it is dimensionally accurate.

JOINING PROCESSES

The term _welding_ can be used to encompass all metal-joining processes that use some source of heat, such as a torch or arc. The term, however, specifically refers to a process in which the base metal itself is melted and allowed to flow together, forming a single piece as the molten weld pool solidifies. During welding, additional filler metal may or may not be added to assist in forming a weld bead. Welding is most commonly accomplished using one of the following processes:

- Oxyfuel (OFW, OAW, or gas welding). A group of welding processes that produce welds in workpieces by heating them with an oxy–fuel gas flame. The processes are used with or without the addition of filler metal (Figure 2-37).
- Shielded metal arc welding (SMAW or stick welding). An arc welding process with an arc between a covered electrode and the weld pool. The process is used with shielding from the decomposition of the electrode covering, with filler metal being added from the electrode core metal (Figure 2-38).

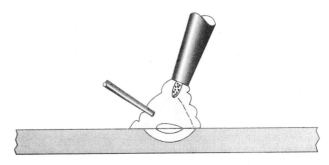

FIGURE 2–37 _Oxyfuel welding._

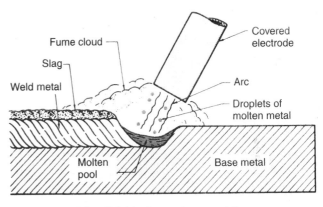

FIGURE 2–38 _Shielded metal arc welding._

- Gas metal arc welding (GMAW or MIG). An arc welding process that uses an arc between a continously fed filler metal electrode and the weld pool. The process is used with a shielding gas to protect the molten weld pool (Figure 2-39).
- Flux-cored arc welding (FCAW) (Figure 2-40).
- Gas tungsten arc welding (GTAW or TIG). An arc welding process that uses an arc between a nonconsumable tungsten electrode and the weld pool. The process is used with shielding gas to protect the molten weld pool (Figure 2-41).
- Spot welding. A weld made between or upon overlapping members in which a weld occurs between the mating surfaces as the result of electrical resistance. The weld cross section is approximately circular (Figure 2-42).
- Brazing. A group of joining processes that produce a bond between heated surfaces when the liquid filler metal is drawn into the joint by capilary action. The process may or may not require a flux, and the filler metal melts above 840°F (450°C) (Figure 2-43).
- Soldering. A group of joining processes that produce a bond between heated surfaces and a liquid filler metal. The process may or may not require a flux, and the filler metal melts below 840°F (450°C) (Figure 2-44).

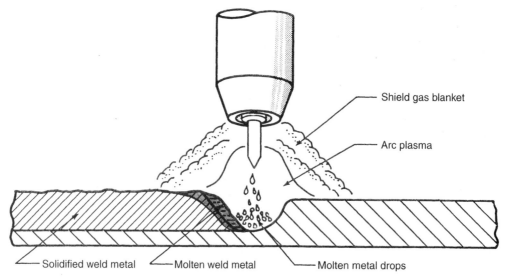

FIGURE 2-39 *Gas-shielded metal arc welding (GMAW).*

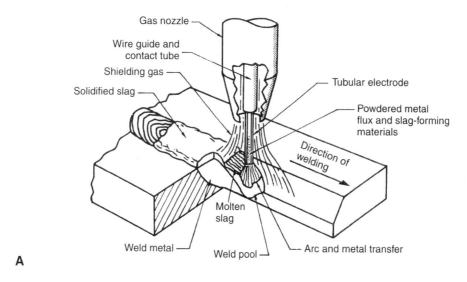

A

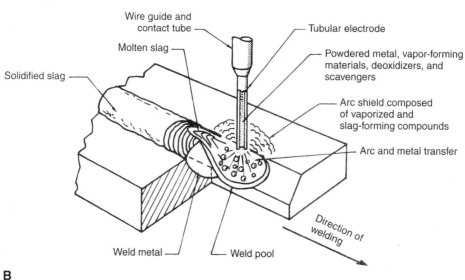

B

FIGURE 2-40 (A) Gas-shielded flux-cored arc welding. (B) Self-shielded flux-cored arc welding. (Courtesy of the American Welding Society)

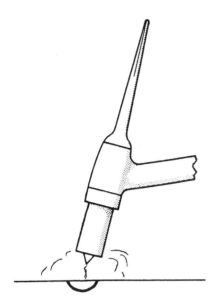

FIGURE 2-41   GTA welding torch.

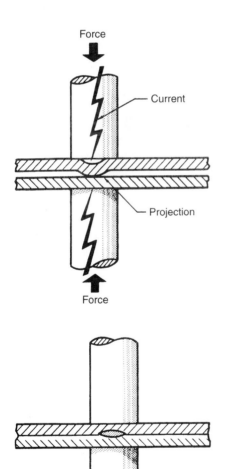

FIGURE 2-42   Spot or projection welding.

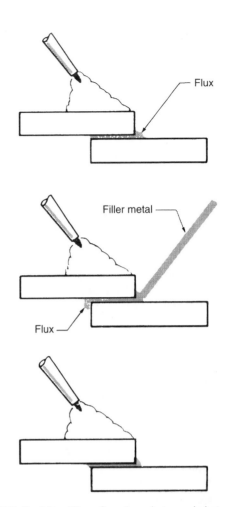

FIGURE 2-43   Flux flowing into a joint reduces oxides to clean the surfaces and gives rise to a capillary action that causes the filler metal to flow behind it.

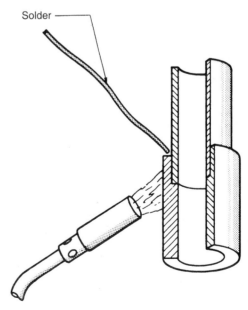

FIGURE 2-44   Soldering copper fitting to copper pipe. (Courtesy of Praxair, Inc.)

- Braze welding. A group of joining processes that produce a bond between heated surfaces when the liquid filler metal is not drawn into the joint by capillary action. The process may or may not require a flux, and the filler metal melts above 840°F (450°C), (Figure 2-45).

The term *welding* is also sometimes used to refer to cutting processes such as:

- Oxyfuel cutting (OFC, OAC, or gas cutting). A group of thermal cutting prosesses that severs or removes metal by means of the chemical reaction between oxygen and the base metal at an elevated temperature. The necessary temperature is maintained by the heat from an arc, an oxyfuel gas flame (Figure 2-46).
- Plasma arc cutting. An arc-cutting process that uses a constricted arc and removes the molten metal with a high-velocity jet of ionized gas issuing from the constricting orifice (Figure 2-47).

SECTIONING BODY RAILS

Virtually all front and rear rails are closed sections. There are two distinct types of closed sections. One is called a closed section; it comes from the factory or the recycler with all four sides intact. Sometimes it is referred to as a *box section* (Figure 2-48). The other type comes as an open hat channel and is closed on the fourth side by being joined to some other component in the body structure (Figure 2-49).

Structural members such as rocker panels, A- and B-pillars, and body rails are closed, both to increase their strength and to prevent internal rusting. Rusting

could significantly reduce the component's strength if it were not sealed completely to prevent moisture from entering. Sealing these areas can be accomplished by using an approved caulking, by producing a contiguous weld around the entire joint, or by both welding and caulking. This type of a weld is referred

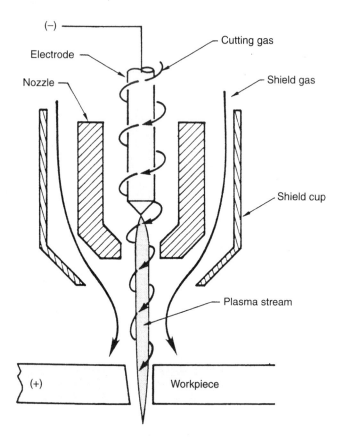

FIGURE 2-47 *The cutting gas can swirl around the electrode to produce a tighter plasma column. (Courtesy of the American Welding Society)*

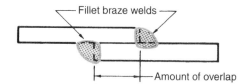

FIGURE 2-45 *A braze-welded lap joint.*

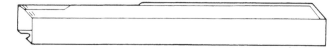

FIGURE 2-48 *A closed rail.*

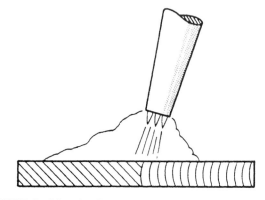

FIGURE 2-46 *Oxyfuel cutting.*

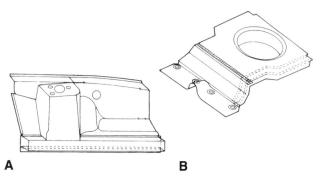

FIGURE 2-49 *Typical hat channels: (A) front rail and (B) rear rail.*

to as a *seal weld*. Welding is the preferred method of insuring that these components remain water-tight, because caulked joints can loosen over time as a result of weathering of the caulk material, improper initial application of the sealant, aging of the caulk material, and mechanical damage.

The butt joint with an insert or backing is the welding procedure for repairing a closed section rail (Figure 2-50). Most rear rails, plus various makes of front rails, are of the hat channel type. Some of the hat channel closures are vertical, such as a front rail joined to a side apron. Some of them are horizontal, such as a rear rail joined to a trunk floor.

In most cases when sectioning the open hat channel type of rail, you should use a lap joint with plug welds in the overlapped areas and a continuous lap weld along the edge of the overlap (Figure 2-51). Whenever sectioning front or rear rails, always remember that they contain crush zones that must be avoided when making cuts.

WELDING

For a seal weld in thin material, an insert or backing plate must be used (Figure 2-52). When closing the job with a butt weld, leave a gap that is wide enough to allow the weld to thoroughly penetrate of the insert. The width of the gap depends on the thickness of the metal, but ideally it should not be less than 1/16 inch (1.5 mm) nor more than 1/8 inch (3 mm) (Figure 2-53). Be careful to remove the burrs from the cut edges before welding. Otherwise, the parts will not fit tightly together, and this can create a flawed weld. A poor weld can result in stress concentration that can cause cracks and weaken the joint (Figure 2-54). The joint is assembled using standard 5/16 in (8 mm)–diameter plug welds so that the edges

of the members contact each other. Large spaces or gaps left between the adjoining members will result in the need to make a larger weld in order to bridge or fill up the gap. This increased weld size will significantly increase the difficulty of this repair.

To insure that a weld bonds all of the layers of metal together, you must make a *full penetration weld*. This is a weld that melts all the way through the metal. When you make a full penetration weld, you should see a slight raise in the surface on the backside of the weld (Figure 2-55).

The metal of a vehicle is so thin that most welding results in the entire thickness of the material becoming liquid, unlike welding on thicker types of metal, where the molten weld pool only penetrates partially

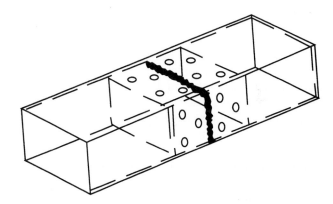

FIGURE 2–52   Butt with insert. (Courtesy of I-Car)

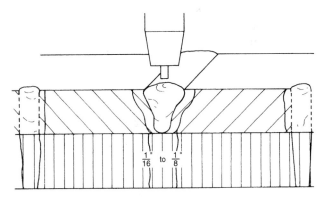

FIGURE 2–53   Butt weld gap.

FIGURE 2–50   A rear rail with a butt joint and insert.

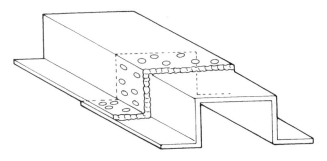

FIGURE 2–51   Joining open rails.

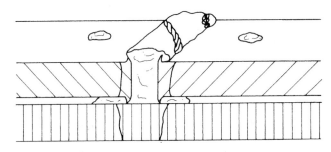

FIGURE 2–54   Burrs in the joint will weaken the weld.

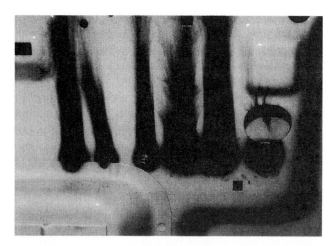

FIGURE 2-55 One hundred percent joint penetration can be seen on the backside of these spot welds. (Courtesy of Larry Maupin)

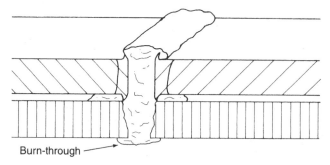

FIGURE 2-56 A burn-through.

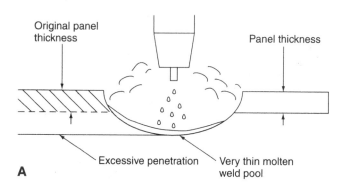

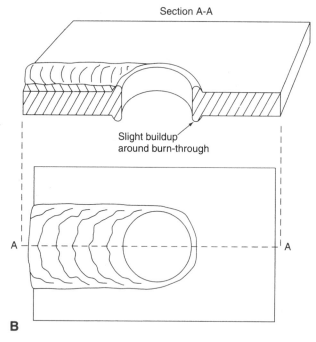

FIGURE 2-57 (A) When surface tension can no longer hold the weld pool together, a burn-through occurs. (B) The edge of a burn-through is thicker than the surrounding metal.

through the material being welded. Welding on thin sections therefore requires that the auto body technician control the weld size, keeping it as small as possible while still maintaining full penetration. Small weld beads are also much easier to finish as part of the post-weld cleanup.

Because of the very thin material, even the slightest increase in weld size can result in a burn-through (Figure 2-56). A burn-through occurs when the molten weld pool becomes so large that surface tension will no longer hold the thin film of molten metal together across the gap (Figure 2-57). The surface tension of the molten weld metal is similar to that of a soap bubble that allows the liquid soap solution to extend the thin film across a small opening. However, as the opening increases, the film becomes less stable, as exhibited by its increased flexibility. When the gap becomes too large, the soap film ruptures.

Burn-through can occur as a result of any welding process. It is a more commonly occurring problem on thinner metals and with larger weld pools.

If a burn-through does occur, it must be repaired using welding. Stop the repair immediately once the burn-through has occurred, and allow the metal to cool. After the metal has cooled for as little as a minute, the welding repair can begin. As the burn-through occurred, surface tension held the molten weld metal to the side, where it solidified, thus making the edge surrounding the hole slightly thicker (Figure 2-57B). Although this thickening is extremely slight, it still supplies enough additional metal to allow for an easier repair than a drilled hole. If the burn-through is no larger than one-quarter inch (6 mm) as measured in any direction, the repair can be made all at one time in the following manner:

1. If the burn-through occurred vertically, begin welding at the top and progress downward. If

the burn-through occurred in the flat or overhead positions, begin welding at the most convenient point.

2. Strike the arc at a point just outside of the burn-through perimeter on solid metal (Figure 2-58), concentrating the heat of the arc and the weld directly on the slightly thickened side of

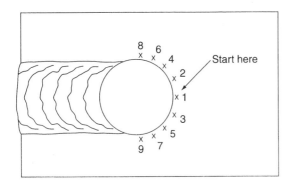

FIGURE 2–58 *Alternate welding from side to side in a zigzag pattern.*

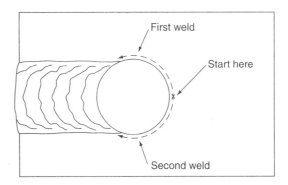

FIGURE 2–59 *Repairing large burn-throughs.*

the burn-through. Because of its increased mass, this area can best handle the weld's heat without significantly increasing in size.

3. Once the arc has stabilized and a very small molten weld pool has been established approximately 1/8 inch (3 mm) in diameter, momentarily stop the weld by releasing the gun's trigger. When the molten weld pool has solidified and cooled only slightly to a dull red color, in about 1 or 2 seconds, strike the arc and begin the weld. This is just like stitch welding but with much less arc-on-time.

4. Again momentarily stop the weld when its size increases to approximately 1/8 inch (3 mm) allowing it to cool as in step 3 before beginning to restart the weld.

5. Continue this process of starting and stopping until the weld has progressed across the entire burn-through.

6. It may be necessary to zig zag—move back and forth from side to side—in order to produce a weld that will be wide enough to close the entire burn-through (Figure 2-58).

Burn-throughs that are larger than 1/4 inch (6 mm) in diameter can be repaired by welding them in the following manner:

1. If the burn-through has occurred in the vertical position, begin welding at the top of the burn-through; however, if the burn-through occurred in the flat or overhead position, welding can begin in any convenient location.

2. Strike the arc just outside of the burn-through perimeter on solid metal.

3. When the weld pool has been established and increases to approximately 1/8 inch in size (3 mm), momentarily stop the arc by releasing the gun trigger.

4. Continue starting and stopping the weld and establish a weld along an entire side of the burn-through (Figure 2-59).

5. Allow the metal to cool for several minutes and then begin a new weld bead as before along the top side of the burn-through.

6. Following the second weld again, allow the metal to cool for several minutes before repeating the process.

7. Continue this welding process, alternating from side to side, until the entire burn-through has been closed.

It is important when using this process of starting and stopping the molten weld pool that good fusion between the weld metal and base metal occurs. To insure that fusion occurs, do not stop the welding process for any longer than the few seconds that is required for the weld to cool from the liquid state to a dull red color. Longer periods of time that allow the base metal to cool below a glowing temperature may result in inadequate fusion between the weld nuggets.

Burn-throughs are easily prevented by learning the warning signs that will occur moments before burn-through occurs. Carefully watch the surfaces of the molten weld pool as its size increases. You will notice in the flat position that the weld begins to sag slightly and may actually become somewhat depressed below the surface of the surrounding metal. If welding is immediately stopped at this point, the weld metal should solidify before the burn-through actually occurs.

On a vertical up weld, the indication that burn-through is possibly going to occur is that the weld will become teardropped in shape (Figure 2-60). Gravity pulls the large molten weld pool downward, resulting in a thinning of the top surface to a point where a burn-through hole suddenly appears. If welding is immediately stopped where this teardrop shape first occurs, the molten weld pool should solidify without resulting in a burn-through.

For vertical down welds, burn-through is not as frequent a problem. It can occur, however. Burn-through occurs on a vertical down weld as a result of the weld pool being pulled down with gravity. The weld's metal

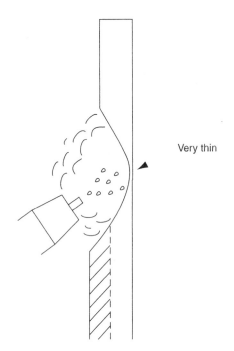

FIGURE 2–60 Burn-through about to occur in a vertical weld.

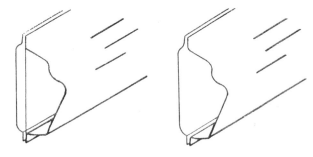

FIGURE 2–61 Rocker panel profiles.

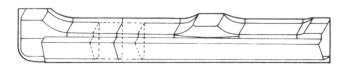

FIGURE 2–62 Joining the rocker panel section with a butt joint and insert.

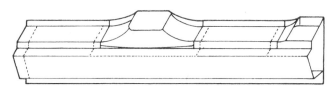

FIGURE 2–63 Replacing the outer rocker panel piece.

may be held up by the force of the arc forming a small ridge near the middle of the molten weld pool. The area above the ridge that has not solidified will become thin enough to eventually rupture, forming the burn-through hole if welding is immediately stopped. When this ridge has first appeared, burn-through can be prevented.

SECTIONING ROCKER PANELS

Before replacing a rocker panel, check the vehicle underbody for dimensional accuracy and make any repairs necessary for proper panel alignment. Once the panel is aligned, check all nearby seams for separation and cracking of the sealant. Any damaged seam must be repaired, since it could create a corrosion problem.

Rocker panels come in two-piece or three-piece designs, depending on the make and model of the unibody (Figure 2-61). In both cases, the rocker panel might contain reinforcements, and the reinforcements can be intermittent or continuous. Depending on the nature of the damage, the rocker panel can be replaced with or without the B-pillar.

To section and repair the rocker panel, a straight-cut butt joint with an insert can be used (Figure 2-62); or the outside piece of the rocker panel can be cut and the repair piece installed with overlapping joints (Figure 2-63). The butt joint with an insert is generally used when installing a recycled rocker panel with the B-pillar attached and when installing a recycled quarter panel.

To do a butt joint with an insert, cut straight across the panel. An insert is fashioned out of one or more pieces cut from the excess length on the repair panel or from the end of the damaged panel. The insert should be 6 to 12 inches (150 to 300 cm) long and should be cut lengthwise into two to four pieces, depending on the rocker panel configuration (Figure 2-64). Remove the pinch weld flange so that the insert will fit inside the rocker panel. With the insert in place, secure it with plug welds. For structural sectioning, 5/16-inch (8 mm) plug weld holes are needed to achieve an adequate nugget and acceptable weld strength. This 5/16-inch (8 mm) hole requires a circular motion of the gun to properly fuse the edge of the hole to the base metal (Figure 2-65).

When installing an insert in a closed section, whether it is a rocker panel, A- or B-pillar, or body rail, make sure the closing weld fully penetrates the insert. When closing the job with a butt weld, leave a gap that is wide enough to allow thorough penetration of the insert (Figure 2-66).

In general, use the lap joint procedure on a rocker panel when installing only the outer rocker or a portion of it. Leave the inner piece intact and cut only the outer piece. One way to make an lap joint is to make the cut in the front door opening and allow for an overlap when measuring. When making this cut, stay several inches away from the base of the

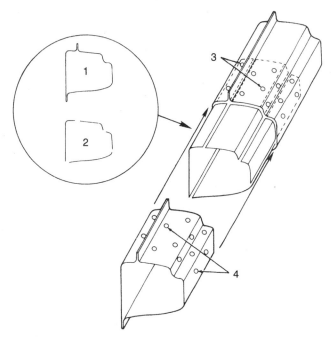

FIGURE 2-64   Cutting an insert to fit a rocker panel.

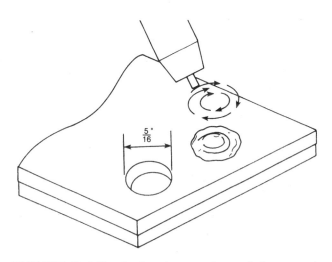

FIGURE 2-65   A circular motion of the gun is required in this 5/16-inch hole.

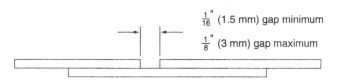

$\frac{1}{16}$" (1.5 mm) gap minimum

$\frac{1}{8}$" (3 mm) gap maximum

FIGURE 2-66   Root gap to aid in penetration.

B-pillar to avoid cutting any reinforcement underneath it (Figure 2-67).

1. Cut around the bases of the B- and C-pillars, leaving overlapping areas around each (Figure 2-68).
2. Cut out the new outer rocker panel so that it overlaps the bases of the pillars and the

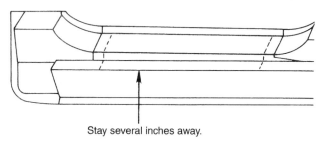

Stay several inches away.

FIGURE 2-67   Overlapping the outer rocker panel section.

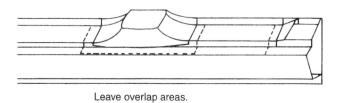

Leave overlap areas.

FIGURE 2-68   An overlap joint around a pillar.

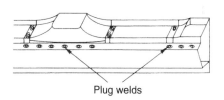

Plug welds

FIGURE 2-69   Plug-weld flanges.

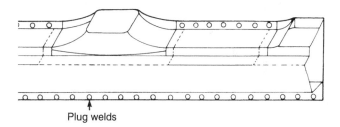

Plug welds

FIGURE 2-70   Plug-weld overlaps.

original piece of the outer rocker still affixed to the car.
3. In the flange weld, use plug welds to replace the factory spot welds (Figure 2-69).
4. Plug-weld the overlapping area around the B- and C-pillars, using approximately the same spacing as in the pinch flange weld (Figure 2-70).
5. Lap-weld the edges with about a 30 percent intermittent seam; about 1/2 inch (25 mm) of weld in every 1-1/2 inches (37 mm) of overlapped edge (Figure 2-71).
6. Put plug welds in the overlapping area in the door opening and lap-weld around the edges to close the joint (Figure 2-72).

Of course, it might be necessary to reverse this procedure due to the nature of the hit. Make the

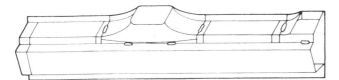

FIGURE 2–71   Intermittent lap weld seams.

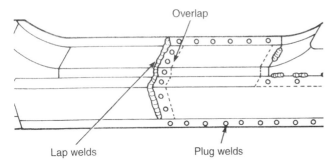

FIGURE 2–72   A plug weld and lap weld overlapping section.

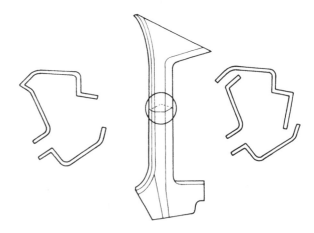

FIGURE 2–73   Two- and three-piece A-pillar sections.

overlap cut in the rear door opening and cut out and overlap around the bases of the A- and B-pillars. Use this same basic technique to replace the entire outer rocker. In this version, cut around the bases of all three pillars and overlap all three bases in the same way as before.

SECTIONING A-PILLARS

A-pillars can be either two-piece or three-piece components (Figure 2-73). They can be reinforced at the upper end or the lower end or both. They are not likely to be reinforced in the middle. Therefore, A-pillars should be cut near the middle to avoid cutting through any reinforcing pieces. It is also the easiest place to work.

To section an A-pillar, use a straight-cut butt joint with an insert or an offset butt joint without an insert. The cut joint with an insert is made in the same manner as already described for the rocker panel. The A-pillar insert should be 4 to 6 inches (100 to 150 mm) in length. After cutting the insert lengthwise and removing any flanges, tap the pieces into place. Secure the insert with plug welds and close all around the pillar with a continuous butt weld (Figure 2-74).

To make the offset butt joint, cut the inner piece of the pillar at a different point than the other piece was cut, creating the offset (Figure 2-75). Whenever possible, try to make the cuts between the factory spot welds so that it will not be difficult to drill them out, and make the cuts no closer to each other than 2 to 4 inches. Butt the sections together and continuous-weld them all around.

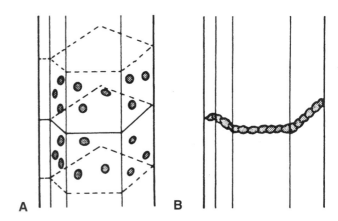

FIGURE 2–74   Welding the butt joint with an insert in an A-pillar using (A) plug welds and (B) butt welds.

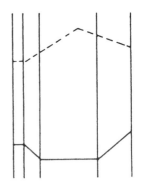

FIGURE 2–75   An offset butt joint.

SECTIONING B-PILLARS

For sectioning B-pillars, two types of joints can be used: the butt joint with an insert and a combination offset cut and overlap. The butt joint with an insert is usually easier to align and fit up when the B-pillar is a simple two-piece cross section without a lot of internal reinforcing members. The insert provides additional strength (Figure 2-76).

Be sure to cut far enough below the D-ring mount to avoid cutting through the D-ring anchor reinforcement. The majority of B-pillars have them. In the case of the B-pillar, use a channel insert in only the outside piece of the pillar. The D-ring anchor reinforcement welded to the inside piece prevents the installation of an insert there.

Begin by overlapping the new inside piece on the existing one, rather than butting them together, and lap-weld the edge (Figure 2-77A). Then, secure the insert with plug welds and close the joint with a continuous butt weld around the outer pillar (Figure 2-77B).

On occasion it is practical to obtain a recycled B-pillar and rocker panel assembly and replace them as a unit, because any time a B-pillar is hit so hard that it needs to be replaced, the rocker panel is almost invariably damaged, too. Install the upper end of the B-pillar with either of the two approved types of joints and make a butt joint with an insert in the rocker panel in the manner already shown. If the main damage is in the rear door opening, make the butt joint with an insert in the front door opening and install the other end of the rocker in its entirety. If the main damage is in the front door opening, reverse the procedure.

Generally speaking, the combination offset and overlap joint (Figure 2-78) is used more often when installing new parts and when working with separate inside and outside pieces.

1. Cut a butt joint in the outside piece above the level of the D-ring anchor reinforcement.
2. Make an overlap cut in the inside piece below the D-ring anchor reinforcement.
3. Install the inside piece with the new segment overlapping the existing segment.
4. Lap-weld the edge (Figure 2-79A).
5. Put the outside pieces in place, make plug welds in the flanges, and close the section with a continuous weld at the butt joint (Figure 2-79B).

Usually, it is advantageous to use the offset and overlap joint on a B-pillar with three or more pieces in its cross section. In fact, sometimes the offset and overlap procedure is mandatory because it is actually not possible to install an insert.

SECTIONING FLOOR PANS

When sectioning a floor pan, do not cut through any reinforcements, such as seat belt anchorages. The rear section should overlap the front section by at

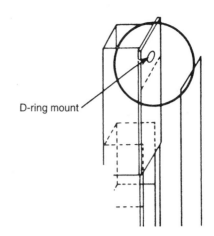

FIGURE 2-76 Two-piece B-pillar.

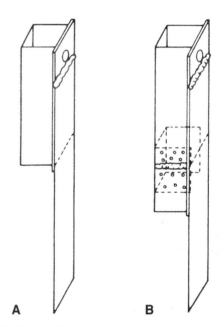

FIGURE 2-77 (A) Lap-weld the inner panel and (B) plug- and butt-weld the outer panel.

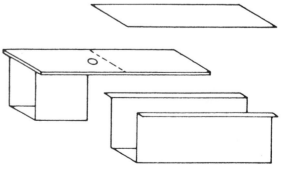

FIGURE 2-78 A combination offset and overlap joint.

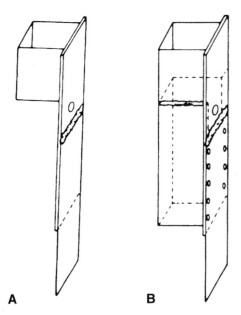

FIGURE 2–79 Creating the combination offset and overlap joint: (A) lap welding inside and (B) plug and lap welding outside.

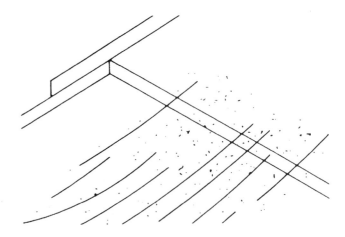

FIGURE 2–80 The rear section overlap shields the joint from wind.

least 2 inches (50 mm) so the edge of the bottom piece under the car is always pointing toward the rear. This is so that road splash, which moves from the front to the rear, streams past the bottom edge and does not strike it head-on (Figure 2-80).

1. Join all floor pan sections with an overlap.
2. Punch or drill plug weld holes in the top section.
3. If the mating surfaces are ungalvanized metal, apply a zinc-rich weld-through primer to them.
4. Fit the sections together and clamp them.
5. Plug-weld the overlap (Figure 2-81).
6. On the bottom side, lap-weld the underlapping edge with a continuous bead.
7. Caulk the forward edge of the top side with a flexible body caulk.
8. Cover the lap weld with a primer, a seam sealer, and a topcoat. The primer helps the sealer hold better, and the topcoat completes the protection. This assures that there will be no carbon monoxide intrusion through the joint into the passenger compartment.

SECTIONING TRUNK FLOORS

When a trunk floor is being sectioned, follow the basic procedures just described for the floor pan with some variations. Also, you should note that in a collision that necessitates sectioning of the trunk floor, the rear rail usually requires sectioning as well.

1. There is generally some kind of a cross member under the trunk floor in the vicinity of

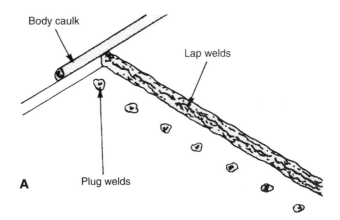

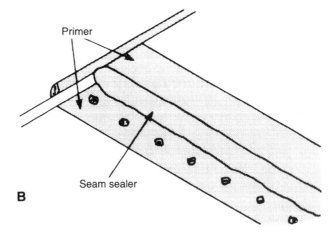

FIGURE 2–81 Plug-weld from the top and lap-weld, seal, and prime the bottom edge.

the rear suspension. Whenever possible, section the trunk floor above the cross member's rear flange. Section the rail just to the rear of the cross member.

2. Plug-weld the trunk floor overlap joint to the cross member, again putting the plugs in from the topside, downward, as in a floor pan.

3. Caulk the topside forward edge just like a floor pan seam.
4. A lap weld on the underlapping edge of the bottom side is not necessary because of the strength provided by the cross member. However, on cars in which the trunk floor section is not above a cross member, the lower edge must be lap welded.

In both cases, cover the bottom-side seam with a primer, a seam sealer, and a topcoat. With the trunk floor, sealing against carbon monoxide intrusion is critical because of the proximity of the tailpipe.

FULL-BODY SECTIONING

To do a complete vehicle reconstruction by full-body sectioning, the front portion of one vehicle is joined to the rear portion of another. This type of repair is generally done on a vehicle that has sustained severe rear-end damage. In such a case, sectioning is more practical and cost effective than conventional methods of repairing severely damaged unitized automobiles. This method reduces the extensive time and labor of dismantling the vehicle and disturbs less of the corrosion protection.

Front full-body sectioning is not recommended, since the vehicle identification number (VIN) would not match the registration, and the power train warranty could be affected. Also, the matching of mileage, condition, and accessories would be difficult.

The techniques and procedures just described for sectioning a rocker panel, an A-pillar, and a floor pan are repeated in a full-body section. You can fully section a vechicle by sectioning both A-pillars, both rocker panels, and the floor pan.

When the individual components are properly sectioned, aligned, and welded using the proper techniques and procedures, full-body sectioning is a suitable and satisfactory procedure. Vehicles repaired by full-body sectioning are completely crashworthy. This has been tested and proven over time. Keep in mind, however, that full-body sectioning is not a frequently required procedure, and full disclosure should be given to the car owner before repairs are started.

A discussion between the insurance representative, car owner, and repairer must be conducted with the following points covered:

- All repair procedures, including alignment and welding, must be fully explained to the car owner.
- The recycled sections—both body and mechanical—must be of like kind and quality. Always verify that all VIN code identifications and EPA emission control requirements are met and

that all suspension, braking, and steering components are in proper working order.
- Carefully inspect the front and rear sections for proper alignment before cutting. If either is out of alignment, proper fit and lineup of the section joints will be difficult, if not impossible, to achieve.

A compact car can have a full body sectioned in a commercial repair shop. In one case, the undamaged front half of one car was joined to the undamaged rear half of another. Butt joints and inserts were used in the middle of the A-pillars, and in the two rocker panels, and an overlap joint was used in the floor pan. The rocker panel and floor pan cuts were made in the middle of the front door opening to avoid any brackets or reinforcements in the A- and B-pillars.

Remember that the floor pan might have reinforcements and brackets that need to be removed before sectioning. Reinforcements can be left on the replacement rear half to aid in alignment. Proper corrosion protection must be restored when replacing brackets and reinforcements.

JOINING THE FULL-BODY SECTIONS

After the front and rear sections have been trimmed to fit, drilled for plug welds, and primed with weld-through, follow these steps to join the sections:

1. Install the rocker and pillar inserts. Clamp them in place with sheet metal screws.
2. Place the A-pillar inserts in the upper or lower portion of the windshield pillar, depending on the angle and contour of the windshield.
3. Fit the two halves together by first joining the rocker panels and then the A-pillars. Clamp the rocker and pillar flanges to prevent the sections from pulling apart.
4. Check the windshield and door openings for proper dimension, using a tram gauge or a steel rule. If possible, install the doors and windshield to verify proper alignment.
5. When proper alignment is achieved, secure the overlapping areas with sheet metal screws to pull the seam areas together and hold the sections together during welding.
6. Using centerline gauges and a tram gauge, double-check the vehicle dimensions and section alignment before welding the sections together.
7. Weld the sections together using the techniques already described in this chapter for joining rocker panels, A-pillars, and the floor pan.

REVIEW QUESTIONS

1. Why must all structural members of a collision-damaged vehicle be restored or replaced to bring them to their original structural integrity?
2. After the front and rear sections have been trimmed to fit and drilled for plug welds, list the steps to join together the sections.
3. What does HSLA mean?
4. When a component must be replaced, what are the two basic procedures used to accomplish a quality restoration?
5. List the steps to removing paint in order to see spot welds.
6. What do you do if the spot weld positions are not visible after the paint is removed?
7. What are the two types of cutting bits that can be used?
8. What are the advantages of removing only those panel sections that need replacing while leaving the remaining sections intact?
9. What is the advantage to using a multiple-hole drill?
10. What is the slowest method used to remove spot welds?
11. What is brazing?
12. What is shielded metal arc welding?
13. What is a spot weld?
14. Define soldering.
15. Explain plasma arc cutting.
16. What is a full penetration weld?
17. When does a burn-through occur?
18. What needs to be checked before replacing a rocker panel?
19. Why is front full-body sectioning not recommended?
20. What points need to be discussed between the insurance representative, the car's owner, and the repairer before proceeding with full-body sectioning?

Chapter 3

MIG Welding Theory

Objectives

After reading this chapter, you should be able to:

- Name the three main welding categories.
- List five characteristics of welding.
- Explain why MIG welding is recommended for all automotive structural collision repairs.
- List and describe the various shielding gases used with MIG welding.
- Explain the process of short-circuiting metal transfer.
- Name the basic components found in any MIG welding setup.
- Explain the slope, current rating, and duty cycle of the MIG power supply.
- List the beneficial characteristics of the flux-cored arc welding process.
- Describe the setup of MIG equipment.
- Explain how variables such as welding current, arc voltage, and tip-to-base metal distance affect a welding job.

The two basic methods of joining metal in automobile assembly are:

- Mechanical, including nuts and bolts and riveting (Figure 3-1)
- Welding (Figure 3-2)

A *weld* is defined by the American Welding Society (AWS) as "a localized coalescence (the fusion or growing together of the grain structure of the materials being welded) of metals or nonmetals produced

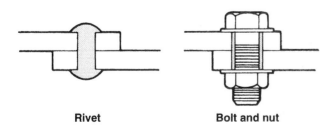

Rivet   **Bolt and nut**

FIGURE 3–1 *Mechanical joining methods.*

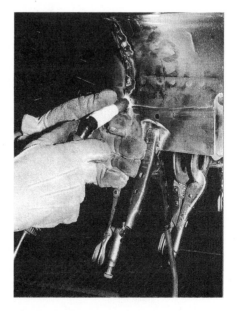

FIGURE 3–2 MIG welding commonly used in auto body repair. (Courtesy of Larry Maupin)

either by heating the materials to the required welding temperatures, with or without the application of pressure, or by the application of pressure alone, and with or without the use of filler materials." *Welding* is defined as "a joining process that produces coalescence of materials by heating them to the welding temperature, with or without the application of pressure or by the application of pressure alone, and with or without the use of filler metal." In less technical language, a weld is made when separate pieces of material to be joined combine and form one piece when heated to a temperature high enough to cause softening or melting and flowing together. Pressure may or may not be used to force the pieces together. In some instances, pressure alone may be sufficient to force the separate pieces of material to combine and form one piece. Filler material is added when needed to form a completed weld in the joint. It is also important to note that the word *material* is used because today welds can be made from a growing list of materials including plastic, glass, and ceramics.

3.1 MIG WELDING

MIG welding uses a welding wire that is fed automatically at a constant speed as an electrode. An arc is generated between the base metal and the wire, and the resulting heat from the arc melts the welding wire and joins the base metals together (Figure 3-3). This method is called a semiautomatic arc welding process, because wire is fed automatically at a constant rate and the welder provides gun movement. During the welding process, a shielding gas protects the weld from the atmosphere and prevents the oxidation of the base metal. The type of shielding gas used depends on the base material to be welded.

This process gets its name from the fact that originally the process only used inert gases for shielding, so the name *metal inert gas* (MIG) welding applied. Today many different gases are used. Some are inert (nonreactive under all conditions), and others are reactive (can combine under some conditions). Because of the changes in the gas shielding, the term *gas metal arc welding* (GMAW) was adopted by the American Welding Society for this process. However, in the collision repair industry, the term MIG is still the one most commonly used. The process has had other names through the years, such as wire welding, but whatever the name, the process is the same.

The advantages of MIG welding (Figure 3-4), over conventional electrode type arc (stick) welding (Figure 3-5), are numerous. Car manufacturers, insurance companies, I-CAR, and governmental regulations require or recommend that MIG be used in virtually all welding repairs. The advantages of MIG welding are:

- MIG welding is easy to learn. The typical welder can learn to use MIG equipment with just a few hours of instruction and practice. More time may be required to master the adjusting of the equipment.
- MIG welding can produce higher quality welds faster and more consistently than conventional stick electrode welds.

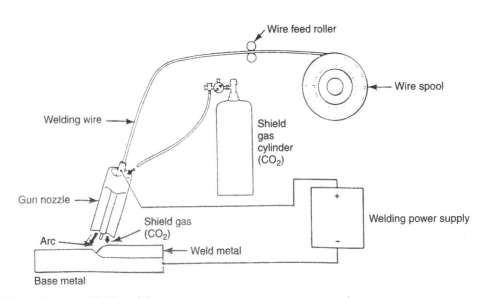

FIGURE 3–3 *The principle of MIG welding.*

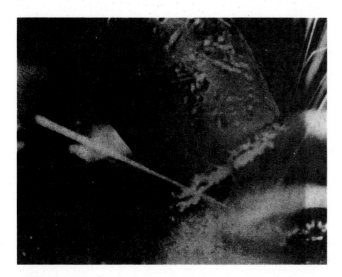

FIGURE 3–4 MIG welding.

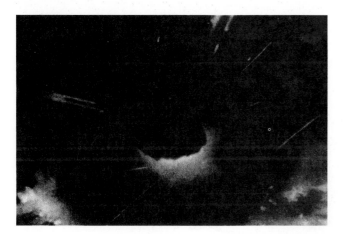

FIGURE 3–5 Stick welding.

- Low current can be used to MIG-weld thin metals.
- Fast welding speeds and low currents prevent heat damage to adjacent areas that can cause strength loss and warping.
- The small molten weld pool is easily controlled (Figure 3-6).
- MIG welding is tolerant of gaps and misfits. Gaps can be welded by making several spot welds on top of each other.
- Almost all auto body steels can be MIG welded with one common type of weld wire.
- Metals of different thicknesses can be MIG welded with the same diameter of wire.
- With MIG welding, vertical and/or overhead welding is possible because the weld pool is small and the metal is molten for a very short time.

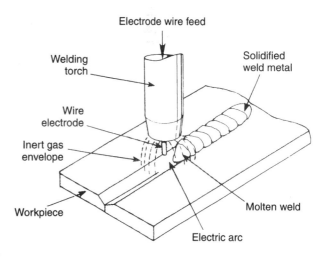

FIGURE 3–6 Basic MIG welding process.

- MIG welds are easily started in the correct spot because the wire is not energized until the gun trigger is depressed.
- With MIG welding, there is minimum waste of welding consumables.

MIG welding became popular when auto manufacturers began using high-strength steel (HSS) and high-strength, low-alloy (HSLA) steel. These materials are strong enough to be used in much thinner gauges than had been used in the past. The only correct way to weld HSS, HSLA, and other thin-gauge steel is with MIG. Welding a rear quarter panel with an oxyacetylene welder takes about four hours; a MIG welder can do the same job in about half an hour.

MIG welding is not limited to body repairs alone. It is also ideal for exhaust pipe welding, repairing mechanical supports, and installing trailer hitches and bumpers. Almost any welding that would be done with either an arc or gas welder can be done faster with MIG. In addition, it is possible to weld aluminum sheet and aluminum castings such as cracked transmission cases, cylinder heads, and intake manifolds.

3.2 — **EQUIPMENT**

The basic MIG equipment consists of the gun, electrode (wire) feed unit, electrode (wire) supply, power source, shielding gas supply with flowmeter/regulator, control circuit, and related hoses, liners, and cables, (Figures 3-7 and 3-8). The system should be portable. In some cases, the system can be used for more than one process. The power sources can be switched over for other uses.

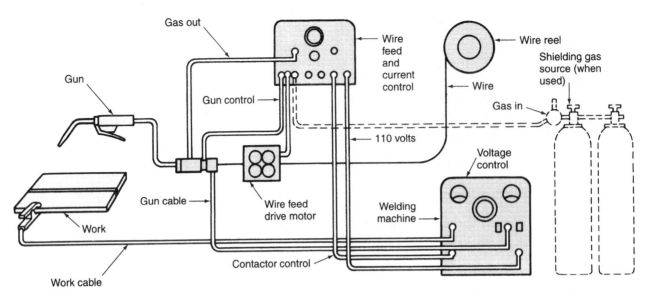

FIGURE 3–7 Schematic of equipment setup for GMA welding. (Courtesy of Hobart Brothers Company)

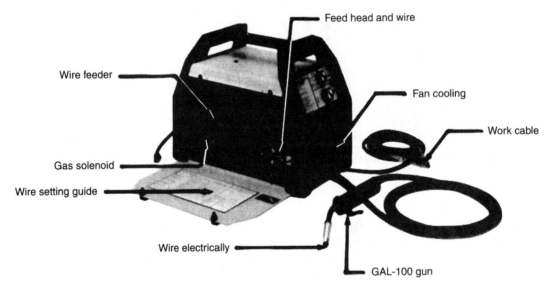

FIGURE 3–8 GMA welding setup. (Courtesy of Hobart Brothers Company)

POWER SOURCE

The power source has a transformer and rectifier. It produces a DC welding current ranging from 40 amperes to 600 amperes with 10 volts to 40 volts, depending upon the machine. In the past, some MIG processes used AC welding current, but DCEP is used exclusively for auto body work. Typical power supplies are shown in Figure 3-9.

Because of the long periods of continuous use, MIG welding machines have a 100 percent duty cycle. This allows the machine to be run continuously without damage.

WIRE FEED UNIT

The purpose of the wire feeder is to provide a steady and reliable supply of wire to the weld. Slight changes in the rate at which the wire is fed have distinct effects on the weld.

The motor used in a feed unit can be continuously adjusted over the desired range. Figures 3-10 and 3-11 show typical wire feed units.

Push-Type Feed System

In a push-type feed system, the wire rollers are clamped securely against the wire to provide the necessary friction to push the wire through the conduit to the gun. The pressure applied on the wire can be adjusted. A groove is provided in the roller to aid in alignment and to lessen the chance of slippage. Most manufacturers provide rollers with smooth or knurled U-shaped or V-shaped grooves (Figure 3-12). Knurling (a series of ridges cut into the groove) helps the roller to grip larger-diameter wires so that they can be pushed along more easily. Soft wires, such as aluminum, are easy to damage if knurled rollers are used. Aluminum wires are best used with U-grooved rollers. Even V-grooved rollers can distort the surface of soft wire, causing problems. V-grooved rollers are best suited for hard wires, such as mild steel and stainless steel. It is also important to use the correct size grooves in the rollers.

In the push-type system, the electrode must have enough strength to be pushed through the conduit without kinking. Mild steel and stainless steel can be

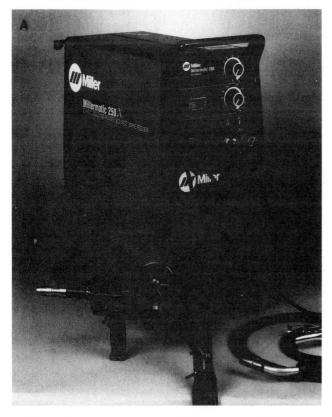

FIGURE 3–9 An all-in-one 250-ampere MIG constant voltage power supply (A), and a 650-ampere constant-voltage and constant-current power supply (B) for multipurpose GMAW applications. (Courtesy of Miller Electric)

A

B

FIGURE 3–10 (A) A 90-ampere power supply with built-in wire feeder for welding sheet steel with carbon dioxide shielding. (B) Modern wire feeder with digital preset and readout of wire-feed speed and closed-loop control. (Courtesy of Miller Electric)

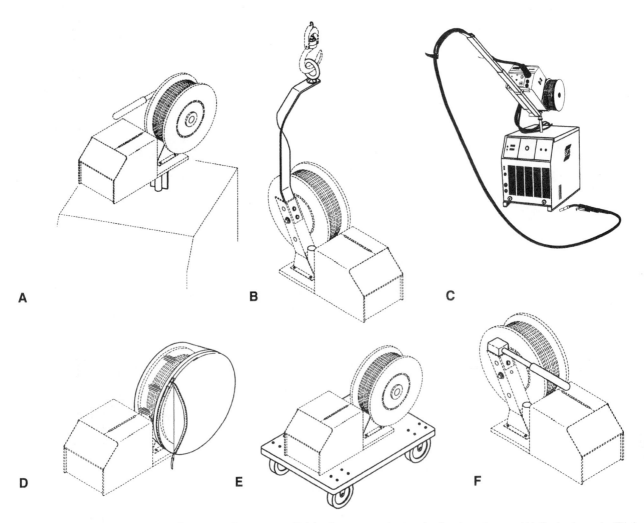

A   **B**   **C**

D   **E**   **F**

FIGURE 3–11 A variety of accessories are available for most electrode feed systems: (A) Swivel post, (B) boom hanging bracket, (C) counterbalance mini boom, (D) spool feeder, (E) wire feeder wheel cart, (F) carrying handle. (Courtesy of ESAB Welding and Cutting Products)

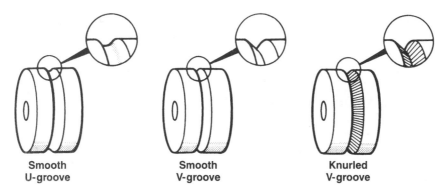

FIGURE 3–12 Feed rollers.

readily pushed 15 ft (4 m) to 20 ft (6 m), but aluminum is much harder to push beyond 10 ft (3 m).

Pull-Type Feed System

In pull-type systems, a smaller but higher-speed motor is located in the gun to pull the wire through the conduit. With this system, it is possible to move even soft wire over great distances. The disadvantages are that the gun is heavier and more difficult to use, rethreading the wire takes more time, and the operating life of the motor is shorter.

SPOOL GUN

A spool gun is a compact, self-contained system consisting of a small drive system and a wire supply (Figure 3-13). This system allows the welder to move freely around a job with only a power lead and shielding gas hose to manage. The major control system is usually mounted on the welder. The feed rollers and motor are found in the gun just behind the nozzle and contact tube. Because of the short distance the wire must be moved, very soft wires (aluminum) can be used. A small spool of welding wire is located just behind the feed rollers. The small spools of wire required in these guns are often very expensive. Although the guns are small, they feel heavy when being used.

ELECTRODE CONDUIT

The electrode conduit or liner guides the welding wire from the feed rollers to the gun. It may be encased in a lead that contains the shielding gas.

Power cable and gun switch circuit wires are contained in a conduit that is made of a tightly wound coil having the needed flexibility and strength. The steel conduit may have a nylon or Teflon® liner to protect

FIGURE 3–13 Spool gun for GMA welding. (Courtesy of Miller Electric)

soft, easily scratched metals, such as aluminum, as they are fed.

If the conduit is not an integral part of the lead, it must be firmly attached to both ends of the lead. Failure to attach the conduit can result in misalignment, which causes additional drag or makes the wire jam completely. If the conduit does not extend through the lead casing to make a connection, it can be drawn out by tightly coiling the lead (Figure 3-14). Coiling will force the conduit out so that it can be connected. If the conduit is too long for the lead, it should be cut off and filed smooth. Too long a lead will bend and twist inside the conduit, which may cause feed problems.

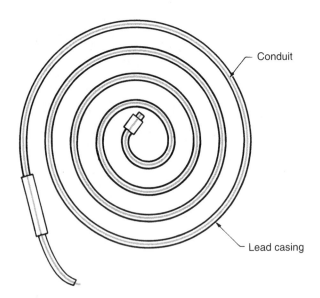

FIGURE 3–14 *Tightly coiled lead casing will force the liner out of the gun.*

FIGURE 3–15 *A typical GMA welding gun used for most welding processes with a heat shield attached to protect the welder's glove hand from intense heat generated when welding with high amperages. (Courtesy of Lincoln Electric, Cleveland, OH)*

WELDING GUN

The welding gun attaches to the end of the power cable, electrode conduit, and shielding gas hose (Figure 3-15). It is used by the welder to produce the weld. A trigger switch is used to start and stop the weld cycle. The gun also has a contact tube that is used to transfer welding current to the electrode moving through the gun, and a gas nozzle that directs the shielding gas onto the weld.

TABLE 3–1:   AWS FILLER METAL SPECIFICATIONS FOR DIFFERENT BASE METALS

Base Metal Type	AWS Filler Metal Specification
Aluminum and aluminum alloys	A5.10
Stainless steel	A5.9
Steel (HSS and HSLA)	A5.18

TABLE 3–2:   FILLER METAL DIAMETERS AND AMPERAGE RANGES

Base Metal	Electrode Diameter Inch	Millimeter	Amperage Range
Carbon Steel	0.023	0.6	35–190
	0.030	0.8	40–220
	0.035	0.9	60–280
Stainless Steel	0.023	0.6	40–150
	0.030	0.8	60–160
	0.035	0.9	70–210

3.3 FILLER METAL SPECIFICATIONS

MIG welding filler metals are available for a variety of base metals (Table 3-1). The most frequently used filler metals are AWS specification A5.18 for carbon steel and AWS specification A5.9 for stainless steel. These filler metals are available in diameters ranging from 0.023 inch (0.6 mm) to 0.125 inch (3.2 mm). Table 3-2 lists the most common sizes for these electrodes and their amperage ranges. The amperage will vary depending on the method of metal transfer, type of shielding gas, and base metal thickness. The small-diameter wires can be used at low currents and voltages, thus greatly reducing heat input to the base material. Most MIG welding for auto body work is done with filler metals from 0.023 inch (0.6 mm) to 0.035 inch (.9 mm) in diameter. Because of the flexibility of the MIG welding process, a single-sized filler metal can be used on various thicknesses of material. It is possible to use a smaller-diameter filler metal to produce satisfactory welds on thicker material, but it is not possible to use a larger-diameter filler wire on thinner material. In other words, it is most practical for a collision auto repair facility to use 0.023 inch filler wire because it can be used to make welds on virtually any thickness of auto body metal. Table 3-3 lists the most common filler metal sizes used in collision repair work and the thickness limits that these wires can be used on.

3.4 WIRE MELTING AND DEPOSITION RATES

The wire melting rates, deposition rates, and wire-feed speeds of the wire welding process are all affected by the same variables. Before discussing them, however, these terms need to be defined. The wire melting rate, measured in inches per minute (mm/sec) or pounds/hr (kg/hr), is the rate at which the arc consumes the wire. The deposition rate, the measure of weld metal deposited, is nearly always less than the melting rate because not all of the wire is converted to weld metal. Some is lost as slag, spatter, or fumes. The amount of weld metal deposited as compared to the wire used is called the deposition efficiency.

Deposition efficiencies depend on the process, on the gas used, and even on how the auto body technician sets welding conditions. With efficiencies of approximately 98 percent, solid wires with argon shields are best.

Auto body technicians can control the deposition rate by changing the current, the electrode extension, and the diameter of the wire. To obtain higher melting rates, they can increase the current or wire extension or decrease the wire diameter. Knowing the precise constants is unimportant. However, it is important to know that current greatly affects melting rate and that the extension must be controlled if results are to be reproducible.

3.5 SHIELDING GAS

The shielding gas selected for a weld will have a definite effect on the weld produced. The items that can be affected include the method of metal transfer,

TABLE 3–3: OPTIMUM WIRE SIZE FOR PANEL THICKNESS USING Ar75% + CO₂ 25%

Wire Size	U.S. Sheet Metal Gauge
0.023″ (0.6 mm) to 0.030″ (0.8 mm)	25 gauge 0.020″ (0.50 mm) 24 gauge 0.023″ (0.58 mm) 23 gauge 0.026″ (0.66 mm) 22 gauge 0.029″ (0.73 mm) 21 gauge 0.032″ (0.81 mm)
0.030″ (0.8 mm) to 0.035″ (0.9 mm)	20 gauge 0.035″ (0.88 mm) 19 gauge 0.041″ (1.04 mm) 18 gauge 0.047″ (1.19 mm) 17 gauge 0.053″ (1.34 mm) 16 gauge 0.059″ (1.49 mm)

TABLE 3–4: GMAW SHIELDING GASES AND METALS

Metal	Shielding Gas	Chemical Reaction
Aluminum	Argon	Inert
	Argon + helium	Inert
Steel, carbon	Argon + oxygen	Slightly oxidizing
	Argon + carbon dioxide	Oxidizing
	Carbon dioxide	Oxidizing
Steel, low-alloy	Argon + oxygen	Slightly oxidizing
	Argon + helium + CO_2	Slightly oxidizing
	Argon + carbon dioxide	Oxidizing
Steel, stainless	Argon + oxygen	Slightly oxidizing
	Argon + helium + CO_2	Slightly oxidizing

welding speed, weld contour, arc cleaning effect, and viscosity of the molten weld pool.

In addition to the effects on the weld itself, the metal to be welded must be considered in selecting a shielding gas. Aluminum must be welded with an inert gas such as argon. Most auto body metals weld more favorably with reactive gases such as carbon dioxide or mixtures of inert gases and reactive gases such as argon and oxygen or argon and carbon dioxide (Table 3-4). The most commonly used shielding gases are 75 percent argon plus 25 percent carbon dioxide, argon plus 1 to 5 percent oxygen, and carbon dioxide (Figure 3-16).

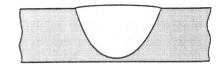

Argon + oxygen   Argon + Carbon dioxide   Carbon dioxide

FIGURE 3–16 *Effect of shielding gas on weld bead shape.*

- **75 percent argon plus 25 percent carbon dioxide.** This mixture is used for the short-circuiting metal transfer process on carbon steels, HSS, and HSLA steels. It produces welds with good wetting characteristics, little spatter, high welding speeds, and low distortion.
- **Argon plus 1 to 5 percent oxygen.** This mixture is used for both short-circuiting metal transfer and the axial spray transfer method. It produces welds with good wetting, arc stability, little undercut, high welding speeds, and minimum distortion.
- **Carbon dioxide (CO_2).** This gas is used for the short-circuiting metal transfer process on carbon steel. CO_2 is not approved for welding on HSS alloys. It produces welds with deep penetration, high welding speeds, and noticeable spatter.

The most commonly used shielding gas for auto body welding is 75 percent argon and 25 percent CO_2. This blend of gases can be used for all of the steels used to manufacture vehicles. Aluminum must be welded with 100 percent argon. Because aluminum is so easily oxidized, reactive gas mixtures cannot be used.

3.6 TYPES OF METAL TRANSFER

The MIG welding process can be divided into 5 different groups according to the method of metal transfer. Metal transfer refers to the process by which the filler wire is melted by the arc and transferred to the base plate in order to form the weld bead. These methods of metal transfer are axial spray, pulsed arc, varied arc, and short-circuiting. Short-circuiting transfer or short arc is the most commonly used method of metal transfer for auto body welding.

SHORT-CIRCUITING TRANSFER

Low currents allow the liquid metal at the electrode tip to be transferred by direct contact with the molten weld pool. A close interaction between the wire feeder and the power supply is required. This technique is called short-circuiting transfer.

The transfer mechanisms in this process are quite simple, as shown schematically in Figure 3-17. To start, the wire is in direct contact with the molten weld pool (Figure 3-17A). Once the electrode touches the molten weld pool, the arc and its resistance are removed. Without the arc resistance, the welding amperage quickly rises as it begins to flow freely through the tip of the wire into the molten weld pool. The resistance to current flow is highest at the point where the electrode touches the molten weld pool. The resistance is high because both the electrode tip and weld pool are very hot. The higher the temperature, the higher the resistance to current flow. A combination of high current flow and high resistance causes a rapid rise in the temperature of the electrode tip.

As the current flow increases, the interface between the wire and the molten weld pool is heated until it explodes into a vapor (Figure 3-17B), establishing an arc. This small explosion produces sufficient force to depress the molten weld pool. A gap between the electrode tip and the molten weld pool (Figure 3-17C) immediately opens. With the resistance of the arc reestablished, the voltage increases as the current decreases.

The low current flow is insufficient to keep melting the electrode tip off as fast as it is being fed into the arc. As a result, the arc length rapidly decreases (Figure 3-17D) until the electrode tip once again contacts the molten weld pool (Figure 3-17A). The liquid formed at the wire tip during the arc-on interval is transferred by surface tension to the molten weld pool, and the cycle begins again with another short circuit.

If the system is properly tuned, the rate of short circuiting can reach hundreds per second, causing a characteristic buzzing sound. The spatter is low and the process easy to use. The low heat produced by MIG short-circuit welding makes it easier to use in all positions on auto body sheet metal from a thickness range of 25 gauge (0.02 in.; 0.5 mm) to 12 gauge (0.1 in.; 2.6 mm). The short-circuiting process does not produce enough heat to make

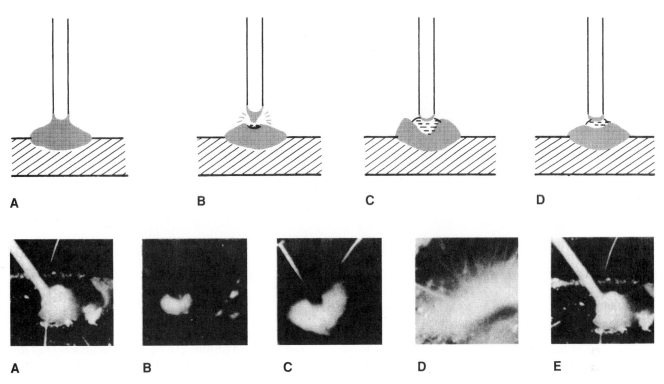

A   B   C   D

A   B   C   D   E

FIGURE 3–17   *Schematic of short-circuiting transfer.*

quality welds in sections much thicker than 1/4 inch. Pure argon as a shield is not as effective because its arc tends to be sluggish and not very fluid. However, a mixture of 25 percent carbon dioxide and 75 percent argon produces a less harsh arc and a flatter, more fluid and desirable weld profile.

The power supply is most critical. It must have a constant potential output and sufficient inductance to slow the time rate of current increase during the short-circuit interval. Too little inductance causes spatter due to high current surges. Too much inductance causes the system to become sluggish. The short-circuiting rate decreases enough to make the process difficult to use. Also, the power supply must sustain an arc long enough to permit the electrode tip to transfer material at recontact with the weld pool.

AXIAL SPRAY METAL TRANSFER

The unique mode of metal transfer called axial spray metal transfer is free from spatter (Figure 3-18). This process is identified by the pointing of the wire tip from which very small drops are projected axially across the arc gap to the molten weld pool. There are hundreds of drops per second crossing from the wire to the base metal. These drops are propelled by arc forces at high velocity in the direction the wire is

pointing. In virtually all cases, the molten weld pool is too large for any auto body welding.

PULSED-ARC METAL TRANSFER

By automatically varying the voltage and amperage, the pulsed-arc process changes from spray arc to globular transfer. The pulsation within a very narrow current range and the globular transfer occur at the rate of only a few drops per second, and a controlled spray transfer at significantly lower average currents is achievable. The time interval below the transition current is short enough to prevent a drop from developing. About 0.1 second is needed to form a globule, so no globule can form at the electrode tip if the time interval at the low base current is about 0.01 second. Actually, the energy produced during this time is very low—just enough to keep the arc alive.

This process requires a special welding machine and is not practical for auto body welding.

BURIED-ARC TRANSFER

In buried-arc transfer, a large drop, partially supported by arc forces, forms on the end of the wire (Figure 3-19). As it becomes heavy enough to overcome those forces, it drops into the pool. This process is best suited for mechanized welding.

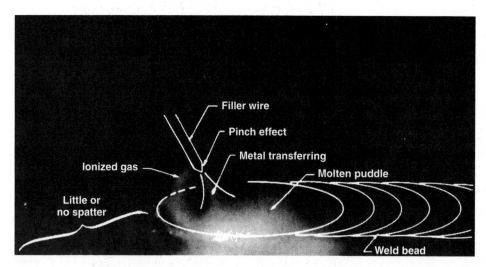

FIGURE 3–18 *Axial spray metal transfer. Note the pinch effect of filler wire and the symmetrical metal transfer column.*

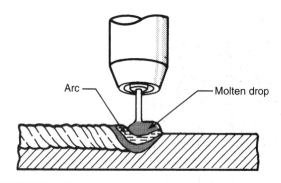

FIGURE 3–19 *Buried-arc metal transfer. Large drop is supported by arc forces.*

WELGING POWER SUPPLIES
3.7

To better understand the terms used to describe the different welding power supplies, you need to know the following electrical terms:

- *Voltage* or *volts (V)* is a measurement of electrical pressure, in the same way that pounds per square inch is a measurement of water pressure.
- *Electrical potential* means the same thing as voltage and is usually expressed by using the term *potential (P).* The terms *voltage, volts,* and *potential* can all be interchanged when referring to electrical pressure.

- *Amperage* or *amps (A)* is the measurement of the total number of electrons flowing, in the same way that gallons is a measurement of the amount of water flowing.
- *Electrical current* means the same thing as amperage and is usually expressed by using the term *current (C).* The terms *amperage, amps,* and *current* can all be interchanged when referring to electrical flow.

MIG welding power supplies are constant-voltage, constant-potential (CV, CP) machines, unlike stick arc welding power supplies, which are constant-current (CC) type. It is impossible to make acceptable welds using the wrong type of power supply. Constant-voltage power supplies are available as transformer-rectifiers or as motor-generators (Figure 3-20). Some newer machines use electronics to enable them to supply both types of welding power by the flip of a switch.

The relationships between current and voltage with different combinations of arc length or wire-feed speeds are called volt-ampere characteristics. The volt-ampere characteristics of arcs in argon with constant arc lengths or constant wire-feed speeds are shown in Figure 3-21. To maintain a constant arc length while increasing current, it is necessary to increase voltage. For example, with a 1/8-inch arc length, increasing current from 150 to 300 amperes requires a voltage increase from about 26 to 31 volts. The current increase illustrated here results from increasing the wire feed-speed from 200 to 500 inches per minute.

FIGURE 3–20 (A) Motor-generator welding power supply. (B) Transformer-rectifier welding power supply. (**A** Courtesy of Lincoln Electric, Cleveland, OH. **B** Courtesy of Miller Electric)

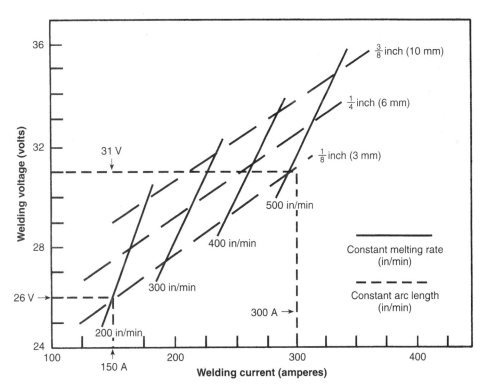

FIGURE 3–21 The arc length and arc voltage are affected by the welding current and wire-feed speed. (0.045-inch [1.43 mm] wire; one-inch (25 mm) electrode extension.)

The amperage can be determined by knowing the wire-feed speed, or the wire-feed speed can be determined by knowing the voltage. This relationship of amperage to wire-feed speed results from the need for a higher amperage to melt the larger quantity of filler metal being introduced to the arc.

POWER SUPPLIES FOR SHORT-CIRCUITING TRANSFER

Although the MIG power source is said to have a constant potential (CP), it is not perfectly constant. The graph in Figure 3-22 shows that there is a slight decrease in voltage as the amperage increases within the working range. The rate of decrease is known as slope. Its effects are shown in Table 3-5. It is expressed as the voltage decrease per 100 ampere increase; for example, 10 V/100 A. For short-circuiting welding, some power supplies are equipped to allow changes in the slope by steps or by continuous adjustment.

The slope, which is called the volt-ampere curve, is often drawn as a straight line because it is fairly straight within the working range of the machine. Whether it is drawn as a curve or a straight line, the slope can be found by finding two points. The first point is the set voltage as read from the voltmeter when the gun switch is activated but no welding is being done. This is referred to as the open circuit voltage. The second point is the voltage and amperage as read during a weld. The voltage control is not adjusted during the test, but the amperage can be changed. The slope is the voltage difference between the first and second readings. The difference can be found by subtracting the second voltage from the first voltage. Therefore, for settings over 100 amperes, it is easier to calculate the slope by adjusting the wire feed so that you are welding with 100 amperes, 200 amperes, 300 amperes, and so on. In other words, the voltage difference can be simply divided by 1 for 100 amperes, 2 for 200 amperes, and so forth.

The duty cycle of a particular machine indicates how many minutes out of ten a welding machine can run without overheating. Figure 3-23 shows the duty cycle chart for a welder. When the heat set-

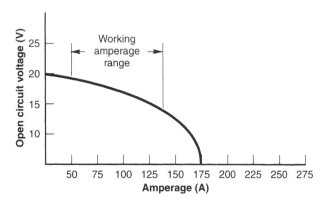

FIGURE 3–22 Constant-potential welder slope.

TABLE 3–5: EFFECT OF SLOPE

Flatter	← Slope →	Steeper
Buildup	Decreases	Increases
Depth of Fusion	Increases	Decreases
Spatter	Increases	Decreases
Shorting	Violently cleared	May not clear

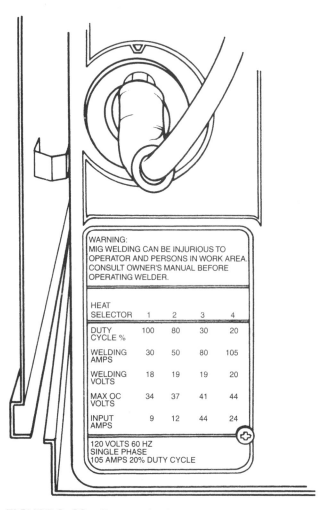

WARNING:
MIG WELDING CAN BE INJURIOUS TO OPERATOR AND PERSONS IN WORK AREA. CONSULT OWNER'S MANUAL BEFORE OPERATING WELDER.

HEAT SELECTOR	1	2	3	4
DUTY CYCLE %	100	80	30	20
WELDING AMPS	30	50	80	105
WELDING VOLTS	18	19	19	20
MAX OC VOLTS	34	37	41	44
INPUT AMPS	9	12	44	24

120 VOLTS 60 HZ
SINGLE PHASE
105 AMPS 20% DUTY CYCLE

FIGURE 3–23 Duty cycle chart.

ting is 30 amps, this machine has a 100 percent duty cycle, so it can weld 10 minutes. The machine only has a 20 percent duty cycle when the heat setting is 105 amps; it is only safe to weld 2 minutes out of 10.

All machines should have a duty cycle displayed either in the owner's manual or on the machine. The higher the duty cycle, the more expensive the machine. Very little auto body welding requires long periods of continuous welding; therefore, a typical auto body welder does not need to have a 100 percent duty cycle rating.

Some machines also display a welding guide (Figure 3-24). A typical chart identifies the type of base metal, thickness, and the recommended machine settings for solid wires and small-diameter flux-cored wires. It is a great convenience when setting the machine.

WELDING GUIDE
Settings are approximate. Adjust as required.

Material	Thickness	Process	Wire Class	Wire Size	Gas Type	Gas Flow	Polarity	Stickout	Welding Voltage	Wire Speed Control
Carbon Steel	24 ga	GMAW	ER-70S-6 (HB 28)	0.024	CO$_2$ or C$_{25}$	20 CFH	DCEP	1/4	1	5.5–6
	18 ga							1/4–5/16	1	6–7
	16 ga							5/16–1/2	2	6.5–7
	10 ga							5/16–1/2	3 (4)	7 (8)
	3/16"							1/2	4	8
	18 ga			0.030				5/16	2	5.5–6.5
	16 ga							5/16–1/2	2	6–6.5
	10 ga							1/2	3	6.5
	3/16"							1/2	4	7–7.5
	18 ga			0.035				5/16	2	5–5.5
	16 ga							5/16	2	5.5
	10 ga							1/2	3	6
	3/16"							1/2	4	6.5
Stainless Steel	10 ga		ER-308L 308L Stainless	0.030	C$_{25}$			1/2	4	7.5
Carbon Steel	18 ga	FCAW	E-717-11 Fabshield 21B	0.045	None		DCEN	1/2–3/4	2	5
	16 ga							1/2–3/4	2	5.5
	10 ga							3/4	3	6
	3/16"							3/4	4	6

24 ga = 0.022", 18 ga = 3/64", 16 ga = 1/16", 10 ga = 1/8"
CO$_2$ = Carbon dioxide
CO$_{25}$ = 25% Carbon dioxide + 75% Argon
DCEP = DC Volts Wire Positive
DCEN = DC Volts Wire Negative

FIGURE 3–24 _Typical welding guide._

3.8 SPOT WELDING

MIG can be used to make high-quality arc spot welds. Welds can be made using standard or specialized equipment. The arc spot weld produced by MIG welding differs from electric resistance spot welding. The MIG welding spot weld starts on one surface of one member and burns through to the other member (Figure 3-25). Fusion between the members occurs, and a small nugget is left on the metal surface. This type of spot weld is not recommended for auto body work. It can be used for tack welding and fitup work.

MIG spot welding has some advantages for such things as the following: (1) welds can be made in thin-to-thick materials, (2) the weld can be made when only one side of the material to be welded is accessible, and (3) the weld can be made when there is paint on the interfacing surfaces. The arc spot weld can also be used to assemble parts for welding to be done at a later time.

Thin metal can be attached to thicker sections using an arc spot weld. If a thin-to-thick butt, lap, or tee joint is to be welded with complete joint penetration, often the thin material will burn back, leaving a hole, or there will not be enough heat to melt the thick section. With an arc spot weld, the burning back of the thin material allows the thicker metal to be melted. As more metal is added to the weld, the burn-through is filled (Figure 3-25).

The MIG spot weld is produced from only one side. Therefore, it can be used on awkward shapes and in cases where the other side of the surface being welded should not be damaged. In addition, because the metals are melted and the molten weld pool is agitated, some thin films of paint between the members being joined need not be removed.

CAUTION: Safety glasses and/or flash glasses must be worn to protect the eyes from flying sparks.

Specially designed nozzles provide flash protection, part alignment, and arc alignment (Figure 3-26). As a result, for some small jobs it may be possible to perform the weld without a welding helmet. Welders can shut their eyes and turn their head during the weld.

CAUTION: This is not advisable for any work requiring more than just a few spot welds. Prolonged exposure to the reflected ultraviolet light will cause skin burns.

The optional control timer provides weld time and burn-back time. To make a weld, the amperage, voltage, and length of welding time must be set correctly. The burn-back time is a short period at the end of the weld when the wire feed stops but the current does not. This allows the wire to be burned back so it does not stick in the weld.

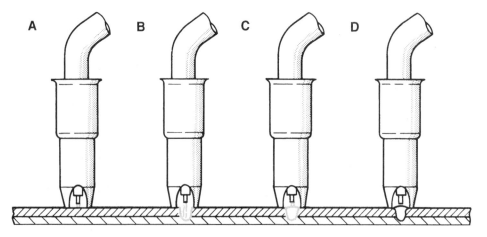

FIGURE 3–25   GMA spot weld: (A) the arc starts, (B) a hole is burned through the first plate, (C) the hole is filled with weld metal, and (D) the wire feed stops and the arc burns the electrode back.

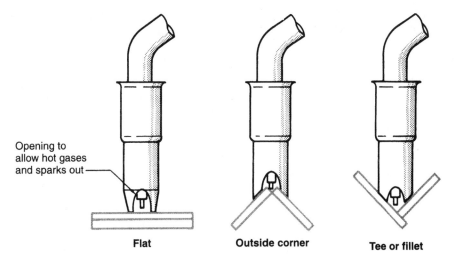

Opening to allow hot gases and sparks out

Flat   **Outside corner**   **Tee or fillet**

FIGURE 3-26 *Special nozzles are available for different types of welds.*

3.9 FLUX-CORED ARC WELDING

Flux-cored arc welding (FCAW) is an electric arc welding process similar to MIG that uses a tubular wire with flux inside. With 0.030-inch (0.8 mm) self-shielded flux-cored wire, the flux-cored welding process has proven to be a valuable tool for working on thin-gauge steels. The FCAW process uses the same type of constant potential power source as MIG welding. It also uses the electrode feed system, contact tube, electrode conduit, welding gun, and many other pieces of equipment that are used in MIG welding.

Not all FCAW requires an external shielding gas. As the flux within the wire vaporizes in the heat of the arc, the gases created shield the molten weld pool, stabilize the arc, help to control penetration, and reduce porosity. The flux also mixes with the impurities on the metal surface and brings them to the top of the weld, where they solidify as slag. The slag must then be chipped or brushed away.

Two very important advantages of the FCAW process over MIG welding are its ability to tolerate surface impurities (thus, it requires less precleaning) and its ability to stabilize the arc. Other beneficial characteristics of the process include the following:

- High deposition rate.
- Efficient electrode metal use.
- Requires little edge preparation.
- Welds in any position.
- Welds a wide range of metal thicknesses with one size of electrode.

- Produces high-quality welds.
- Molten weld pool is easily controlled, and its surface appearance is smooth and uniform even with minimal operator skill.
- Produces a weld with less porosity than MIG welding when welding galvanized steels.

The nozzle can be removed when using self-shielded wires to improve visibility.

While the FCAW process has a number of advantages over MIG welding, it has the following drawbacks:

- FCAW wires are more expensive than MIG welding hard wires.
- Spatter is worse when using flux-cored wires.
- Only ferrous metals can be welded.

Like MIG welding, FCAW has certain variables that must be carefully controlled to deliver quality welds. These include

- Welding voltage (heat)
- Welding current (wire speed)
- Travel speed
- Gun angle
- Contact tip height (electrical stickout)

Optimum voltage and current value for 0.030-inch flux-cored wire is approximately 100 amps at 16 volts. With the variety of machines in use today, it is best to check the owner's manual for setting recommendations. Settings for 0.030-inch hard wire are similar to settings for 0.030-inch flux-cored wire. Remember that voltage and current are determined primarily by the thickness of the base metal. Secondary considerations are the type of joint, the position in which the welding is to be done, and the skill of the operator.

REVIEW QUESTIONS

1. Define welding.
2. When is filler material added?
3. What is the purpose of a shielding gas?
4. What does MIG stand for?
5. What are the advantages of MIG welding?
6. When did MIG welding become popular?
7. Name some of the areas that can be welded using MIG besides body repairs.
8. What does the basic MIG equipment consist of?
9. What is the purpose of the wire feeder?
10. What type of grooved roller is best suited for hard wires such as stainless steel and mild steel?
11. What happens when there is a failure to attach the conduit?
12. What is a welding gun?
13. What are the most frequently used filler metals?
14. How is the wire melting rate measured?
15. What are the methods of metal transfer?
16. Define voltage.
17. The duty cycle of a particular machine indicates what?
18. What is the burn-back time?
19. What is FCAW?
20. What are the advantages of the FCAW process over MIG welding?
21. What is the optimum voltage and current value for 0.030-inch flux-cored wire?

Chapter
4

MIG Welding Techniques

Objectives

After reading this chapter, you should be able to:

- Name the six basic welding techniques used with MIG equipment and describe how each is accomplished.
- Identify butt, lap or flange, plug, spot, and stitch welds.
- Describe the problems unique to MIG welding of galvanized metals.
- Distinguish between weld discontinuities and weld defects.
- Describe how to perform penetrant visual and ultrasonic inspection.
- Identify common MIG weld defects and their causes.

Performing a satisfactory MIG weld requires more than just manipulative skill. The setup, voltage, amperage, electrode extension, and welding angle, as well as other factors, can dramatically affect the weld produced. The very best welding conditions are those that will allow an auto body technician to produce the largest quantity of successful welds in the shortest period of time. Because this is a semiautomatic or automatic process, increased productivity can be obtained. This does not mean that you will work harder, but rather that you will work more productively, resulting in a greater cost efficiency.

The more cost efficient you can be, the more competitive your companies become. This can make the difference between getting a job or losing work.

CHOOSING THE PROPER EQUIPMENT

It is important to select the proper size MIG welder for your welding situation. While smaller welders, such as

the one in Figure 4-1, are fine for welding layers of standard 18- to 25-gauge sheet metal, some high-production shops may require a larger capacity welder.

Using a welder with enough capacity also promotes rapid fusion of the panels, thus reducing the chances of the adjoining metal being affected by the heat zone. Keep in mind that most welds in structural areas are performed on a flanged, U-shaped panel; the factory makes all the original spot welds on the outer flanges of this panel. The strength is derived from the 90-degree bends in the panel and the welds attached to the outer flanges. The role of the welds is to hold the rigid panel in position and maintain structural integrity, both during normal driving and in the event of a collision. It is therefore crucial that the heat be contained in the flange and not allowed to enter the bent areas of the panel, since this is where the structural strength is obtained. To do this, heat control is crucial; the shorter the duration of the weld, the better the heat is contained.

Always disconnect the battery terminals before doing any MIG or arc welding on the vehicle. Also, remove the computer module (if applicable) and store it

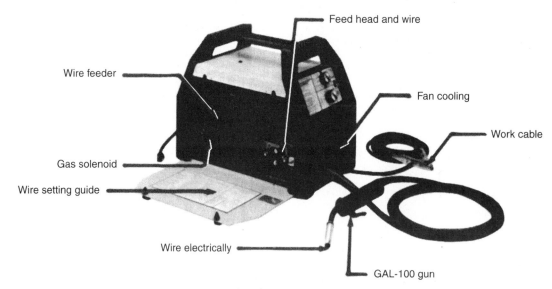

FIGURE 4–1 *Small welders are suitable for welding single layers of sheet metal. (Courtesy of Hobart Brothers Company)*

outside the immediate welding area. The magnetic fields created by welding machines have been known to affect on-board computers.

ARC VOLTAGE AND AMPERAGE CHARACTERISTICS

The arc-voltage and amperage characteristics of MIG welding are different from those in most other welding processes. The voltage is set on the welder, and the amperage is set by changing the wire feed speed. At a voltage setting, the amperage required to melt the wire as it is fed into the weld must change. It requires more amperage to melt the wire the faster it is fed and less amperage the slower the wire is fed.

Because changes in the wire feed speed directly change the amperage, it is possible to set the amperage by using a chart and measuring the length of wire fed per minute (Table 4-1). The voltage and amperage required for a specific metal transfer method will be different for various wire sizes, shielding gases, and metals.

The voltage and amperage settings will be specified for most welding that is done according to the vehicle manufacturer's specifications. Welding that is done on older vehicles is not done to any specific set of standards, and therefore no specific settings exist. For that reason, and because even when specifications are given there is a range of acceptable settings, it is important that you learn to make the necessary adjustments to allow you to produce quality welds.

TABLE 4–1: WELDING SPEED	
Panel Thickness	**Welding Speed (in./min.)**
1/32"	41-11/32 to 45-9/32
More Than 1/32"	39-3/8
3/64"	35-7/16 to 39-3/8
1/16"	31-1/2 to 33-15/32

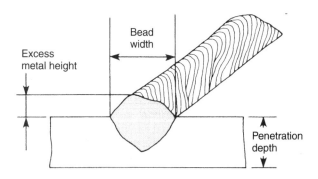

FIGURE 4–2 *Penetration depth, excess metal height, and bead width.*

WELDING CURRENT

The welding current affects the base metal penetration depth (Figure 4-2), the speed at which the wire is melted, arc stability, and the amount of weld spatter. As the current is increased, the penetration depth, excess metal height, and bead width also increase (Table 4-2).

TABLE 4–2:   ADJUSTMENTS IN WELDING VARIABLES AND TECHNIQUES

Welding Variables to Change	Desired Changes							
	Penetration		Deposition Rate		Bead Size		Bead Width	
	Increase	Decrease	Increase	Decrease	Increase	Decrease	Increase	Decrease
Current and Wire Feed Speed	Increase	Decrease	Increase	Decrease	Increase	Decrease	No effect	No effect
Voltage	Little effect	Little effect	No effect	No effect	No effect	No effect	Increase	Decrease
Travel Speed	Little effect	Little effect	No effect	No effect	Decrease	Increase	Increase	Decrease
Stickout	Decrease	Increase	Increase	Decrease	Increase	Decrease	Decrease	Increase
Wire Diameter	Decrease	Increase	Decrease	Increase	No effect	No effect	No effect	No effect
Shield Gas Percent CO_2	Increase	Decrease	No effect	No effect	No effect	No effect	Increase	Decrease
Torch Angle	Backhand to 25°	Forehand	No effect	No effect	No effect	No effect	Backhand	Forehand

ARC VOLTAGE

Good welding results depend on a proper arc length. The length of the arc is determined by the arc voltage. When the voltage is set properly, a continuous light hissing or crackling sound is emitted from the welding area. When the voltage is high, the arc length increases, the penetration is shallow, and the bead is wide and flat. When the voltage is low, the arc length decreases, penetration is deep, and the bead is narrow and dome-shaped.

Since the length of the arc depends on the amount of voltage, voltage that is too high will result in an overly long arc and an increase in the amount of weld spatter (Figure 4-3). A sputtering sound and no arc means that the voltage is too low.

4.3 SETUP

The basic MIG welding installation consists of the following: welding gun, gun switch circuit, electrode conduit–welding contractor control, electrode feed unit, electrode supply, power source, shielding gas supply, shielding gas flowmeter regulator, shielding gas hoses, and both power and work cables. The equipment setup in this chapter is similar to most equipment built by all manufacturers, which means that any skills developed can be transferred easily to other equipment.

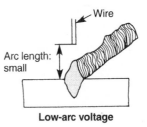

Low-arc voltage

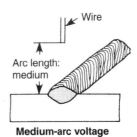

Medium-arc voltage

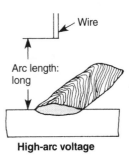

High-arc voltage

FIGURE 4–3 Arc voltage and bead shape.

PRACTICE 4–1

MIG Welding Equipment Setup

For this practice, you will need a MIG welding power source, welding gun, electrode feed unit, electrode supply, shielding gas supply, shielding gas flowmeter regulator, electrode conduit, power and work leads, shielding gas hoses, assorted hand tools, spare parts, and any other required materials. In this practice, you will properly set up a MIG welding installation.

If the shielding gas supply is a cylinder, it must be chained securely in place before the valve protection cap is removed (Figure 4-4). Standing to one side of the cylinder, quickly crack the valve to blow out any dirt in the valve before the flowmeter regulator is attached (Figure 4-5). Attach the correct hose from the regulator to the "gas-in" connection on the electrode feed unit or machine.

Install the reel of electrode (welding wire) on the holder and secure it (Figure 4-6). Check the feed roller size to ensure that it matches the wire size (Figure 4-7). The conduit liner size should be checked to be sure that it is compatible with the wire size. Connect the conduit to the feed unit. The conduit or an extension should be aligned with the groove in the roller and set as close to the roller as possible without touching it (Figure 4-8). Misalignment at this point can contribute to a "bird's nest" (Figure 4-9). Bird nesting

FIGURE 4–5 *Attach the flowmeter regulator. Be sure the tube is vertical.*

FIGURE 4–4 *Make sure the gas cylinder is chained securely in place before removing the safety cap.*

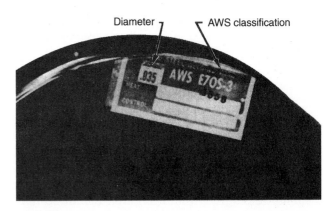

FIGURE 4–6 *When installing the spool of wire, check the label to be sure that the wire is the correct type and size.*

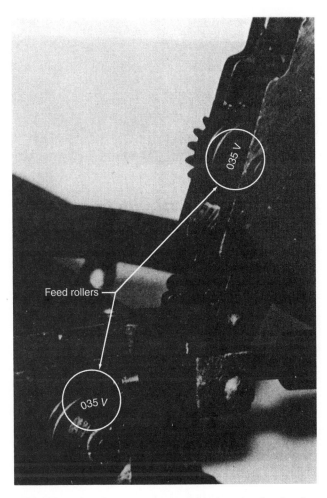

FIGURE 4–7 Check to be certain that the feed rollers are the correct size for the wire being used.

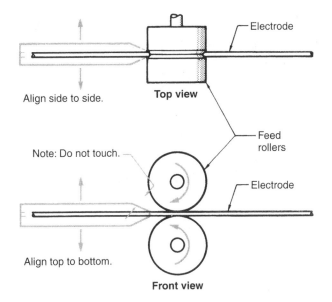

FIGURE 4–8 Feed roller and conduit alignment.

FIGURE 4–9 "Bird's nest" in the filler wire at the feed rollers.

of the electrode wire results when the feed roller pushes the wire into a tangled ball because the wire would not go through the outfeed-side conduit. It looks like a bird's nest.

Be sure the power is off before attaching the welding cables. The electrode and work leads should be attached to the proper terminals. The electrode lead should be attached to electrode or positive (+). If necessary, it is also attached to the power cable part of the gun lead. The work lead should be attached to work or negative (−).

The shielding "gas-out" side of the solenoid is then also attached to the gun lead. If a separate splice is required from the gun switch circuit to the feed unit, it should be connected at this time (Figure 4-10). Check to see that the welding contractor circuit is connected from the feed unit to the power source.

The welding gun should be securely attached to the main lead cable and conduit (Figure 4-11). There should be a gas diffuser attached to the end of the conduit liner to ensure proper alignment. A contact tube (tip) of the correct size to match the electrode wire being used should be installed (Figure 4-12). A shielding gas nozzle is attached to complete the assembly.

Recheck all fittings and connections for tightness. Loose fittings can leak; loose connections can cause added resistance, reducing the welding efficiency.

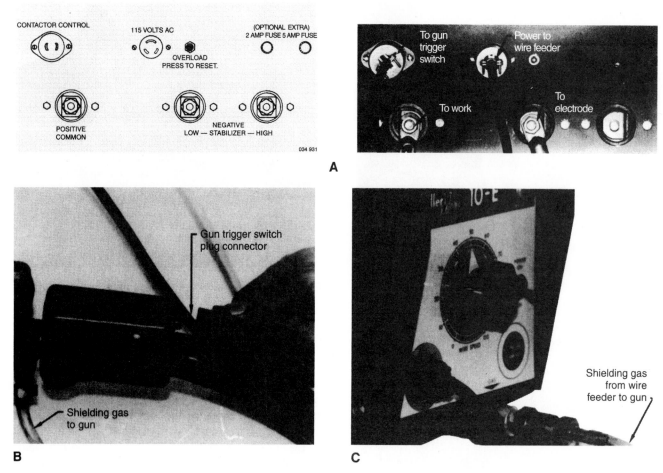

FIGURE 4–10   Connect the leads and other lines as shown in the owner's manual.

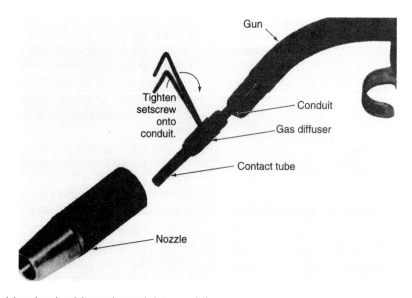

FIGURE 4–11   Attaching lead cable and conduit to welding gun.

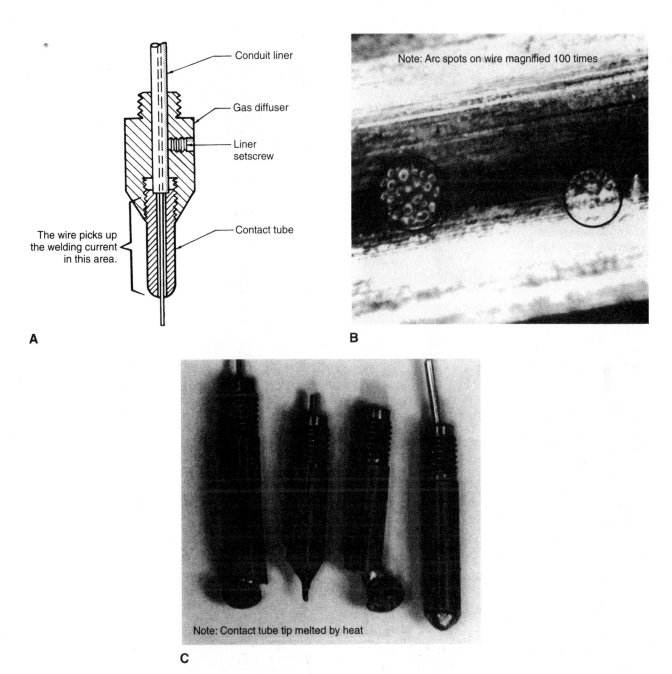

A

The wire picks up the welding current in this area.

Conduit liner

Gas diffuser

Liner setscrew

Contact tube

B

Note: Arc spots on wire magnified 100 times

C

Note: Contact tube tip melted by heat

FIGURE 4–12 *(A) The contact tube must be the correct size. Too small a contact tube will cause the wire to stick. (B) Too large a contact tube can cause arcing to occur between the wire and tube. (C) Heat from the arcing can damage the tube.*

Some manufacturers include detailed setup instructions with their equipment (Figure 4-13).

PRACTICE 4–2

Threading MIG Welding Wire

Using the MIG welding machine that was properly assembled in Practice 4-1, you will turn the machine on and thread the electrode wire through the system.

Check to see that the unit is assembled correctly according to the manufacturer's specifications. Switch on the power and check the gun switch circuit by depressing the switch. The power source relays, feed relays, gas solenoid, and feed motor should all activate.

Cut the end of the electrode wire free. Hold it tightly so that it does not unwind. The wire has a natural curve that is known as its cast. The cast is measured by the diameter of the circle that the wire would make if it were loosely laid on a flat

1. Open side using easy "swell" latches.

2. Remove wire spool nut. Unload wire spool, remove protective packaging, reload wire spool with free end unreeling from bottom, left to right. Reattach wire spool nut.

3. Release upper feed roller.

4. Thread wire through guide between rollers and into wire cable.

5. Clamp upper feed roller.

6. For spec. 7144-1, plug unit into a 120 VAC grounded wall outlet.

7. Turn input switch on. Set weld voltage range switch to "1." Advance wire to end of cable with gun trigger. **Warning:** Wire is electrically hot when trigger is pulled.

8. Set voltage and wire feed speed using wire settings guide on inside of case.

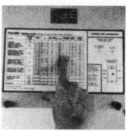

9. Attach work cable clamp to work to be welded.

10. Connect gas to coupling at rear of case.

11. For gas-shielded solid wire, be sure B cable is attached to negative (−) terminal and A cable to positive (+). Unit is shipped this way. To use handler with self-shielding tubular wire, reverse polarity by connecting A to negative (−) and B to positive (+) terminal. Refer to guide inside case. See photo 8.

12. **Always wear proper safety equipment.** Pull trigger and weld.

FIGURE 4–13   *Example of manufacturer's setup instructions. (Courtesy of Hobart Brothers Company)*

surface (Figure 4-14). The cast helps the wire to make a good electrical contact as it passes through the contact tube (Figure 4-15). However, the cast can be a problem when threading the system. To make threading easier, straighten about 12 inches (305 mm) of the end of the wire and cut any kinks off.

Separate the wire feed rollers, and push the wire first through the guides, then between the rollers, and finally into the conduit liner (Figure 4-16). Reset the rollers so that there is a slight amount of compression on the wire (Figure 4-17). Set the wire feed speed control to a slow speed. Hold the welding gun so that the electrode conduit and cable are as straight as possible.

Press the gun switch. The wire should start feeding into the liner. Watch to make certain that the wire feeds smoothly, and release the gun switch as soon as the end comes through the contact tube.

***

CAUTION: If the wire stops feeding before it reaches the end of the contact tube, stop and check the system. If no obvious problem can be found, mark the wire with tape and remove it from the gun. It then can be held next to the system to determine the location of the problem.

***

With the wire feed running, adjust the feed roller compression so that the wire reel can be stopped easily by a slight resistance. Too light a roller pressure will cause the wire to feed erratically. Too high a pressure can turn a minor problem into a major

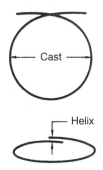

FIGURE 4–14   Cast of electrode wire.

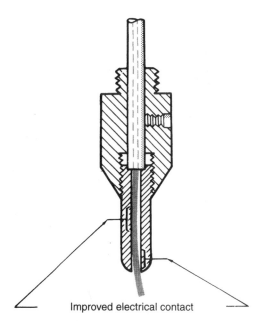

FIGURE 4–15   Cast forces the wire to make better electrical contact with the tube.

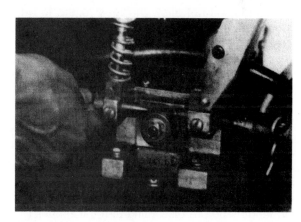

FIGURE 4–16   Push the wire through the guides by hand.

FIGURE 4–17   Adjust the wire feed tensioner.

disaster. If the wire jams when the roller pressure is high, the rollers will keep feeding the wire, causing it to bird nest and possibly short out. With a light roller pressure, the wire can stop, preventing bird nesting. This is very important with soft wires. The other advantage of a light pressure is that the feed will stop if something like clothing or gas hoses are caught in the reel.

With the feed running, adjust the spool drag so that the reel stops when the feed stops. The reel should not coast to a stop, because the excess wire that rolls off can be snagged easily. Also, when the feed restarts, a jolt occurs when the slack in the excess wire is taken up. This jolt can be enough to momentarily stop the wire, possibly causing a discontinuity in the weld.

When the test runs are completed, the wire can either be rewound or cut off. Some wire feed units have a retract button. This allows the feed driver to reverse and retract the wire automatically. To rewind the wire on units without this retract feature, release the rollers and turn them backward by hand. If the machine will not allow the feed rollers to be released without upsetting the tension, you must cut the wire.

CAUTION: Do not discard pieces of wire on the floor. They present a hazard to safe movement around the machine. In addition, a small piece of wire can work its way into a filter screen on the welding power source. If the piece of wire shorts out inside the machine, it could become charged with high voltage, which could cause injury or death. Always wind the wire tightly into a ball or cut it into short lengths before discarding it in the proper waste container.

4.4 GAS DENSITY AND FLOW RATES

Density is the chief factor determining how effective a gas is for arc shielding. The lower the density of a gas, the higher the flow rate required for arc protection. The flow rates, however, are not in proportion to the densities. Helium, with about one-tenth the density of argon, requires only twice the flow for equal protection.

PRACTICE 4-3

Setting Gas Flow Rate

Using the equipment setup as described in Practice 4-1 and the threaded machine as described in Practice 4-2, you will set the shielding gas flow rate.

The exact flow rate required for a job will vary depending upon welding conditions. This experiment will help you to determine how those conditions affect the flow rate. You will start by setting the shielding gas flow rate at 35 cfh (16 L/min).

Turn on the shielding gas supply valve. If the supply is a cylinder, the valve is opened all the way. With the machine power on and the welding gun switch depressed, you are ready to set the flow rate. Slowly turn in the adjusting screw and watch the float ball as it rises in a tube on a column of gas. The faster the gas flows, the higher the ball will float. A scale on the tube allows you to read the flow rate. Different scales are used with each type of gas being used. Since various gases have different densities (weights), the ball will float at varying levels even when the flow rates between gases are the same (Figure 4-18). The line corresponding to the flow rate may be read as it compares to the top, center, or bottom of the ball, depending upon the manufacturer's instructions. There should be some marking or instruction on the tube or regulator to tell a person how it should be read (Figure 4-19).

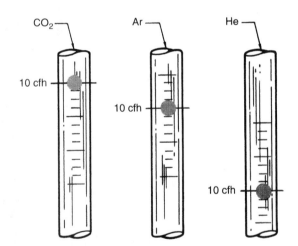

FIGURE 4-18 *Each of these gases is flowing at the same cfh (L/min) rate. Because helium (He) is the least dense, its indicator ball is the lowest. Be sure that you are reading the correct scale for the gas being used.*

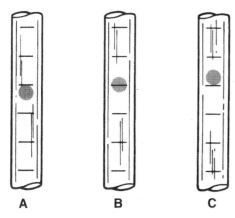

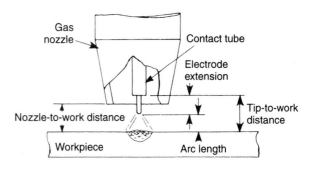

FIGURE 4-20 Tip-to-base metal distance.

FIGURE 4-19 Three methods of reading a flowmeter: (A) top of ball, (B) center of ball, and (C) bottom of ball.

Release the welding gun switch; the gas flow should stop. Turn off the power and spray the hose fittings with a leak-detecting solution.

When stopping for a period of time, the shielding gas supply valve should be closed and the hose pressure released.

4.5 — OPERATOR VARIABLES

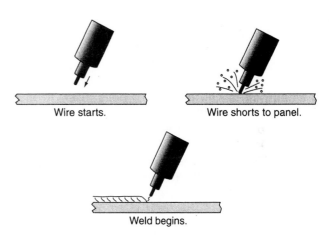

FIGURE 4-21 The beginning of a MIG weld.

TIP-TO-BASE METAL DISTANCE

The tip-to-work distance (Figure 4-20) is also an important factor in obtaining good welding results. The standard distance is approximately 1/4 to 5/8 inch (6 mm to 9 mm).

If the tip-to-base metal distance is too long, the length of wire protruding from the end of the gun increases and becomes preheated, which decreases the heat input to the weld and decreases weld penetration. Also, the shield gas effect will be reduced. If the tip-to-base metal distance is too short, the penetration increases, visibility decreases, and the tip can short to the work.

WELDING SPEED

If you weld at a rapid pace, the penetration depth and bead width will decrease, and the bead will be dome-shaped. If the speed is increased even more, undercutting can occur. Welding at too slow a speed can cause burn-through holes. Ordinarily, welding speed is determined by base panel thickness and/or voltage of the welding machine.

WIRE SPEED

An even, high-pitched buzzing sound indicates the correct wire-to-heat ratio producing the correct welding temperature. A steady, reflected light should be observed; it will start to fade in intensity as the arc is shortened and wire speed is increased.

If the wire speed is too slow, a hissing and plopping sound will be heard as the wire melts away from the molten weld pool and deposits the molten droplet. There will be a much brighter reflected light. Too much wire speed will choke the arc; more wire is being deposited than the heat and molten weld pool can absorb. The result is that the wire melts into tiny balls of molten metal (spatter) that fly away from the weld. There is a strobe light arc effect.

When the trigger is first activated, a solid steel wire makes its initial contact with a solid steel plate. The wire has been charged with current and the gas flow has been started prior to contact (Figure 4-21). The first contact produces tiny sparks being burned off the wire and base metal.

Immediately after the sparks appear, tiny molten droplets form as the wire melts. Once the heat creates the molten weld pool, the spatter reduces greatly.

TABLE 4–3:   WELDING VARIABLES

Weld Acceptability	Voltage	Amperage	Electrode Extension	Contact Tube-to-Work Distance	Bead Shape
Poor	20	100	1 in (25 mm)	1 1/4 in (31 mm)	Narrow, high with little penetration

Electrode Diameter	.035 in (0.9 mm)
Shielding Gas	75% Ar + 25% CO_2
Welding Direction	Forehand

After the arc transfer has been started, an on-off arc cycling action occurs in rapid sucession. It cycles on and off at approximately 1 to 230 times per second. Every time the metal is deposited a small pop is heard; when it pulls away, a hiss is heard. Speeding it up to approximately 200 plops and hisses per second creates a smooth buzzing sound.

Decreasing the wire speed causes the droplets to burn farther back at about 150 transfers per second. At 150 transfers per second, the sound is a lower buzz and the light is brighter than the high-pitched buzz and dimmer light at 230. The more time the droplets have to burn back, the more time the arc has to strike the molten weld pool. Therefore, a flatter weld is produced with a slower wire speed.

When you are welding overhead, the danger of having too large a molten weld pool and droplets is obvious. The droplets can be pulled by gravity down onto the contact tip or into the gas nozzle, where they can create serious problems. Therefore, overhead welding should always be done at a high wire speed, with a low voltage-to-amperage ratio. This is done to keep the molten weld pool smaller and more easily controlled.

Normal buildup of spatter in the gas nozzle area must always be removed before it falls inside and shorts out the nozzle. More spatter buildups in the nozzle will occur during overhead welding.

Table 4-3 lists several variables that affect weld characteristics. Later you will add your own practice data to Table 4-3.

4.6 GUN NOZZLE

The nozzle used on MIG welding guns controls the shielding gas so that it will provide proper gas protection for the molten weld pool. These nozzles are made out of copper or copper alloys. Copper and its alloys provide excellent heat resistance and easy spatter removal.

An insulator separates the nozzle from the gas diffuser/contact tip. If the insulation is bridged by spatter, the power intended for the wire is transferred to the gas nozzle. This can cause the nozzle to arc to the work. Welding on dirty or rusty material can cause heavy spatter buildup in the nozzle.

The basic adjustment procedure for a gas nozzle is as follows:

- *Arc Generation.* Position the tip of the gun near the base metal. When the gun switch is activated, the wire is fed at the same time as the shield gas. Bring the end of the wire in contact with the base metal and create an arc. If the distance between the tip and the base metal is shortened a little, it will be easier to create the arc (Figure 4-22). If the end of the wire forms a large ball, it will be difficult to generate an arc, so cut off the end of the wire with a pair of wire cutters (Figure 4-23).

CAUTION: Hold the tip of the gun away from your face when cutting off the end of the wire.

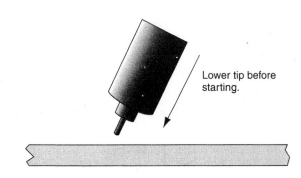

FIGURE 4–22 Creating an arc.

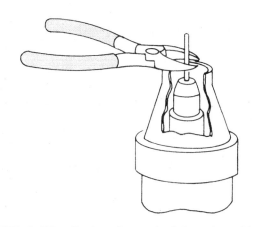

FIGURE 4–23 Cutting the end of the wire with wire cutters.

- *Spatter Treatment.* Remove weld spatter periodically. If it is allowed to build up excessively it can adhere to the end of the nozzle, impeding the shield gas flow and resulting in a poor weld. Antispatter compounds are available that reduce the amount of spatter on the nozzle.
- *Contact Tip Conditions.* To ensure a stable arc, the tip should be replaced if it has become worn. For a good current flow and a stable arc, keep the tip properly tightened.

PRACTICE 4–4

Setting the Current

Using a properly assembled MIG welding machine, proper safety protection, and one piece of mild steel plate approximately 12 inches long by 1/4 inch thick,

you will change the current settings and observe the effect on MIG welding.

On a scale of 0 to 10, set the wire feed speed control dial at 5, or halfway between the low and high settings of the unit. The voltage is also set at a point halfway between the low and high settings. The shielding gas should be a 25 percent CO_2, 75 percent argon mixture. The gas flow should be adjusted to a rate of 35 cfh (16 L/min).

Hold the welding gun at a comfortable angle, lower your welding hood, and pull the trigger. As the wire feeds and contacts the plate, the weld will begin. Move the gun slowly along the plate. Note the following welding conditions as the weld progresses: voltage, amperage, weld direction, metal transfer, spatter, molten weld pool size, and penetration. Stop and record your observations in Table 4-3. Evaluate the quality of the weld as acceptable or unacceptable.

Reduce the voltage somewhat and make another weld, keeping all other weld variables (travel speed, stickout, direction, amperage) the same. Observe the weld, and upon stopping, record the results. Repeat this procedure until the voltage has been lowered to the minimum value indicated on the machine. Near the lower end the wire may stick, jump, or simply no longer weld.

Return the voltage indicator to the original starting position and make a short test weld. Stop and compare the results to those first observed. Then slightly increase the voltage setting and make another weld. Repeat the procedure of observing and recording the results as the voltage is increased in steps until the maximum machine capability is obtained. Near the maximum setting the spatter may become excessive. Care must be taken to prevent the wire from fusing to the contact tube.

Return the voltage indicator again to the original starting position and make a short test weld. Compare the results observed with those previously obtained.

Lower the wire feed speed setting slightly and use the same procedure as before. First lower and then raise the voltage through a complete range and record your observations. After a complete set of test results is obtained from this amperage setting, again lower the wire feed speed for a new series of tests. Repeat this procedure until the amperage is at the minimum setting shown on the machine. At low amperages and high voltage settings, the wire may tend to pop violently as a result of the uncontrolled arc.

Return the wire feed speed and voltages to the original settings. Make a test weld and compare the

TABLE 4–4:   GRAPH FOR MIG MACHINE SETTINGS

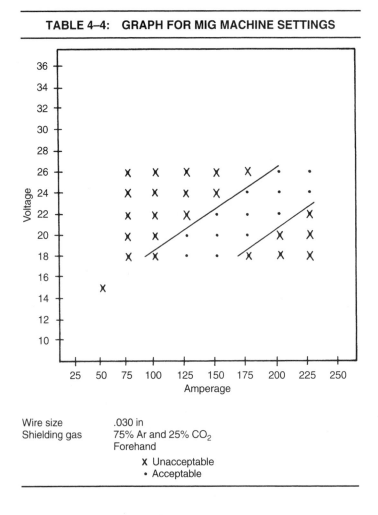

Wire size   .030 in
Shielding gas   75% Ar and 25% CO$_2$
                   Forehand

                   X Unacceptable
                   • Acceptable

results with the original tests. Slightly raise the wire speed and again run a set of tests as the voltage is changed in small steps. After each series, return the voltage setting to the starting point and increase the wire feed speed. Make a new set of tests.

All of the test data can be gathered into an operational graph for the machine, wire type, size, and shielding gas. Use Table 4-4 to plot the graph. The acceptable welds should be marked on the lines that extend from the appropriate voltages and amperages. Upon completion, the graph will give you the optimum settings for the operation of this particular MIG welding setup. The optimum settings are along a line in the center of the acceptable welds.

Experienced welders will follow a much shorter version of this type of procedure any time they are starting to work on a new machine or testing for a new job.

This experiment can be repeated using different types of wire, wire sizes, shielding gases, and weld directions. Turn off the welding machine and shielding gas supply, and clean up your work area when you are finished welding.

ELECTRODE EXTENSION

Because of the constant potential (CP) power supply, the welding current will change as the distance between the contact tube and the work changes. Although this change is slight, it is enough to affect the weld being produced. The longer the electrode extension, the greater the resistance to the welding current flowing through the small welding wire. This results in some of the welding current being changed to heat at

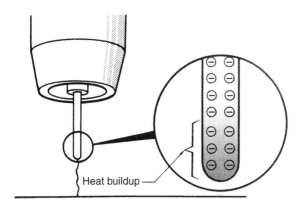

FIGURE 4-24 *Heat buildup due to the extremely high current for the small conductor (electrode).*

the tip of the electrode (Figure 4-24). With a standard SMA welding CC power supply, this would also reduce the arc voltage, but with a CP power supply, the voltage remains constant and the amperage increases. If the electrode extension is shortened, the welding current decreases.

The increase in current does not result in an increase in penetration, because the current is being used to heat the electrode tip and is not being transferred to the weld metal. Penetration is reduced and buildup increased as the electrode extension is reduced in length. Penetration is increased and buildup decreased as the electrode extension is lengthened. Being able to control the weld penetration and buildup by changing the electrode will help during welding. It will also help you to better understand what may be happening if a weld starts out correctly but begins to change as it progresses along the joint. You may be changing the electrode extension without noticing the change.

PRACTICE 4-5

Electrode Extension

Using a properly assembled MIG welding machine, proper safety protection, and a few pieces of mild steel, each about 12 inches long and ranging in thickness from 16 to 24 gauge, you will observe the effect of changing the electrode extension on the weld.

Start at a low current setting. Using the graph developed in Practice 4-4, set both the voltage and the amperage. The settings should be equal to those on the optimum line established for the wire type and size being used with the same shielding gas.

Holding the welding gun at a comfortable angle and height, lower your helmet, and start to weld. Make a weld approximately 2 inches long. Then reduce the distance from the gun to the work while continuing to weld. After a few inches, again shorten the electrode extension. Keep doing this in steps until the nozzle is as close as possible to the work. Stop and return the gun to the original starting distance.

Repeat the process just described, but now increase the electrode extension in steps of an 1/8 of an inch. Keep increasing the electrode extension until the weld will no longer fuse or the wire becomes impossible to control.

Change the plate thickness and repeat the procedure. When the series has been completed with each plate thickness, raise the voltage and amperage to a medium setting and repeat the process again. Upon completing this series of tests, adjust the voltage and amperage upward to a high setting. Make a full series of tests using the same procedures as before.

Record the results after each series of tests. The final results can be plotted on a graph, as is shown in Table 4-5, to establish the optimum electrode extension for each thickness, voltage, and amperage. Turn off the welding machine and shielding gas supply, and clean up your work area when you are finished welding.

WELDING GUN ANGLE

The term *welding gun angle* refers to the angle between the MIG welding gun and the work as it relates to the direction of travel. Backhand welding or a dragging angle (Figure 4-25) produces a weld with deep penetration and higher buildup. Forehand welding or a pushing angle (Figure 4-26) produces a weld with shallow penetration and little buildup.

Slight changes in the welding gun angle can be used to control the weld as the groove spacing changes. A narrow gap may require more penetration,

TABLE 4–5: ELECTRODE EXTENSION GRAPH

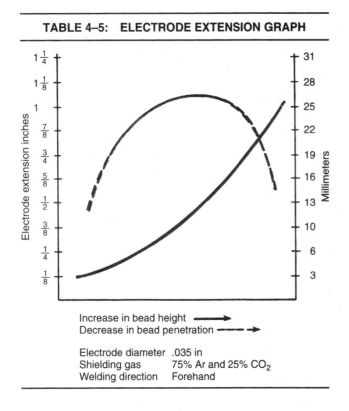

Increase in bead height →
Decrease in bead penetration ---→

Electrode diameter .035 in
Shielding gas 75% Ar and 25% CO_2
Welding direction Forehand

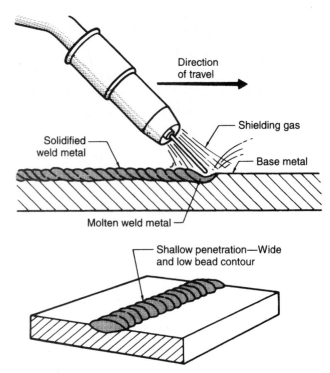

FIGURE 4–26 Forehand welding or pushing angle.

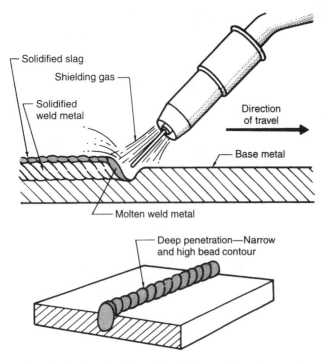

FIGURE 4–25 Backhand welding or dragging angle.

but as the gap spacing increases, a weld with less penetration may be required. Changing the electrode extension and welding gun angle at the same time can result in a quality weld being made with less than ideal conditions.

PRACTICE 4–6

Welding Gun Angle

Using a properly assembled MIG welding machine, proper safety protection, and some pieces of mild steel, each approximately 12 inches long and ranging in thickness from 16 to 24 gauge, you will observe the effect of changing the welding gun angle on the weld bead.

Starting with a medium current setting and a 16-gauge plate, hold the welding gun at a 30-degree angle to the plate in the direction of the weld (Figure 4-27). Lower your welding hood and depress the trigger. When the weld starts, move in a straight line and slowly pivot the gun angle as the weld progresses. Keep the travel speed, electrode

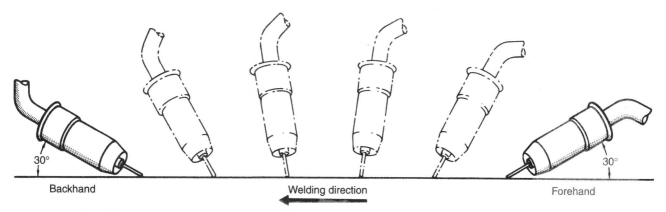

FIGURE 4-27 *Welding gun angle.*

extension, and weave pattern (if used) constant so that any change in the weld bead is caused by the angle change.

The pivot should be completed within the 12-inch weld. You will proceed from a 30-degree pushing angle to a 30-degree dragging angle. Repeat this procedure using different welding currents and plate thicknesses.

After the welds are complete, note the differences in width and reinforcement along the welds. Turn off the welding machine and shielding gas supply, and clean up your work area when you are finished welding.

4.8 EFFECT OF SHIELDING GAS ON WELDING

Shielding gases in the gas metal-arc process are used primarily to protect the molten metal from oxidation and contamination. Other factors must be considered, however, in selecting the right gas for a particular application. Shielding gas can influence arc and metal-transfer characteristics, weld penetration, width of fusion zone, surface shape patterns, welding speed, and undercut tendency. Inert gases such as argon and helium provide the necessary shielding because they do not form compounds with any other substance and are insoluble in molten metal. When used as pure gases for welding ferrous metals, argon and helium may produce an erratic arc action and may promote undercutting and other flaws.

It is therefore usually necessary to add controlled quantities of reactive gases to achieve good arc action and metal transfer with these materials. Adding oxygen or carbon dioxide to the inert gas tends to stabilize the arc, promote favorable metal transfer, and minimize spatter. As a result, the penetration pattern is improved, and undercutting is reduced or eliminated.

Oxygen or carbon dioxide are often added to argon. The amount of reactive gas required to produce the desired effects is quite small. As little as a one-half of one percent of oxygen will produce noticeable change. One to five percent oxygen is common in shielding gases. Carbon dioxide may be added to argon in the range of 20 to 30 percent. Mixtures of argon with less than 10 percent carbon dioxide may not have enough arc voltage to give the desired results.

Adding oxygen or carbon dioxide to an inert gas causes the shielding gas to become oxidizing. This in turn may cause porosity in some ferrous metals. In such cases, a filler wire containing suitable deoxidizers should be used.

Pure carbon dioxide has not become widely used as a shielding gas for MIG welding in collision repair. The chief drawback in the use of pure carbon dioxide is the less-steady arc characteristics that make it difficult to maintain a very short, uniform arc length. Also, it is not approved for auto body welding.

4.9 MIG WELDING TECHNIQUES

There are six basic welding techniques employed with MIG equipment, as shown in (Figure 4-28).

- *Tack Weld.* The tack weld is exactly that: a tack, or relatively small, temporary MIG weld that is used instead of a clamp or sheet metal screw to hold the fitup in place while proceeding to make a permanent weld. Tack welds are used like clamps or sheet metal screws, and are only a temporary holding device. The length of the tack weld is determined by the thickness of the panel. Ordinarily, a length of 15 to 30 times the thickness of the panel is appropriate (Figure 4-29). Temporary welds must not interfere with additional assembly and welding, so they must be done accurately.

- *Continuous Weld.* In a continuous weld, an uninterrupted bead is laid down in a steady, ongoing movement. The gun must be supported securely so that it does not wobble. Use the forward method, moving the gun continuously at a constant speed, carefully watching the welding bead. The gun should be inclined between 10 and 15 degrees to obtain the optimum bead shape and shield gas effect (Figure 4-30).

  Maintaining the proper tip-to-base metal distance and correct gun angle are both important when making a continuous weld. If the weld is not progressing well, stop and make the necessary adjustment. Welds that do not penetrate the metal adequately must be removed. Handling the gun in a smooth and even manner will help make the bead consistent with a uniform, closely spaced ripple.

- *Plug Weld.* In a MIG plug weld, a hole is drilled or punched through the outside piece (or pieces) of metal, the arc is directed through the hole to penetrate the inside piece, and the hole is filled with molten metal (Figure 4-31).

- *Spot Weld.* In a MIG spot weld, the arc is directed to penetrate both pieces of metal, while triggering a timed impulse of wire feed.

- *Lap Spot Weld.* In the MIG lap weld technique, the arc is directed to penetrate the bottom piece

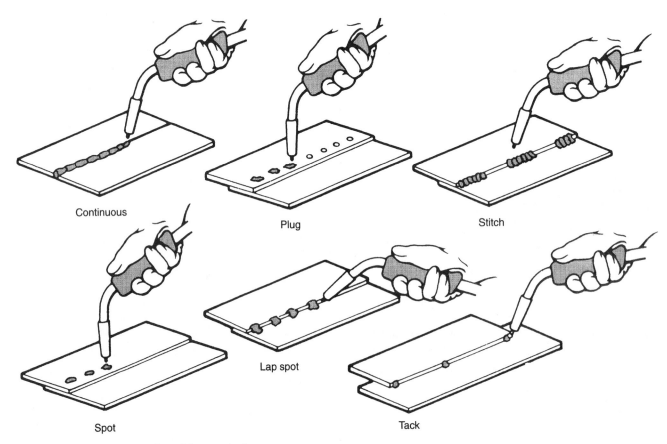

Continuous   Plug   Stitch

Spot   Lap spot   Tack

FIGURE 4–28 Basic MIG welding techniques.

of metal, and the molten weld pool is allowed to flow into the edge of the top piece (Figure 4-32).

- *Stitch Weld.* A stitch weld is a series of connecting or overlapping spot welds that create a continuous seam. Each weld that makes up the stitch weld can vary in length, depending on the metal thickness. Each weld can be longer on thicker metal and must be shorter on thinner sections.

4.10 BASIC WELDING JOINTS

The MIG welds used for the repair or reattachment of damaged or replacement sections are butt welds, lap welds, flange welds, and plug or spot welds. Each type of joint can be welded by several different techniques, depending mainly on the manufacturer's recommendations. Other factors include location application, the thickness or thinness of the metal, the

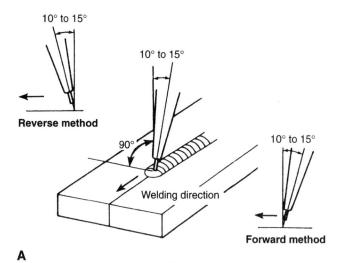

A

B

FIGURE 4–30 (A) Continuous welding. (B) Vertical down continuous weld. (Courtesy of Larry Maupin)

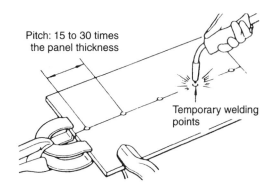

A

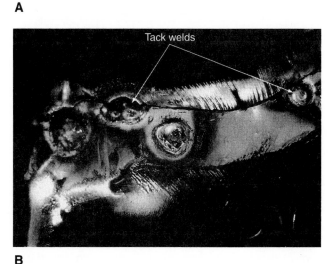

B

FIGURE 4–29 (A) Temporary or tack welding. (B) Tack welds on plug-welded lap joint. (Courtesy of Larry Maupin)

FIGURE 4–31 Plug weld. (Courtesy of Larry Maupin)

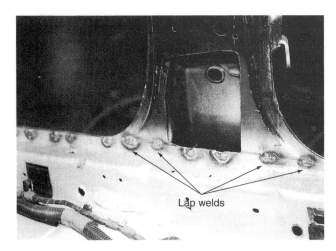

FIGURE 4–32 *Lap weld. (Courtesy of Larry Maupin)*

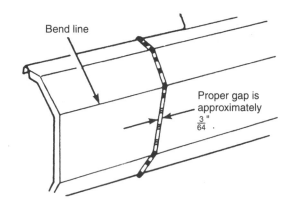

FIGURE 4–33 *Tack-welding panels prevents warpage. (Courtesy of Nissan Motor Corp.)*

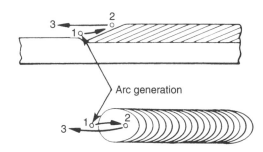

FIGURE 4–34 *If the gun handling is smooth and even, the bead will be of consistent height and width with a uniform, closely spaced ripple.*

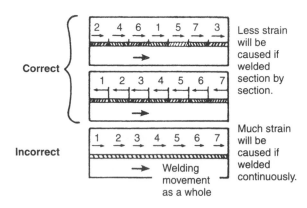

FIGURE 4–35 *Right and wrong welding sequences. (Courtesy of Nissan Motor Corp.)*

condition of the metal, the amount of gap, if any, between the pieces to be welded, and the welding position. For instance, the butt joint can be welded using the continuous technique or the stitch technique. It can be tack-welded at various points along the joint to hold the fitup in place while completing the joint with a permanent continuous weld or a stitch weld. Lap and flange joints can be made using all six of the previously discussed welding techniques.

BUTT WELDS

Butt welds are formed by fitting two edges of adjacent panels together and welding along the mating or butting edges. When butt welding, especially on thin panels, it is wise not to weld more than 3/4 of an inch (19 mm) at one time. Closely watch the melting of the panel, the welding wire, and the continuity of the bead. At the same time, be sure the end of the wire does not wander away from the butted portion of the panels.

If the butt weld is to be long, it is a good idea to tack-weld the panels in several locations to prevent panel warpage (Figure 4-33). Figure 4-34 illustrates how to generate an arc a short distance ahead of the point where the weld ends and then immediately move the gun to the point where the bead should begin. The bead width and height should be consistent at this time, with a uniform, closely spaced ripple—provided the gun handling is smooth and even.

When butt welding, establish a sequence to allow the weld area to cool naturally before the next area is welded (Figure 4-35). While butt welds of outer panels are far less sensitive, the same sequencing procedure should be used to prevent warpage and distortion from temperature buildup. To fill the spaces between intermittently placed beads, first grind the beads along

the surface of the panel using a sander or grinder, then fill the space with metal (Figure 4-36). If weld metal is placed without grinding the surface of the beads, blowholes can result.

When butt welding panels that are 1/32 inch (1 mm) or less, an intermittent or stitch welding technique is a must to prevent burn-through. The combination of a proper gun angle and correct cycling techniques will achieve a satisfactory weld bead

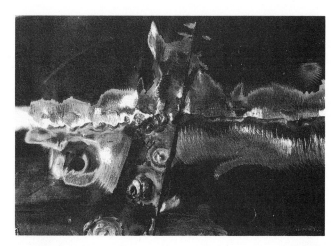

FIGURE 4–36   Butt weld ground to remove weld reinforcement to allow better fitup of the lap joint. (Courtesy of Larry Maupin)

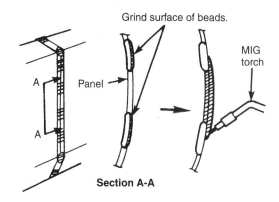

FIGURE 4–37   Filling the space between intermittently spaced beads.

(Figure 4-37). The reverse welding method should be used for moving the gun because it is easier to aim at the bead.

Figure 4-38 shows a typical butt welding procedure for installing a replacement panel. If the desired results are not obtained using this method, the cause for the problem may be that the distance between the tip of the gun and the base metal is too great. Remember that weld penetration decreases as the distance between the tip and the base metal increases. Try holding the tip of the gun at several different distances away from the base metal until the proper distance that gives the desired results is found (Figure 4-39).

Moving the gun too fast or too slowly (Figure 4-40) will yield poor results. Gun movement that is too slow will cause melt-through. Conversely, gun movement that is too fast will cause shallow penetration and inferior weld strength.

Even if a proper bead is formed during butt welding, panel warpage can result if the weld is started at or near the edge of the metal (Figure 4-41A). Therefore, to prevent panel warpage, disperse the heat into the base metal by starting the weld in the center and frequently changing the location of the weld area (Figure 4-41B). A good rule of thumb is this: the thinner the panel thickness, the shorter the bead length.

When you weld a butt joint, be sure the weld penetrates all the way through to the backside. Where the metal thickness at a butt joint is 1/16 inch (1.5 mm) or more, a gap should be left to assure full penetration. If it is not practical to leave a gap, grind a V-groove in the joint so the weld can penetrate to the backside.

Do not build up butt welds more than 1/8 inch (3 mm). Reinforcement can create a weaker weld because of stress or structural buildup at the area of reinforcement.

LAP AND FLANGE WELDS

Both lap and flange welds are made with identical MIG techniques. They are formed by melting the two surfaces to be joined at the edge of the top of the two overlapping surfaces. This is similar to butt welding, except only the top surface has an edge. Lap and flange welds should only be made in repairs where they replace identical original factory welds, or where outer panels and not structural panels are involved. The same technique used for temperature control in butt welding should be followed for lap and flange welding. Welds should never be made continuously but should be sequenced to allow for natural cooling; this will prevent temperature buildup in the weld area.

Because of this difference in heating rate, the gun must be directed in the center or on the bottom plate (Figure 4-42). The arc force and surface tension will pull the molten weld pool up, so it is not necessary to put metal on the top plate. Keep in mind that if the weld pool becomes too large, surface tension will pull the molten pool back from the joint. When this happens, a notch will form. Stop for a moment to allow the weld to cool slightly before restarting. When this is done properly, the weld appearance and strength will not be affected.

PLUG WELDS

The plug weld is the body shop alternative to the original equipment manufacturer (OEM) resistance spot welds made at the factory. It can be used anywhere in the body structure where a resistance spot weld was

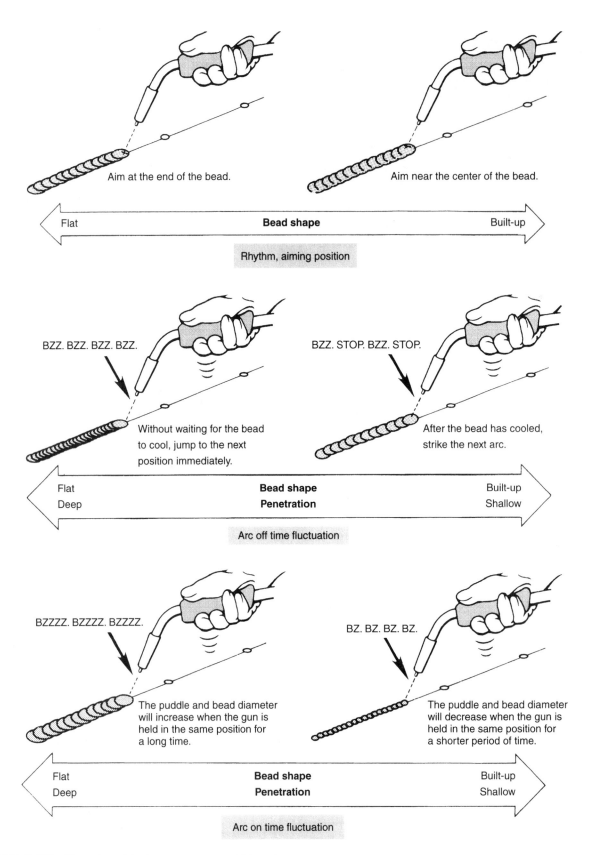

FIGURE 4–38 *Steps in achieving a proper weld bead.*

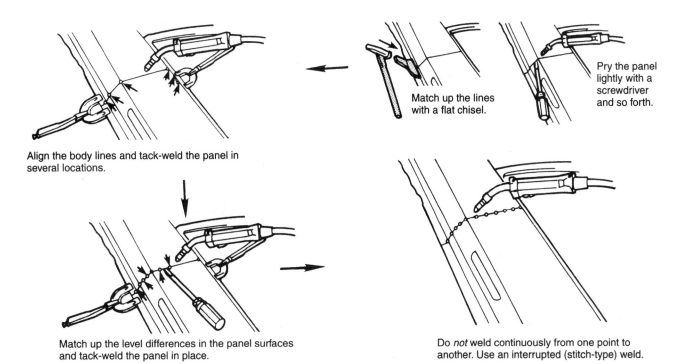

Align the body lines and tack-weld the panel in several locations.

Match up the lines with a flat chisel.

Pry the panel lightly with a screwdriver and so forth.

Match up the level differences in the panel surfaces and tack-weld the panel in place.

Do *not* weld continuously from one point to another. Use an interrupted (stitch-type) weld.

FIGURE 4–39 Procedure for butt welding replacement panels.

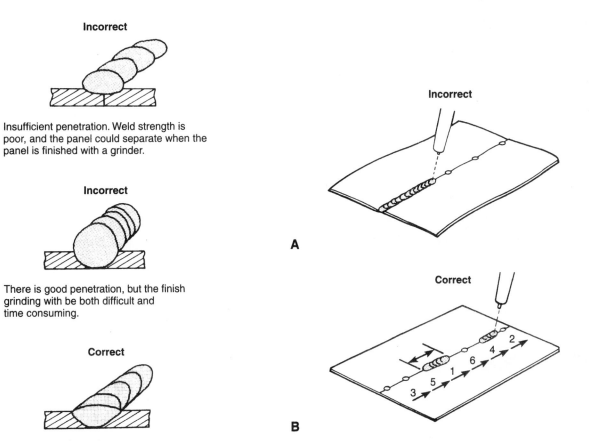

Incorrect

Insufficient penetration. Weld strength is poor, and the panel could separate when the panel is finished with a grinder.

Incorrect

There is good penetration, but the finish grinding with be both difficult and time consuming.

Correct

There is good penetration, and the bead will be easy to grind.

FIGURE 4–40 *Analyzing bead cutaway shapes. The relation of gun speed to bead shape.*

Incorrect

A

Correct

B

FIGURE 4–41 *(A) Starting a butt weld at or near the edge of the panel causes warpage. (B) Instead, always start in the center and frequently change the location of the weld area.*

FIGURE 4–42 Heat is conducted away faster in the bottom plate, resulting in the top plate melting more quickly.

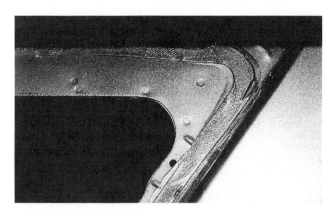

FIGURE 4–43 Factory spot welds. (Courtesy of Larry Moupin)

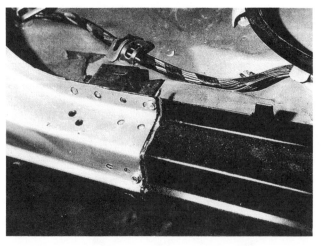

FIGURE 4–44 Plug welds to be used for joining rocker panel to B-pillar. (Courtesy of Larry Maupin)

used (Figure 4-43). A plug weld has ample strength for welding load-bearing structural members, and it can also be used on cosmetic body skins and other thin-gauge sheet metal (Figure 4-44).

FIGURE 4–45 Plug welds. (Courtesy of Larry Maupin)

Plug welding (Figure 4-45) is similar in function to spot welding; it is, however, welded through a hole. That is, a plug weld is formed by drilling or punching a hole in the outer panel being joined (Figures 4-46 and 4-47). The materials should be tightly clamped together. Holding the torch at right angles to the surface, put the electrode wire in the hole, trigger the arc briefly, then release the trigger. Move the gun in a circle inside the hole so that the molten weld pool fills the hole and solidifies.

When plug welding, try to duplicate both the number and the size of the original factory spot welds. The hole that is punched or drilled should not be larger in diameter than the factory weld nugget. The 5/16-inch (8 mm) hole customarily used for plug welding cosmetic panels is not sufficient for plug welding structural members. Some structural members require a 5/16- to 3/8-inch hole to achieve acceptable weld strength. For larger holes, move the gun slowly in a circular motion around the edges of the hole (Figure 4-48) to fill it in. For smaller holes, it is best to aim the gun at the center of the hole and keep it stationary. A flat, gently raised bead gives a nice appearance and reduces the grinding or sanding necessary.

Proper wire length is an important factor in obtaining a good plug weld. If the wire protruding out of the end of the gun is too long, the wire will not melt properly, resulting in inadequate weld penetration. The weld will improve if the gun is held closer to the base metal. Be sure the weld penetrates into the lower panel; there should be a round or dome-shaped protrusion on the underside of the metal as a good indicator of proper weld penetration (Figure 4-49).

The weld area should be allowed to cool naturally before any adjacent welds are made. Areas around the weld should not be force-cooled using water or air; it is important that they are allowed to cool naturally. Slow, natural cooling without the use of water or

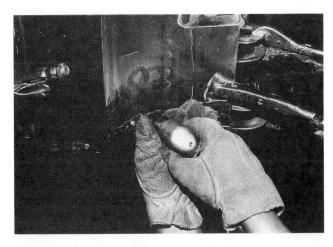

FIGURE 4–46 Drilling for plug welds. (Courtesy of Larry Maupin)

FIGURE 4–48 Plug welding. (Courtesy of Larry Maupin)

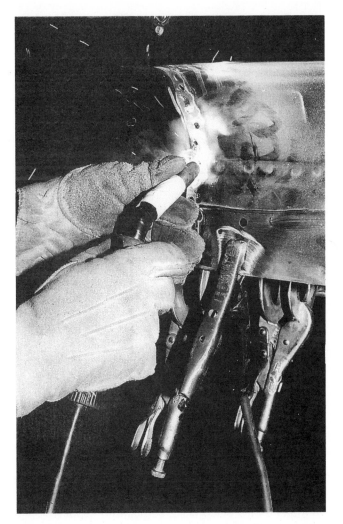

FIGURE 4–47 Making a MIG plug weld. (Courtesy of Larry Maupin)

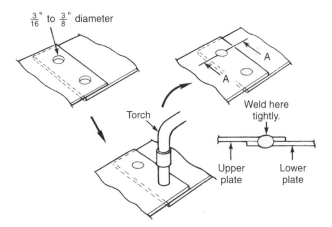

FIGURE 4–49 Steps in making a plug weld. (Courtesy of Nissan Motor Corp.)

air will minimize any panel distortion and maintain the strength designed into the panels.

Plug welds can also be used to join more than two panels together. When this is being done, a hole is punched in every panel except the bottom one (Figure 4-50). The diameter of the plug weld hole in each additional panel being joined should be smaller than the diameter of the hole above it. Likewise, if panels of different thicknesses are being joined, a larger hole is punched in the thinner panel to assure that the thicker panel is melted into the weld first. In addition, make sure the thinner panel is on top (Figure 4-50).

A plug weld can be accomplished in a minimum amount of time using a MIG welder, creating minimal temperature buildup in adjacent panels. Adjacent welds should not be made immediately; the area being welded will cool in a very short time.

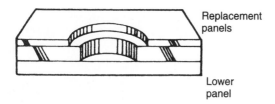

FIGURE 4–50 *Welding two or more panels using the plug welding technique.*

SPOT WELDS

It is possible to spot weld with MIG equipment (Figure 4-51). In fact, most of the better MIG machines now available and designed for collision repair work have built-in timers that shut off the wire feed and welding arc after the time required to weld one spot. Some MIG equipment also has a burnback time setting that is adjusted to prevent the wire from sticking in the molten weld pool. The setting of these timers depends on the thickness of the workpiece. This information can be found in the machine's owner's manual.

Since spot welding usually requires a higher current setting than making a continuous weld bead, make a test weld first to select the right welding settings. Check the penetration of the spot weld by pulling the test weld apart. A good weld will tear a small hole out of the bottom piece, while a weak weld will break off at the surface. To get more penetration, increase the weld time or heat. To reduce penetration, reduce the time or heat.

For MIG spot welding, a special nozzle (Figure 4-52) must replace the standard nozzle. Once the gun is in place and the spot timing, welding heat, and backburn times are set for the situation, the spot nozzle is held against the weld site and the gun triggered. For a very brief period of time, the timed pulses of wire feed and welding current are activated, and the arc melts through the outer layer and penetrates the inner layer (Figure 4-53). Then the automatic shutoff goes into action, and no matter how long the trigger is squeezed, nothing will happen. However, when the trigger is released and then squeezed again, the next spot pulse is obtained.

The quality of a MIG spot weld is difficult to ascertain because of varying conditions. Therefore, the plug weld or resistance spot welding technique described later is the preferred method on load-bearing members. However, the MIG lap spot technique is effective for quick welding of lap joints and flanges on thin-gauge, nonstructural sheets and skins. Set the spot timer, and position the nozzle over the edge of the outer sheet at an angle slightly off 90 degrees. This will allow contact with both pieces of metal at the

FIGURE 4–51 *Controls on a typical MIG welder. (Courtesy of HTP America, Inc.)*

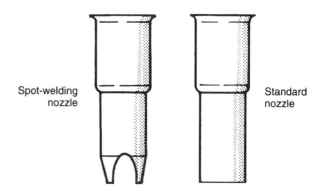

FIGURE 4–52 *Note the notch in the spot welding nozzle to let the welding heat and sparks out.*

same time. The arc melts into the edge and penetrates the lower sheet.

STITCH WELDS

To make a MIG stitch weld, combine the spot welding process with the continuous welding technique and gun travel (Figure 4-54). Use the standard nozzle, not the spot welding nozzle. If you are using a machine with a stitch weld setting, set the automatic timer—either a shutoff or pulsed interval timer, depending on the MIG machine. The spot weld pulses and shutoffs

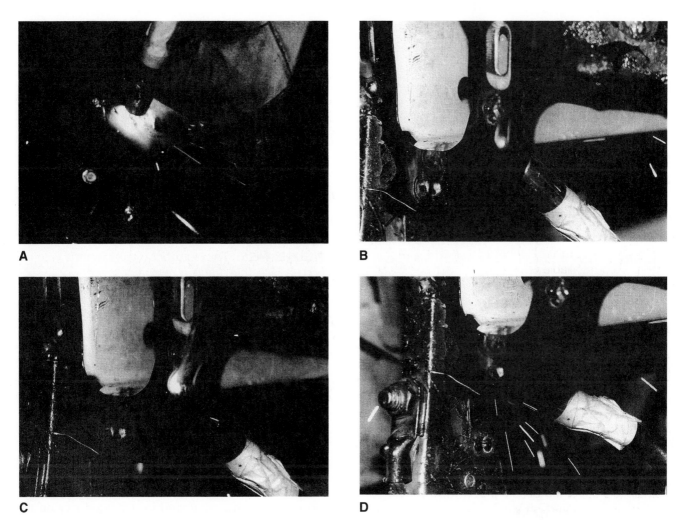

FIGURE 4–53 (A) Weld is started. (B) Back side heats up. (C) 100 percent penetration occurs. (D) Weld reinforcement can be seen as weld cools. (Courtesy of Larry Maupin)

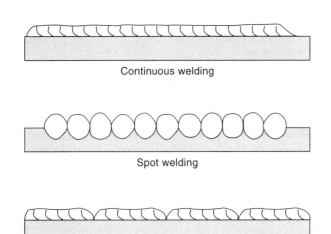

Continuous welding

Spot welding

Stitch welding

FIGURE 4–54 Finished weld looks continuous for both spot and stitch welds if done correctly.

recur with automatic regularity: weld-stop-weld-stop-weld-stop as long as the trigger is held in.

Another way to look at this is weld-cool-weld-cool-weld-cool and so on, because the arc-off period allows the last spot to cool slightly and start to solidify before the next spot is deposited. This intermittent technique means less distortion and less melt-through or burn-through. These characteristics make the stitch weld preferable to the continuous weld for working thin-gauge cosmetic panels.

The intermittent cooling and solidifying of the stitch weld also makes it preferable to continuous welding on vertical joints where distortion is a problem (Figure 4-55). The operator does not have to contend with a continuous molten weld pool that gravity is trying to pull down the joint ahead of the arc. By the same token, stitch welding is also preferable in the overhead position because there is virtually no molten weld pool for gravity to pull.

If the MIG machine does not have the automatic stitch modes, the spot and stitch welds can be made

FIGURE 4–55 *You can still see each weld bead in this lightly ground vertical stitch-welded rocker panel. (Courtesy of Larry Maupin)*

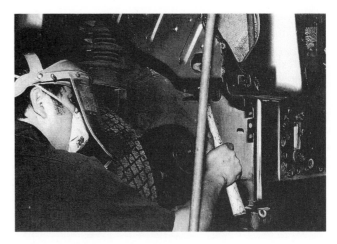

FIGURE 4–56 *This auto body technician is wearing both a face shield and a respirator while working on zinc-coated metal. (Courtesy of Larry Maupin)*

manually. The operator merely has to be capable of triggering the gun on-off, on-off, on-off—the same way the automatic system does. When finish grinding is required after welding, use light pressure and short grinds to keep the heat from building up. Otherwise clean the weld area with a wire brush.

USING NOZZLE SPRAY

Welding spray or paste should never be present in a vaporous state in the nozzle when welding. Instead, nozzle spray is highly preferred because it does not leave a heavy residue inside the nozzle. Nozzle spray should never be put into the nozzle while it is attached to the welding gun. Proper application is with the nozzle off, allowing the spray to dry or "flash off" prior to reinstalling the nozzle on the gun. A straight-blade screwdriver makes an excellent tool for scraping the nozzle.

Care should also be taken never to put the nozzle spray into the contact tip; this can form a coating inside the tip, thus preventing proper electrical contact of the wire with the tip.

MIG WELDING GALVANIZED METALS

There are governmental requirements for the use of coated steels because of their superior durability when compared to uncoated steels. It is very important that the auto body technician be able to weld them. Although this material is called many different names—for example, galvanized, zinc-metallized, zinc-coated—the basic problem remains the same: coated steels can be hard to weld. In fact, the weldability is so poor that some technicians grind off the coating near the weld seam to avoid dealing with it. Of course, this results in premature corrosion of the steel unless a weld-through primer is used.

Some of the problems associated with welding coated steels compared to uncoated include the following:

- Zinc fumes
- Increased spatter
- Erratic arc
- Reduced penetration
- Porosity

Although these problems cannot be totally eliminated, they can at least be controlled to a degree.

CAUTION: Breathing zinc fumes is hazardous to your health. You must have adequate ventilation so that the fumes are drawn away from your face. If there is no practical way to vent the fumes, you must wear a respirator (Figure 4-56).

DEALING WITH FUMES AND SPATTER

The fumes given off while welding galvanized metals are due to the fact that zinc melts at 786° F (419° C) and has a boiling point of 1665° F (907° C). The melting point of steel is about 2600° F (1425° C). Zinc fumes form a white, powdered smoke. The zinc melts at a temperature so much lower than that of steel that as soon as the welding arc is started, zinc fumes appear. These fumes can cause the welding arc to become very erratic, which results in porosity, spatter, and reduced penetration.

The first considerations for anyone welding galvanized metals are safety and health. Inhaling substantial amounts of zinc fumes will usually make a person sick. Always provide adequate ventilation and wear a respirator.

When one is welding zinc-coated steels, sometimes the molten weld pool seems to literally explode, sending globules of hot molten metal everywhere. These globules can cause severe burns, damage the vehicle, and ignite combustibles. Make sure to dress appropriately for welding. Leather cape sleeves are a great help in protecting the upper body and arms from burns. Use a good-quality welding helmet and wear earplugs and a skull cap.

Make sure to protect nearby glass, upholstery, and trim from the spatter and globules. Make sure that combustible materials are moved away from the weld site. A portable welding screen (Figure 4-57) offers excellent protection for other workers and objects that cannot be easily moved out of the general welding area. These screens are flame retardant, so weld spatter falls right off them.

ERRATIC ARCS

As already noted, zinc melts at a lower temperature than steel, and this causes white smoke or residue. When the zinc fumes become vaporized in the heat of the arc, the molten weld pool becomes hard to control. It is very difficult to weld thin-gauge metals with an arc that seems to have a mind of its own, but there are some things the technician can do to stabilize the arc.

First, increase the amperage by five or ten and increase the volts by one. On MIG machines, the increased amperage means an increase in wire speed of two or three short marks on the wire speed adjustment dial. This change in machine setting to a more forceful arc will help control the molten weld pool. This will give the leading edge of the molten weld pool time to vaporize the zinc before the bead actually passes over that area.

Next, move the gun somewhat more slowly than normal. Pulling the gun concentrates the heat of the arc, which aids in burning off the zinc. Pushing the gun angle spreads the heat of the arc sideways, which vaporizes more zinc than necessary.

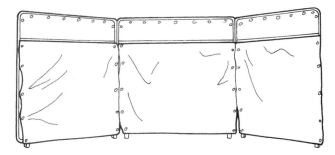

FIGURE 4–57 *A flame-retardant welding screen.*

HEAT INPUT

Because a "hotter" arc is recommended for welding galvanized metals, controlling this added heat input is very important. Heat input is in direct relationship to the machine settings and travel speed. Because of the "hotter" settings and slow travel speed, use a skip welding technique. Make short beads no more than 1-1/2 inches (37 mm) at a time, and allow the heat to dissipate before continuing to weld. Pause several minutes between welds to allow cooling.

Penetration can be reduced slightly when one is welding zinc-coated steels. Wider gaps are usually required for butt joints, which makes penetration control difficult. For this reason, the I-CAR procedure, which uses a backup piece for splices, is the ideal way to weld a butt joint. Otherwise, when the gap is opened up to compensate for the reduced penetration on an open butt joint, the gap becomes so wide that burn-through problems are created. Avoid the open butt joint whenever possible, and use the I-CAR method of inserts on all zinc-coated steel repairs.

The recommended shielding gas for welding zinc-coated steel is 75 percent argon and 25 percent CO_2. A flow of 40 cfh seems to work a bit better on coated steels than the normal flow of 20 or 30 cfh. The ER 70S-6 filler wire is recommended for MIG welding all zinc-coated steel.

DEALING WITH POROSITY

Weld porosity is a common problem when weld beads are made on coated steels. If the porosity holes are no larger than 1/32 inch (1 mm) in diameter and are no more frequent than one hole for every 1 inch (300 mm) of weld, the porosity will probably not impair the quality of the weld or the strength of the joint. If there are three or four holes for every inch of weld, grind them out, then reweld the bad portion. The extent of porosity will depend on amperage, volts, welding speed, and thickness of the coating.

Remember to increase amperage slightly, increase the volts by one, and travel more slowly than when welding uncoated steels. The problem of porosity is explored in further detail later in this chapter.

Welding on coated steels will result in some loss of zinc immediately on each side of the welded seam. This loss usually extends 1/8 to 3/16 of an inch (3 mm to 4 mm) from the weld on either side. This area will create no special problems if proper rustproofing methods are employed to make the repair corrosion resistant.

When you are grinding the weld, be careful not to remove any more coating than is necessary. Also, avoid grinding away too much surface metal; this can result in a reduction in the thickness of the metal.

4.13 EVALUATING MIG WELDS

To ensure the quality, durability, and strength of MIG welds, it is important to periodically evaluate the welding results. Naturally, the extent of any evaluation program depends on the type of service the welded part will be required to withstand. For example, structural members must be welded to much more exacting standards than purely cosmetic parts of the vehicle. Results acceptable in one welding application might not meet the needs of another.

Weld discontinuity is the term used for an interruption of the typical structure of a weld. *Weld defect* is the term used for a discontinuity that exceeds the acceptable standards for the welding. A finished weld should have no defects. When you inspect a weld, note the type, size, and location of discontinuity and defects. These are the most common weld discontinuities, and they occur in most welding processes:

- Porosity
- Inclusions
- Inadequate joint penetration
- Incomplete fusion
- Overlap
- Undercut
- Cracks
- Laminations
- Delaminations
- Lamellar tears
- Arc strikes

POROSITY

Porosity results from gas trapped during the solidification of the weld metal. It is most often caused by improper welding techniques, contamination, or an improper chemical balance between the filler and the base metal. Porosity is either spherical or cylindrical in shape. Cylindrical porosity is also known as wormholes; the rounded edges tend to reduce the stresses around them. Therefore, unless this type of porosity is extensive, there is little or no loss in strength.

The intense heat of a weld can decompose paint, dirt, oil, rust, or other materials, producing hydrogen. This gas can become trapped in the solidifying weld pool, producing porosity. When it causes porosity, hydrogen can also diffuse into the heat-affected zone, producing cracking in some steels.

Porosity can be grouped into the following major types:

- Uniformly scattered porosity (Figure 4-58) is most frequently caused by poor welding techniques or dirty materials.
- Clustered porosity (Figure 4-59) is most often caused by improper starting and stopping techniques.
- Linear porosity (Figure 4-60) is most frequently caused by contamination within the joint, root, or interbead boundaries.

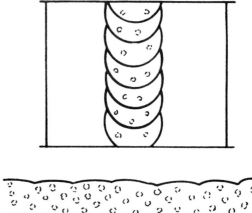
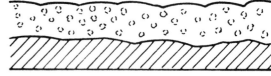

FIGURE 4-58 *Uniformly scattered porosities.*

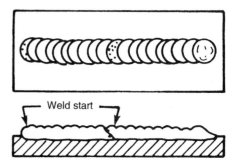

FIGURE 4-59 *A clustered porosity.*

- Piping porosity (Figure 4-61) is most often caused by contamination. This porosity is unique, because its formation depends on the gas escaping from the weld pool at the same rate that the pool is solidifying.

INCLUSIONS

Inclusions are nonmetallic materials (such as slag and oxides) that are trapped in the weld metal, between weld beads, or between the weld and the base metal. Inclusions are sometimes jagged and irregularly shaped and can form in a continuous line. This reduces the structural integrity of the weld.

Inclusions are often found when the following conditions exist:

- Slag and/or oxides do not have enough time to float to the surface of the molten weld pool.
- Sharp notches exist between weld beads or between the bead and the base metal that trap the material so that it cannot float out.
- The joint was designed with insufficient room for the correct manipulation of the molten weld pool.

INADEQUATE JOINT PENETRATION

Inadequate joint penetration means that the depth at which the weld penetrates the joint is less than is needed to fuse through the plate or into the preceding weld (Figure 4-62). A defect results that could reduce the required cross-sectional area of the joint or become a source of stress concentration that leads to fatigue failure.

The major causes of inadequate joint penetration are improper welding technique, insufficient welding current, improper joint fitup, and improper joint design. A root gap that is too small will also keep the weld from penetrating adequately.

INCOMPLETE FUSION

Incomplete fusion is the failure of the molten filler metal and the base metal to mix completely. Even if the base metal melts, a thin layer of oxide can still prevent fusion from occurring. This can be caused by improper welding techniques, improper edge preparation, or improper joint design. Incomplete fusion can also result from insufficient heat to melt the base metal, inadequate agitation to break up oxide layers, or too little space allowed for correct molten weld pool manipulation.

OVERLAP

When the weld metal protrudes beyond the toe, face, or root of the weld without actually being fused to it, the resulting discontinuity is known as _overlap_ (Figure 4-63). It can occur as a result of lack of control of the welding process, improper selection of welding materials, or improper preparation of the materials prior to welding. If there are tightly adhering oxides on the base metal that interfere with fusion, overlap will

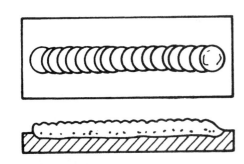

FIGURE 4-60 A linear porosity.

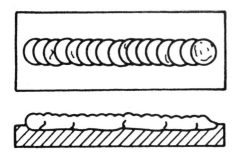

FIGURE 4-61 A piping porosity.

FIGURE 4-62 Example of inadequate joint penetration.

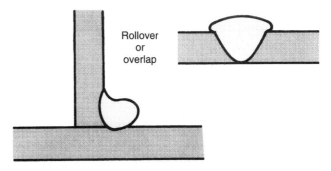

Rollover or overlap

FIGURE 4-63 Rollover or overlap.

result. Overlap is a surface-connected discontinuity that forms a severe notch that can start a crack if the part is subjected to vibration.

UNDERCUT

This discontinuity is generally associated with either improper welding techniques or excessive current or both. It is generally located at the junction of the weld and the base metal (the toe or root) (Figure 4-64). When undercut is controlled within the limits of the specifications and does not constitute a sharp or deep notch, it is not usually considered a weld defect.

CRACKS

If localized stress exceeds the strength of a weld, cracks occur. Cracks are an excellent indicator of discontinuities in either the weld or the base metal. Welding-related cracks are generally brittle, usually due to the high residual stresses present.

Cracks are classified as either hot or cold, depending upon the temperatures at which they form. Hot cracks develop at elevated temperatures and commonly form during the solidification of the metal at temperatures near the melting point (Figure 4-65). Cold cracks develop after the solidification of the metal is complete and are sometimes called *delayed cracks*. Hydrogen embrittlement is usually associated with the formation of cold cracks. Hot cracks form between the grains, while cold cracks form both between and through the grains.

ARC STRIKES

Arc strikes are small, localized points where surface melting occurs away from the joint. These spots can be caused by accidentally striking the arc in the wrong place and/or by faulty ground connections. Even though arc strikes can be ground smooth, they cannot be removed. Keep in mind that these spots can be the starting point for cracking.

DESTRUCTIVE EVALUATION

Most of the destructive testing methods used in industrial welding situations are simply not feasible in an automotive application. Probably the simplest, yet most effective, destructive test for automotive welds is illustrated in Figure 4-66. Use a hammer and chisel to break the weld apart. Once the part has been separated, the weld can be examined. A good weld with adequate fusion will result in a portion of the metal being torn off with the weld. The metal removed can be either a portion of

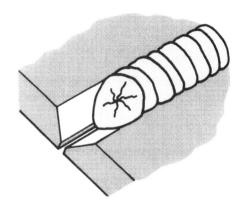

FIGURE 4–65 Crater or star cracks are hot cracks.

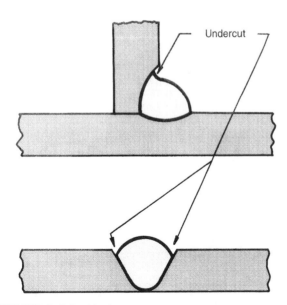

FIGURE 4–64 Undercut.

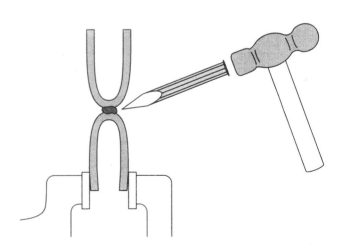

FIGURE 4–66 Using a chisel, a hammer, and a vice to break apart a weld.

the thickness or can result in a hole torn completely out of the plate (Figure 4-67). The area torn out should be approximately equal to the size of the weld. If no metal is torn off with the weld or a portion much smaller than the size of the weld is torn out, inadequate fusion or penetration occurred. This indicates an improper, unacceptable weld.

A plug or spot weld can be tested by twisting the test plate. This is accomplished by first producing a weld in the center of the test plates. Next, bend the test plates, as illustrated in Figure 4-68. Clamp one of the bent tabs in a vise and grasp the other tab with an adjustable wrench (Figure 4-69). Twist the upper plate until the weld separates. Examine the weld to see that it tore out a portion of the base metal. If the torn section is large enough, the weld is acceptable. However, if the torn section is too small or no bonding occurred between the weld and test plates, the weld fails.

A third method of destructive evaluation is used for butt or lap welds. Once the weld is completed, it is bent repeatedly back and forth until it breaks (Figure 4-70). Examine the fracture to see if the weld is acceptable. An acceptable weld results in a portion of the base metal being fractured. An unacceptable weld will result in all or part of the weld pulling free from the base metal without removing any portion of the base metal. This indicates a lack of fusion and is unacceptable.

Because it is impractical to do destructive testing on a vehicle, it is good practice to make test welds from time to time. Before making any welds on the vehicle, secure test panels of the same gauge steel that is to be repaired. Making a few welds on the test panels can insure that the proper settings on the MIG welder are obtained. Making these test welds can also help insure that the quality of the welds on the vehicle is acceptable.

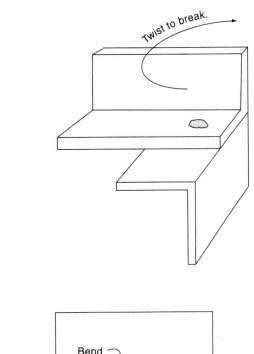

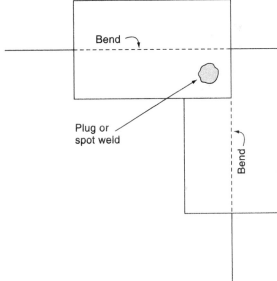

FIGURE 4–68 Bend weld plate for testing.

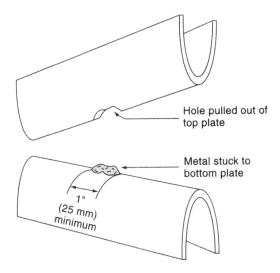

FIGURE 4–67 Good butt weld.

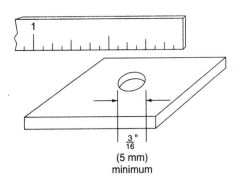

FIGURE 4–69 A good plug weld will pull out a hole in the bottom plate at least 3/16 inch (5 mm) in diameter.

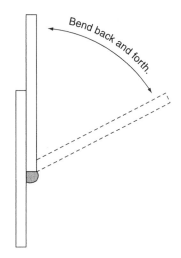

FIGURE 4–70 *Testing a lap joint weld.*

NONDESTRUCTIVE EVALUATION

Nondestructive evaluation is an effective method of checking for surface defects, including cracks, undercuts, arc burns, and lack of penetration. The three methods of nondestructive testing that will be discussed here include visual inspection, penetrant inspection, and ultrasonic inspection.

Visual Inspection

Visual inspection is the most frequently used nondestructive testing method. Many times a weld receives only a visual inspection. Measuring is an important part of visual inspection. This procedure can be easily overlooked when more sophisticated nondestructive testing methods are used. A visual inspection should be done before any other tests to eliminate the obvious problem welds.

A visual inspection schedule can reduce the finished weld rejection rate. Visual inspection can easily be used to check for fitup, interpass acceptance, welder technique, and other variables that ultimately affect weld quality. Minor problems can be identified and corrected before the weld is completed, thus eliminating costly repairs or rejection.

Penetrant Inspection

Penetrant inspection is used to locate very small surface cracks and porosity. There are two types of penetrants now in use, the color-contrast and the fluorescent versions. Color-contrast penetrants contain a colored dye that is visible under ordinary white light. Fluorescent penetrants contain a more effective fluorescent dye that is only visible under black light.

When using a penetrant, proceed as follows:

1. The test surface must be clean and dry. Any flaws must be free of oil, water, and other contaminants.
2. Cover the test surface with a film of penetrant. This can be accomplished by dipping, bathing, spraying, or brushing.
3. Wipe, wash, or rinse the test surfaces free of excess penetrant. Dry with cloths or hot air.
4. Apply developing powder to the test surface. It will act like a blotter and speed the penetrant out of any flaws on the test surface.
5. Depending on the type of penetrant applied, make the inspection under ordinary white light or near-ultraviolet black light. When viewed under the latter, the penetrant fluoresces to a yellow-green color that clearly defines the defect.

Ultrasonic Inspection

Ultrasonics is a fast and relatively low-cost nondestructive testing method. It employs electronically produced, high-frequency sound waves (roughly 25 million cycles per second), which penetrate metal at speeds of several thousand feet per second. The two types of ultrasonic equipment are pulse and resonance. The pulse-echo system, most often employed with welding, uses sound generated in short bursts or pulses. Since high-frequency sound has little ability to travel through air, it must be conducted into the component through a medium such as oil or water.

Sound is directed into the component with a probe held at a preselected angle or direction so that any defects will reflect energy back to the probe. Ultrasonic devices operate very much like depth sounders or fish finders. The speed of sound through a material is a known quantity; ultrasonic devices measure the time required for a pulse to return from a reflective surface. Internal computers calculate the distance and present the information on a cathode ray tube, where the results are interpreted. Sound not reflected by defects continues into the component. Defect size is determined by plotting the length, height, width, and shape.

4.14 ACCEPTING OR REJECTING WELDS

Table 4-6 summarizes common discontinuities and defects and their causes. Keep in mind that it is not intended to be all-inclusive; there are others not

TABLE 4–6:   MIG WELDING DEFECTS AND THEIR CAUSES

Defect	Defect Condition	Remarks	Possible Causes
Pores/Pits		There is a hole made when gas is trapped in the weld metal	There is rust or dirt on the base metal. There is rust or moisture adhering to the wire. Improper shielding action (the nozzle is blocked or wind or the gas flow volume is low). Weld is cooling off too fast. Arc length is too long. Wrong wire is selected. Gas is sealed improperly. Weld joint surface is not clean.
Undercut		Undercut is a condition in which the overmelted base metal has made grooves or an indentation. The base metal's section is made smaller and, therefore, the weld zone's strength is severely lowered.	Arc length is too long. Gun angle is improper. Welding speed is too fast. Current is too large. Torch feed is too fast. Torch angle is tilted.
Improper Fusion		This is an unfused condition between weld metal and base metal or between deposited metals.	Check torch feed operation. Is voltage lowered? Weld area is not clean.
Overlap		Overlap is apt to occur in fillet weld rather than in butt weld. Overlap causes stress concentration and results in premature corrosion.	Welding speed is too slow. Arc length is too short. Torch feed is too slow. Current is too low.
Insufficient Penetration		This is a condition in which there is insufficient deposition made under the panel.	Welding current is too low. Arc length is too long. The end of the wire is not aligned with the butted portion of the panels. Groove face is too small.
Excess Weld Spatter		Excess weld spatter occurs as speckles and bumps along either side of the weld bead.	Arc length is too long. Rust is on the base metal. Gun angle is too severe.
Spatter (short throat)		Spatter is prone to occur in fillet welds.	Current is too great. Wrong wire is selected.

mentioned here, and those that are mentioned will not always conform to these descriptions. When discontinuities or defects do occur, look for ways that the method of operation can be changed to prevent the same mistakes from happening again.

REVIEW QUESTIONS

1. Why do technicians sometimes fail to select the proper size MIG welder for a given welding situation?
2. What is the best way to prevent the welding heat from entering the wrong places?
3. What parts should be removed and/or disconnected before welding?
4. Where can you find the voltage and amperage settings for most welds on newer cars?
5. What determines the length of the arc?
6. What is the result of voltage that is set too high?
7. List the components in a basic MIG welding installation.
8. What steps should be performed if the wire stops feeding before it reaches the end of the contact tube?
9. Why is it important that the pressure be set correctly for the wire feed unit?
10. Why is it important that wire be discarded in the proper container?
11. What is the chief factor determining how effective a gas is for arc shielding?
12. What determines the welding speed?
13. What sound can be heard when the correct wire-to-heat ratio produces the correct welding temperature?
14. What is the best wire speed for overhead welding?
15. What is the basic adjustment procedure for a gas nozzle?
16. What are the reasons for using a backhand or dragging angle and a forehand or pushing angle?
17. List the factors to be considered when selecting the correct shielding gas.
18. List the six types of MIG welds and define them.
19. What determines which type of MIG weld is used?
20. How is panel warpage avoided?

Chapter

5

MIG Welding Practices

Objectives

After reading this chapter, you should be able to:

- **Make MIG butt, lap, and tee welds using the short-circuiting metal transfer method.**
- **Make MIG plug welds using the short-circuiting metal transfer method.**
- **Make MIG spot welds using the short-circuiting metal transfer method.**
- **Pass the I-CAR Automotive MIG welding qualification test.**

The practices in this chapter are grouped according to those requiring similar techniques and setups. For consistent, acceptable MIG welds, the major skill required is the ability to set up the equipment and weld practice plates. Changes such as variations in material thickness, position, and type of joint require changes both in technique and setup. A correctly set up MIG welding unit can, in many cases, be operated with minimum skill as compared to stick welding.

Ideally, only a few tests would be needed for the auto body technician to make the necessary adjustments in setup and technique to achieve a good weld. Any previous welding experiments should have given the auto body technician help in making the correct changes. In addition to keeping the setup data, you may want to keep the sample plates for more accurate comparisons.

The grouping of practices in this chapter will keep the number of variables in the setup to a minimum. Often, the only change required before going on to the next weld is to adjust the power settings.

Figures that are given in some of the practices will give the auto body technician general operating conditions, such as voltage, amperage, and shielding gas and/or gas mixture. These are general values, so the technician will have to make some fine adjustments. Differences in the type of machine being used and the

material thickness will affect the settings. For this reason, it is preferable to use the settings developed during the practices.

 ## 5.1 — METAL PREPARATION

Your practice welds can be made on hot-rolled steel, galvanized steel, or cold-rolled steel. In the following practices, all three of these metals will be referred to generally as sheet metal. Hot-rolled steel can be easily identified because it has an oxide layer that is formed during the rolling process called *mill scale*. Mill scale is a thin layer of dark gray or black iron oxide. This layer is left on hot-rolled steel because it offers some protection from rusting.

Mill scale is not removed for welding, since it does not present a major problem during most auto body practice welding. Filler metals have deoxidizers added to them so that the adverse effects of the mill scale are reduced or eliminated. The most common problem that mill scale causes is porosity. If the porosity becomes a problem, then surfaces within an inch of the weld must be cleaned to bright metal. Cleaning may be either grinding, filing, sanding, or blasting.

Galvanized steel has a thin coating of zinc applied to the surface to prevent rusting. Many automotive body panels and structural members are also zinc-coated to prevent rusting. The zinc coating can be welded through without the need to have it removed. In fact, it is a good idea to leave it on the test plates so that you can learn this welding technique. It is a skill you will need to have later for vehicle repairs.

Welding on zinc-coated metals produces a hazardous fume that must not be breathed. Both additional ventilation and a respirator are required any time zinc-coated metal is being welded.

CAUTION: Never weld on zinc-coated metals in a confined space without both forced ventilation that will blow the fumes away from you and an approved welding respirator.

Cold-rolled steel and some hot-rolled steels have the mill scale removed during manufacturing. These metals have a thin coating of oil to protect them from rusting. If the coating is too thick, it must be removed before welding to prevent weld porosity. The oil coating can be removed with soap and water or an approved solvent. Do not remove the coating from more material than you will use in a single welding practice session; without the coating, the metal can rust quickly.

Most newer automotive panel and structural members range in thickness from 20 gauge to 26 gauge. Older vehicles, trucks, buses, and some recreational vehicles may have similar members constructed from metal as thick as 18 gauge. Some hard points or frames may be 1/8 inch (3 mm) thick or thicker. Welding on thicker metal is easier than welding on thinner sections. For this reason, you should start out welding on the thickest sheet metal, and as you develop your skills move on to thinner metal. Also, hot-rolled and cold-rolled steels are easier to learn on than galvanized metal.

The practice welds can be made either with a continuous weld process or as a continuous series of stitch welds. Stitch welding is the most commonly used method in the repair shops but continuous welding is required on the I-CAR Qualification Test. Practice both methods as you make the following practice welds. You will probably find that the only way you can make some of the welds on thinner sections is to use the stitch welding technique.

5.2 TACK WELDING

PRACTICE 5-1

MIG Tack Welding

Using a properly set up and adjusted MIG welding machine (see Table 5-1 or the manufacturer's operating manual for the welder for settings), welding gloves, safety glasses, appropriate clothing, approved welding respirator, 0.023-inch and/or 0.030-inch (0.6 mm and/or 0.8 mm) diameter wire, and two or more pieces of sheet metal 5 inches (127 mm) long by 3 inches (76 mm) wide and 16 gauge to 24 gauge thick, you will make a series of tack welds to hold plates in the butt, lap, and tee joint positions for welding (Figure 5-1).

Start at one end of the plates with the nozzle resting on the joint at a 45-degree angle and the wire directed into the joint. Pull the trigger on the welding gun for a few seconds. Examine the tack weld for acceptability. The tack weld should be small enough so that it does not interfere with the welding, but large enough to hold the parts together. Finding the correct size will take some practice. If the weld is too large or burns through, use less welding time. If the weld is too small, it will not fuse both plates. You can use

---

TABLE 5–1: MIG WELD DEPOSITION RATES

Weld Deposition Rate Pounds per Hour			
Electrode Diameter			
Amperage	0.35	0.45	0.63
50	2.0		
100	4.8	4.2	
150	7.5	6.7	5.1
200		8.7	7.8
250		12.7	11.1
300			14.4

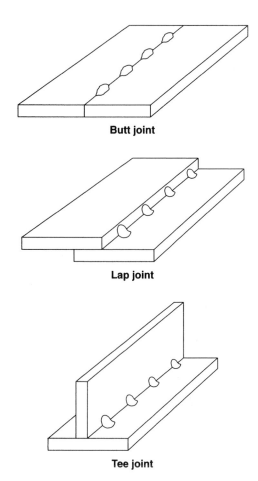

Butt joint

Lap joint

Tee joint

FIGURE 5–1 Butt, lap, and tee joints.

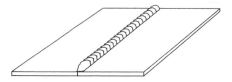

FIGURE 5–2 Butt weld flat position.

operating manual for the welder for settings), welding gloves, safety glasses, appropriate clothing, approved welding respirator, 0.023-inch and/or 0.030-inch diameter wire, and two or more pieces of sheet metal 5 inches (127 mm) long by 3 inches (76 mm) wide and 16 gauge to 24 gauge thick, you will make a stringer bead weld in the flat position (Figure 5-2).

Starting at one end of the plate and using either a pushing or dragging technique, make a weld bead along the entire 5 inches length of the metal. After the weld is complete, check its appearance. Make any needed changes to correct the weld and then record your new welding setup on Tables 5-2 and 5-3. Repeat the weld and make additional adjustments, each time recording the information on the tables. After the machine is set, start to work on improving the straightness and uniformity of the weld.

Keeping the bead straight and uniform can be hard because of the limited visibility due to the small amount of light and the size of the molten weld pool. The welder's view is further restricted by the shielding gas nozzle (Figure 5-3). Even with limited visibility, it is possible to make a satisfactory weld by watching the edge of the molten weld pool, the sparks, and the weld bead produced. Watching the leading edge of the molten weld pool (forehand welding, push technique) will show you the molten weld pool fusion and width. Watching the trailing edge of the molten weld pool (backhand welding, drag technique) will show you the amount of buildup and the relative heat input (Figure 5-4). The backhand method is preferred for welding on coated (galvanized) steels. The quantity and size of sparks produced can indicate the relative location of the filler wire in the molten weld pool. The number of sparks will increase as the wire strikes the solid metal ahead of the molten weld pool. The gun itself will begin to vibrate or bump as the wire momentarily pushes against the cooler, unmelted base metal before it melts. Changes in weld width, buildup, and proper joint tracking can be seen by watching the bead as it appears from behind the shielding gas nozzle.

Repeat the bead as needed on each available type and thickness of metal until consistently good beads are obtained. Turn off the welding machine, shielding gas supply, and clean up your work area when you are finished welding.

more time and try rocking the nozzle back and forth during the tacking process. Make additional tack welds along the joint, spacing them about every one to two inches. Repeat the process on each available type and thickness of metal in each of the joints designated until consistently good tack welds are obtained. Turn off the welding machine and shielding gas supply, and clean up your work area when you are finished welding.

5.3 ___ **FLAT POSITION, 1G AND 1F POSITIONS**

PRACTICE 5–2

Stringer Beads Using the Short-Circuiting Metal Transfer Method in the Flat Position

Using a properly set up and adjusted MIG welding machine (see Table 5-1 or the manufacturer's

TABLE 5–2:   SETTING THE CURRENT

Weld Acceptability	Voltage	Amperage	Spatter	Molten Pool Size	Penetration
Good	20	75	Light	Small	Little

Electrode Diameter   .035 in (0.9 mm)
Shielding Gas CO_2
Welding Direction   Backhand

TABLE 5–3:   ELECTRODE EXTENSION

Weld Acceptability	Voltage	Amperage	Electrode Extension	Contact Tube-to-Work Distance	Bead Shape
Poor	20	100	1 in (25 mm)	1 1/4 in (31 mm)	Narrow, high with little penetration

Electrode Diameter   .035 in (0.9 mm)
Shielding Gas CO_2
Welding Direction   Forehand

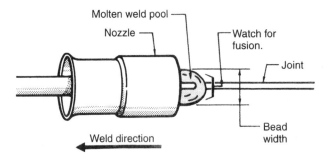

FIGURE 5–3   The shielding gas nozzle restricts the welder's view.

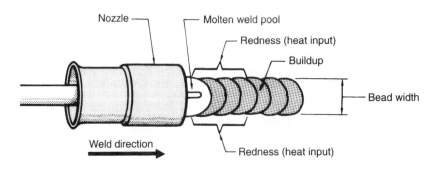

FIGURE 5–4   Watch the trailing edge of the molten weld pool.

PRACTICE 5-3

Flat Position Butt Joint, Lap Joint, and Tee Joint

Using the same equipment, materials, and procedures listed in Practice 5-1, make welded butt joints, lap joints, and tee joints in the flat position (Figure 5-5).

- Tack-weld the sheets together and place them flat on the welding table (Figure 5-6).
- Starting at one end, run a bead along the joint. Watch the molten weld pool and bead for signs that a change in technique may be required.
- Make any needed changes as the weld progresses. By the time the weld is complete, you should be making the weld almost perfectly.
- Using the same technique that was established in the last weld, make another weld. This time, the entire 5-inch weld should be flawless.

Repeat the bead as needed on each available type and thickness of metal until consistently good beads are obtained. Turn off the welding machine, shielding gas supply, and clean up your work area when you are finished welding.

PRACTICE 5-4

Flat Position Butt Joint with a Backing

Using the same equipment, materials, and procedures listed in Practice 5-1, make a welded butt joint with a backing plate (Figure 5-7).

- Tack-weld the sheets together with a root gap of from 1/16 to 3/32 of an inch (1.5 mm to 2 mm) and place them flat on the welding table.
- Starting at one end, run a bead along the joint. Watch the molten weld pool and bead for signs that a change in technique may be required. The weld groove must be filled completely but may not have more than 1/16 inch (1.5 mm) of weld face reinforcement.
- Make any needed changes as the weld progresses. By the time the weld is complete, you should be making the weld almost perfectly.
- Using the same technique that was established in the last weld, make another weld. This time, the entire weld should be flawless.

Repeat the bead as needed on each available type and thickness of metal until consistently good beads are obtained. Turn off the welding machine, shielding gas supply, and clean up your work area when you are finished welding.

PRACTICE 5-5

Flat Position Butt Joint Lap Joint and Tee Joint, All with 100 Percent Penetration

Using the same equipment, materials, and setup listed in Practice 5-1, make a welded joint in the flat position with 100 percent penetration along the entire 5-inch (127-mm) length of the welded joint. Repeat each type of joint as needed until consistently good beads are obtained. Turn off the welding machine, shielding gas supply, and clean up your work area when you are finished welding.

PRACTICE 5-6

Flat Position Butt Joint with a Backing Plate, with 100 Percent Penetration

Using the same equipment, materials, and setup listed in Practice 5-1, make a welded butt joint with a backing plate in the flat position with 100 percent penetration along the entire 5-inch (127-mm) length of the welded joint. Repeat the weld as needed until consistently good beads are obtained. Turn off the welding machine, shielding gas supply, and clean up your work area when you are finished welding.

PRACTICE 5-7

Flat Position Butt Joint, Lap Joint, and Tee Joint, All Welds To Be Tested

Using the same equipment, materials, and setup listed in Practice 5-1, make each of the welded joints in the flat position (Figure 5-8). Each weld joint must pass the bend test. Repeat each type of weld joint until all pass the guided bend test. Turn off the welding machine, shielding gas supply, and clean up your work area when you are finished welding.

PRACTICE 5-8

Flat Position Butt Joint with a Backing Plate, Weld To Be Tested

Using the same equipment, materials, and setup listed in Practice 5-1, make the welded joint with a backing plate in the flat position. The welded joint must pass the guided bend test. Turn off the welding machine, shielding gas supply, and clean up your work area when you are finished welding.

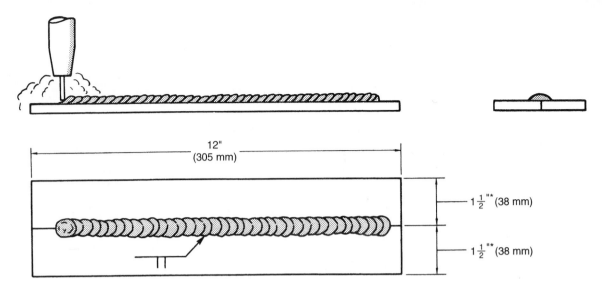

12"
(305 mm)

$1\frac{1}{2}$"* (38 mm)

$1\frac{1}{2}$"* (38 mm)

*This dimension will decrease as the old weld is cut out so the metal can be reused.

A

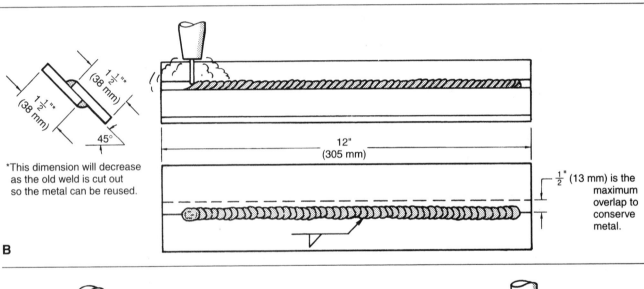

$1\frac{1}{2}$"* (38 mm)

$1\frac{1}{2}$"* (38 mm)

45°

*This dimension will decrease as the old weld is cut out so the metal can be reused.

12"
(305 mm)

$\frac{1}{2}$" (13 mm) is the maximum overlap to conserve metal.

B

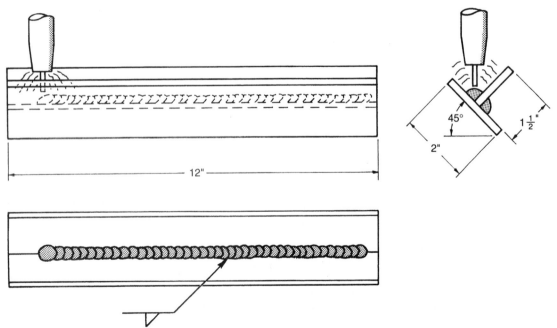

12"

45°

2"

$1\frac{1}{2}$"

C

FIGURE 5–5 (A) Butt joint in the flat position. (B) Lap joint in the flat position. (C) Tee joint in the flat position.

FIGURE 5–6 *Use enough tack welds to keep the joint in alignment during welding. Small tack welds are easier to weld over without adversely affecting the weld.*

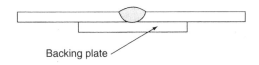

Backing plate

FIGURE 5–7 *Butt joint with backing plate.*

5.4 VERTICAL DOWN 3G AND 3F POSITIONS

The vertical down welding technique can be useful when making auto body welds. The major advantages of this technique are:

- *Speed.* Very high rates of travel are possible.
- *Shallow penetration.* Thin sections or root openings can be welded with little burn-through.
- *Good bead appearance.* The weld has a nice width-to-height ratio and is uniform.

Vertical down welding is the best process to use on thin sheet metals. The combination of controlled penetration and higher welding speeds makes vertical down the best choice for such welds. The most common problem with these welds is lack of fusion or overlap. To prevent these problems, the arc must be kept at or near the leading edge of the molten weld pool.

PRACTICE 5–9

Stringer Bead at a 45-Degree Vertical Down Angle

Using the same equipment, materials, and setup as listed in Practice 5-1, you will make a vertical down stringer bead on a plate at a 45-degree inclined angle.

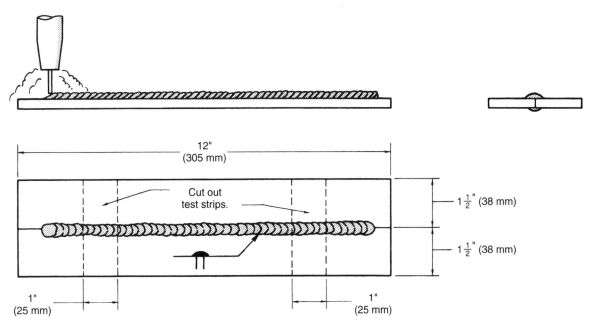

FIGURE 5–8 *Butt joint in the flat position to be tested.*

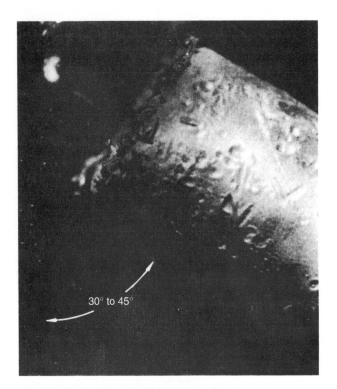

30° to 45°

FIGURE 5–9 *Vertical down position.*

Holding the welding gun at the top of the plate with a slight dragging angle (Figure 5-9) will help to increase penetration, hold back the molten weld pool, and improve the visibility of the weld. Be sure that your movements along the length of the plate are unrestricted.

Lower your hood and start the weld. Watch both the leading edge and the sides of the molten weld pool for fusion. The leading edge should flow into the base metal, not curl over it. The sides of the molten weld pool should also show fusion into the base metal and not be flashed (ragged) along the edges.

The weld may be made with or without a weave pattern. If a weave pattern is used, it should be a very small C pattern. The C should follow the leading edge of the weld. Some changes in the gun angle may help to increase penetration. Experiment with the gun angle as the weld progresses.

Repeat these welds until you have established a rhythm and technique that work well for you. The welds must be straight and uniform and have complete fusion. Turn off the welding machine and shielding gas supply, and clean up your work area when you are finished welding.

PRACTICE 5–10

Stringer Bead in the Vertical Down Position

Repeat Practice 5-9 and increase the angle of the plate until you have developed the skill to repeatedly make good welds in the vertical down position. The weld bead must be straight and uniform and have complete fusion. Turn off the welding machine, shielding gas supply, and clean up your work area when you are finished welding.

PRACTICE 5–11

Butt Joint, Lap Joint, and Tee Joint in the Vertical Down Position

Using the same equipment, materials, and setup as listed in Practice 5-1, you will make vertical down welded joints.

Tack-weld the pieces of metal together and brace them in position. Using the same technique developed in Practice 5-10, start at the top of the joint and weld down the length of the joint. When the weld is complete, inspect it for discontinuities and make any necessary changes in your technique. Repeat each type of joint as needed until consistently good welds are obtained. Turn off the welding machine, shielding gas supply, and clean up your work area when you are finished welding.

PRACTICE 5–12

Butt Joint and Lap Joint in the Vertical Down Position with 100 Percent Penetration

Using the same equipment, materials, and setup as listed in Practice 5-1, you will make welded joints with 100 percent weld penetration.

It may be necessary to adjust the root opening on the butt joint to meet the penetration requirements. The tee joint was omitted from this practice because little additional skill can be developed with it that is not already acquired with the lap joint. Repeat each type of joint as needed until consistently good welds are obtained. Turn off the welding machine, shielding gas supply, and clean up your work area when you are finished welding.

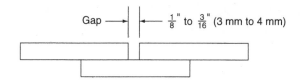

FIGURE 5-10 Butt joint with backing plate for vertical down weld.

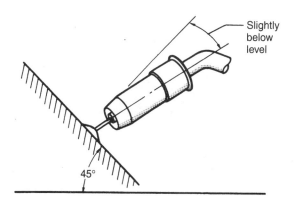

FIGURE 5-11 Forty-five-degree horizontal position.

PRACTICE 5-13

Flat Position Butt Joint with a Backing Plate in Vertical Down Position

Using the same equipment, materials, and procedures listed in Practice 5-1, make a welded butt joint with a backing plate (Figure 5-10).

- Tack-weld the sheets together and place them flat on the welding table.
- Starting at the top, run a bead along the joint. Watch the molten weld pool and bead for signs that a change in technique may be required. Underfill and overfill are both common problems when making this type of weld. No underfill is allowed, and the maximum weld face reinforcement allowed is 1/16 inch (1.5 mm).
- Make any needed changes as the weld progresses. By the time the weld is complete, you should be making the weld almost perfectly.
- Using the same technique that was established in the last weld, make another weld. This time, the entire weld should be flawless.

Repeat the bead as needed on each available type and thickness of metal until consistently good beads are obtained. Turn off the welding machine, shielding gas supply, and clean up your work area when you are finished welding.

PRACTICE 5-14

Butt Joint and Lap Joint in the Vertical Down Position, Welds To Be Tested

Using the same equipment, materials, and setup as listed in Practice 5-1, you will make the welded joints in the vertical down position. Each weld must pass the bend test. Repeat each type of weld joint until they pass the bend test. Turn off the welding machine, shielding gas supply,and clean up your work area when you are finished welding.

PRACTICE 5-15

Butt Joint with a Backing Plate in the Vertical Down Position, Weld To Be Tested

Using the same equipment, materials, and setup as listed in Practice 5-1, you will make the welded joint in the vertical down position. The weld must pass the bend test. Repeat the weld joint until it passes the bend test. Turn off the welding machine, shielding gas supply, and clean up your work area when you are finished welding.

5.5 HORIZONTAL 2G AND 2F POSITIONS

PRACTICE 5-16

Horizontal Stringer Bead at a 45-Degree Angle

Using the same equipment, materials, and setup as listed in Practice 5-1, you will make a horizontal stringer bead on a plate at a 45-degree reclined angle.

Start at one end with the gun pointed in a slightly upward direction (Figure 5-11). You may use a pushing or a dragging leading or trailing gun angle, depending upon the current setting and the penetration desired. Undercutting along the top edge and overlap along the bottom edge are problems with both gun angles. Careful attention must be paid to the weave technique used to overcome these problems.

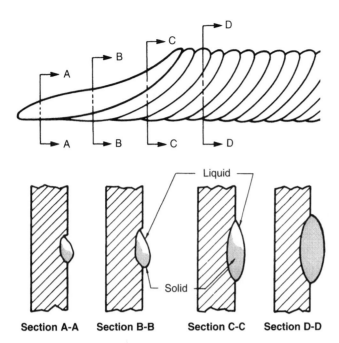

FIGURE 5–12 *The actual size of the molten weld pool remains small along the weld.*

The most successful weave patterns are very small C or J patterns. The J pattern is the most frequently used. The J pattern allows weld metal to be deposited along a shelf created by the previous weave (Figure 5-12). The length of the J can be changed to control the weld bead size. The maximum allowed weld bead width is 3/16 to 3/8 of an inch (5 mm to 10 mm).

Repeat these welds until you have established the rhythm and technique that work well for you. The weld must be straight and uniform and have complete fusion. Turn off the welding machine, shielding gas supply, and clean up your work area when you are finished welding.

PRACTICE 5-17

Stringer Bead in the Horizontal Position

Repeat Practice 5-9 and increase the angle of the plate until you have developed the skill to repeatedly make good horizontal welds on a vertical surface. The weld bead must be straight and uniform and have complete fusion. Turn off the welding machine, shielding gas supply, and clean up your work area when you are finished welding.

PRACTICE 5-18

Butt Joint, Lap Joint, and Tee Joint in the Horizontal Position

Using the same equipment, materials, and setup listed in Practice 5-1, you will make horizontal welded joints.

Tack-weld the pieces of metal together and brace them in position using the same skills developed in Practice 5-16. Starting at one end, make a weld along the entire length of the joint. When making the butt or lap joints, it may help to recline the plates at a 45-degree angle until you have developed the technique required. Repeat each type of joint as needed until consistently good welds are obtained. Turn off the welding machine, shielding gas supply, and clean up your work area when you are finished welding.

PRACTICE 5-19

Butt Joint and Lap Joint in the Horizontal Position with 100 Percent Penetration

Using the same equipment, materials, and setup as listed in Practice 5-1, you will make horizontal joints having 100 percent penetration in the horizontal position.

It may be necessary to adjust the root opening on the butt joint to meet the penetration requirements. Repeat each type of joint as needed until consistently good welds are obtained. Turn off the welding machine, shielding gas supply, and clean up your work area when you are finished welding.

PRACTICE 5-20

Butt Joint with a Backing Plate in the Horizontal Position with 100 Percent Penetration

Using the same equipment, materials, and setup as listed in Practice 5-1, you will make a welded joint with a backing plate having 100 percent penetration in the horizontal position.

It may be necessary to adjust the root opening to meet the penetration requirements. Repeat the weld as needed until consistently good welds are obtained. Turn off the welding machine, shielding gas supply, and clean up your work area when you are finished welding.

PRACTICE 5-21

Butt Joint and Lap Joint in the Horizontal Position, Welds To Be Tested

Using the same equipment, materials, and setup as listed in Practice 5-1, you will make the welded joints in the horizontal position. Each weld must pass the bend test. Repeat each type of weld joint until both pass the bend test. Turn off the welding machine, shielding gas supply, and clean up your work area when you are finished welding.

PRACTICE 5-22

Butt Joint with a Backing Plate in the Horizontal Position, Weld To Be Tested

Using the same equipment, materials, and setup as listed in Practice 5-1, you will make the welded joint in the horizontal position. The weld must pass the bend test. Repeat each type of weld joint until each weld passes the bend test. Turn off the welding machine, shielding gas supply, and clean up your work area when you are finished welding.

5.6 — OVERHEAD 4G AND 4F POSITIONS

There are several advantages to the use of short-circuiting arc metal transfer in the overhead position, including:

- *Small molten weld pool size.* The smaller size of the molten weld pool allows surface tension to hold it in place. Less molten weld pool sag results in improved bead contour with less undercut and icicles (Figure 5-13).
- *Direct metal transfer.* The direct metal transfer method does not rely on other forces to get the filler metal into the molten weld pool. This results in efficient metal transfer and less spatter and loss of filler metal.

PRACTICE 5-23

Stringer Bead, Overhead Position

Using the same equipment, materials, and setup as listed in Practice 5-1, you will make a welded stringer bead in the overhead position.

The molten weld pool should be kept as small as possible for easier control. A small molten weld pool can be achieved by using lower current settings, by traveling faster, or by pushing the molten weld pool. The technique used is the welder's choice. Often a combination of techniques can be used with excellent results.

Lower current settings require closer control of gun manipulation to ensure that the wire is fed into the molten weld pool just behind the leading edge. The low power will cause overlap and more spatter if this wire-to-molten weld pool contact position is not closely maintained.

Weld penetration into the base metal at the start of the bead can be obtained by using a slow start or quickly reversing the weld direction. Both the slow start and the reversal of weld direction put more heat into the weld start to increase penetration (Figure 5-14). The higher speed also reduces the amount of weld distortion by reducing the amount of time that heat is applied to a joint.

The pushing or trailing gun angle forces the bead to be flatter by spreading it out over a wider area as compared to the bead that results from a dragging or backhand gun angle. The wide, shallow molten weld

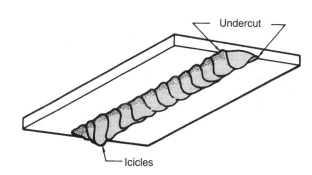

FIGURE 5-13 Overhead weld.

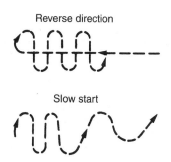

FIGURE 5-14 Two methods of concentrating heat at the beginning of a weld bead to aid in penetration depth.

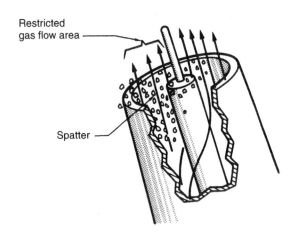

FIGURE 5-15 *Shielding gas flow affected by excessive weld spatter in nozzle.*

pool cools faster, resulting in less time for sagging and the formation of icicles.

When welding overhead, extra personal protection is required to reduce the danger of burns. Leather sleeves or leather jackets should be worn.

Much of the spatter created during overhead welding falls into the shielding gas nozzle. The effectiveness of the shielding gas is reduced (Figure 5-15), and the contact tube may short out to the gas nozzle (Figure 5-16). Turbulence caused by the spatter obstructing the gas may lead to weld contamination. The shorted gas nozzle may arc to the work, causing damage both to the nozzle and to the plate. To control the amount of spatter, a longer stickout and/or a sharper gun-to-plate angle is required to allow most of the spatter to fall clear of the gas nozzle. Applying antispatter to the nozzle won't stop the spatter from building up but will make it much easier to remove.

Make several short weld beads using various techniques to establish the method that is most successful and most comfortable for you. After each weld, stop and evaluate it before making a change. When you have decided on the technique to be used, make a welded stringer bead 5 inches long.

Repeat the weld until it can be made straight, uniform, and free from any visual defects. Turn off the welding machine, shielding gas supply, and clean up your work area when you are finished welding.

PRACTICE 5-24

Butt Joint, Lap Joint, and Tee Joint in the Overhead Position

Using the same equipment, materials, and setup as listed in Practice 5-1, you will make an overhead welded joint.

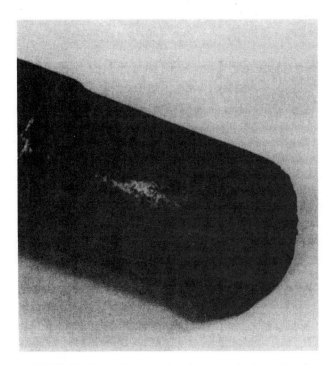

FIGURE 5-16 *Gas nozzle damaged after shorting out against the work.*

Tack-weld the pieces of metal together and secure them in the overhead position. Be sure you have an unrestricted view and freedom of movement along the joint. Start at one end and make a weld along the joint. Use the same technique developed in Practice 5-23.

Repeat the weld until it can be made straight, uniform, and free from any visual defects. Turn off the welding machine, shielding gas supply, and clean up your work area when you are finished welding.

PRACTICE 5-25

Butt Joint and Lap Joint in the Overhead Position with 100 Percent Penetration

Using the same equipment, materials, and setup as listed in Practice 5-1, you will make overhead welded joints having 100 percent penetration.

Tack-weld the metal together. It may be necessary to adjust the root opening to allow 100 percent weld metal penetration. During these welds, it may be necessary to use a dragging or backhand torch angle. When used with a C or J weave pattern, this torch angle helps to achieve the desired depth of penetration. A keyhole just ahead of the molten weld pool is a good sign that the metal is being penetrated (Figure 5-17).

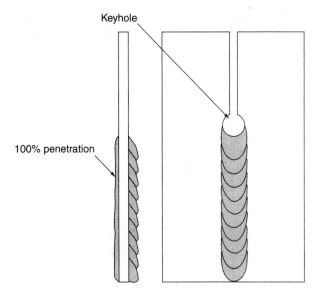

Keyhole

100% penetration

FIGURE 5–17 *A keyhole ahead of the molten weld pool indicates 100 percent penetration.*

Repeat the weld until it can be made straight, uniform, and free from any visual defects. Turn off the welding machine, shielding gas supply, and clean up your work area when you are finished welding.

Butt Joint with a Backing Plate in the Overhead Position with 100 Percent Penetration

Using the same equipment, materials, and setup as listed in Practice 5-1, you will make an overhead welded joint with backing plate having 100 percent penetration.

Tack-weld the metal together. It may be necessary to adjust the root opening to allow 100 percent weld metal penetration. During these welds, it may be necessary to use a dragging or backhand torch angle. When used with a C or J weave pattern, this torch angle helps to achieve the desired depth of penetration. A keyhole just ahead of the molten weld pool is a good sign that the metal is being penetrated.

Repeat the weld until it can be made straight, uniform, and free from any visual defects. Turn off the welding machine, shielding gas supply, and clean up your work area when you are finished welding.

Butt Joint and Lap Joint in the Overhead Position, Welds To Be Tested

Using the same equipment, materials, and setup as listed in Practice 5-1, you will make welded joints in the overhead position. Each weld must pass the bend test. Repeat each type of weld joint until both pass the bend test. Turn off the welding machine, shielding gas supply, and clean up your work area when you are finished welding.

Butt Joint with a Backing Plate in the Overhead Position, Welds To Be Tested

Using the same equipment, materials, and setup as listed in Practice 5-1, you will make a welded joint with backing plate in the overhead position. The weld must pass the bend test. Repeat the weld joint until it passes the bend test. Turn off the welding machine, shielding gas supply, and clean up your work area when you are finished welding.

5.7 — MIG PLUG WELDS

An MIG plug weld is a very common type of weld that is used in making both panel and structural repairs. Making either type of repair requires the same skill. The major difference between the two weld applications is that the structural material is thicker and the plug hole may be larger. The following practices will all use a standard 5/16-inch (8 mm) plug hole unless otherwise noted.

Plug-Welded Lap Joint in the Flat Position

Using a properly set up and adjusted MIG welding machine (see Table 5-1 or manufacturer's operating manual for the welder for settings), welding gloves, safety glasses, appropriate clothing, approved welding respirator, 0.023-inch and/or 0.030-inch (0.6 mm and/or 0.8 mm) diameter wire, and two or more pieces of sheet metal 5 inches (127 mm) long by

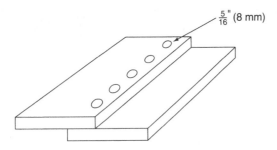

FIGURE 5-18   *Plug-welded lap joint.*

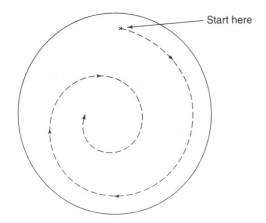

Start here

FIGURE 5-19   *Start near the edge and spiral inward until plug is filled.*

3 inches (76 mm) wide and 18 gauge to 24 gauge thick, you will make a plug weld on a lap joint in the flat position (Figure 5-18).

- Drill or punch two 5/16 inch (8 mm) diameter holes in one of the two plates.
- Clamp the plates together so that they overlap.
- Start welding on the inside edge of the hole and continue all the way around the inside circumference (Figure 5-19). If the plug is not completely filled after the first pass, spiral inward toward the center of the hole. Do not overfill the plug. The maximum buildup should be 1/8 inch (3 mm) or less.
- When the first weld is completed, allow it to cool and examine it for appearance. Make any needed adjustments before making the next weld.
- Repeat this process until consistently high-quality welds can be produced.

Turn off the welding machine, shielding gas supply, and clean up your work area when you are finished welding.

PRACTICE 5-30

Plug-Welded Lap Joint in the Flat Position with 100 Percent Penetration

Using the same equipment, materials, and setup as listed in Practice 5-29, you will make a plug-welded lap joint having 100 percent penetration.

Repeat the weld until it can be made uniform and free from any visual defects. Turn off the welding machine, shielding gas supply, and clean up your work area when you are finished welding.

PRACTICE 5-31

Plug-Welded Lap Joint in the Flat Position, Welds To Be Tested

Using the same equipment, materials, and setup as listed in Practice 5-29, you will make a plug-welded lap joint in the flat position. Each weld must pass the break test (Figure 5-20). Repeat the weld joint until the test is passed. Turn off the welding machine, shielding gas supply, and clean up your work area when you are finished welding.

PRACTICE 5-32

Plug-Welded Lap Joint in the Vertical Position

Using the same equipment, materials, and setup as listed in Practice 5-29, you will make a plug-welded lap joint in the vertical position.

Repeat this process until consistently high-quality welds can be produced. Turn off the welding machine, shielding gas supply, and clean up your work area when you are finished welding.

PRACTICE 5-33

Plug-Welded Lap Joint in the Vertical Position with 100 Percent Penetration

Using the same equipment, materials, and setup as listed in Practice 5-29, you will make a plug-welded lap joint having 100 percent penetration.

Repeat the weld until it can be made uniform and free from any visual defects. Turn off the welding machine, shielding gas supply, and clean up your work area when you are finished welding.

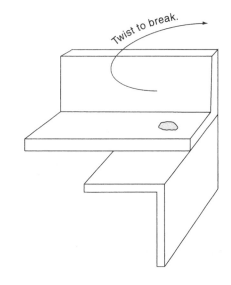

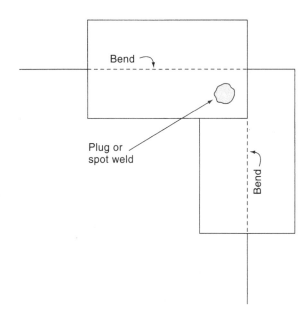

Bend

Plug or
spot weld

Bend

A

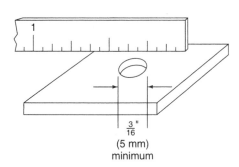

1

$\frac{3}{16}$"
(5 mm)
minimum

B

FIGURE 5–20 *(A) Break-testing a plug weld. (B) A good plug weld will pull out a hole in the bottom plate at least 3/16 of an inch (5 mm) in diameter.*

Plug-Welded Lap Joint in the Vertical Position, Welds To Be Tested

Using the same equipment, materials, and set-up as listed in Practice 5-29, you will make a plug-welded lap joint in the vertical position. Each weld must pass the break test (Figure 5-21). Repeat the weld joint until the test is passed. Turn off the welding machine, shielding gas supply, and clean up your work area when you are finished welding.

Plug-Welded Lap Joint in the Overhead Position

Using the same equipment, materials, and setup as listed in Practice 5-29, you will make a plug-welded lap joint in the overhead position.

Repeat this process until consistently high-quality welds can be produced. Turn off the welding machine, shielding gas supply, and clean up your work area when you are finished welding.

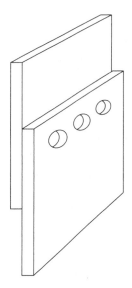

FIGURE 5–21 *Break-testing a plug-welded lap joint.*

PRACTICE 5-36

Plug-Welded Lap Joint in the Overhead Position with 100 Percent Penetration

Using the same equipment, materials, and setup as listed in Practice 5-29, you will make a plug-welded lap joint having 100 percent penetration.

Repeat the weld until it can be made uniform and free from any visual defects. Turn off the welding machine, shielding gas supply, and clean up your work area when you are finished welding.

PRACTICE 5-37

Plug-Welded Lap Joint in the Overhead Position, Welds To Be Tested

Using the same equipment, materials, and setup as listed in Practice 5-29, you will make a plug-welded lap joint in the overhead position. Each weld must pass the break test. Repeat the weld joint until the test is passed. Turn off the welding machine, shielding gas supply, and clean up your work area when you are finished welding.

5.8 I-CAR MIG WELDING QUALIFICATION TEST

The I-CAR welding qualification test is partially based on the American Welding Society's *AWS D1.3 Structural Welding Code for Sheet Metal*. The requirements have been modified to meet the specific needs of the automotive industry. All I-CAR qualification tests will be made on 18-gauge mild steel sheet. Only the vertical and overhead positions are used for the test. These two positions are used because it is believed that if technicians can make satisfactory welds in these positions, then they can also make quality welds in others.

I-CAR testing must be done under the supervision of I-CAR-qualified personnel. Your ability to succeed in the previous practices should enable you to pass the I-CAR testing. Contact I-CAR directly for specific information about taking the test.

The evaluation of the weld test plates is divided into two areas, visual examination and destructive testing. The weld test plates must pass the visual exam before they can be destructively tested.

5.9 VISUAL INSPECTION OF WELDS

During the I-CAR test you are allowed to make 2 weld test plates for each of the 6 required welds, for a total of 12 sample welds. It is then your responsibility to decide which weld from each set is better. For that reason, practicing both the welding and the examination procedures will help you to pass the qualification test.

To make a visual inspection of the test plate, you will need paper and pencil and either a rule, tape measure, or welding gauge (Figure 5-22). The visual inspection acceptance criteria for welds are as follows:

WELD FACE DEFECTS

- No porosity, skips, or voids greater than 1/8 inch (3 mm) as measured in any direction (Figure 5-23A).
- The total size of all porosity, skips, and voids cannot exceed 1/4 inch (6 mm) (Figure 5-23B).
- No underfill (Figure 5-23C).
- No cracks (Figure 5-23D).
- No undercut (Figure 5-23E).

ROOT FACE DEFECTS

- No burned-through holes (Figure 5-24A).
- Joint penetration as evidenced by a slight rise along the weld (Figure 5-24B).
- No excessive penetration higher than 1/16 inch (1.5 mm) (Figure 5-24C).

DIMENSION REQUIREMENTS

Butt and Lap Welds

- The weld must be centered along the length of the joint (Figure 5-25).
- The weld length must be at least 1 inch (25 mm) but no longer than 1 1/2 inches (38 mm).
- The bead width must be at least 3/16 inch (5 mm) but no wider than 3/8 inch (10 mm).
- The bead height must be at least flush with the plate but no higher than 1/8 inch (3 mm).
- No melt-through larger than 3/16 inch (5 mm) in diameter.

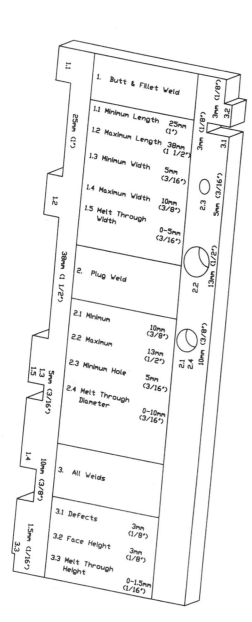

FIGURE 5–22   Welding gauge.

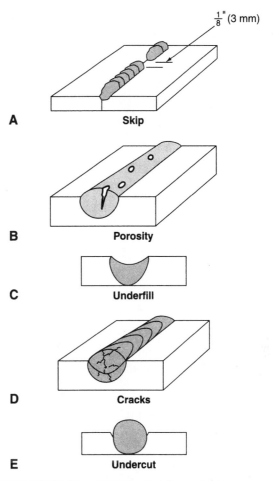

FIGURE 5–23   Weld face defect criteria.

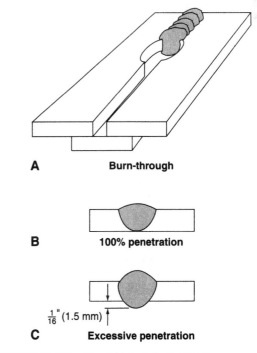

FIGURE 5–24   Root face defect criteria.

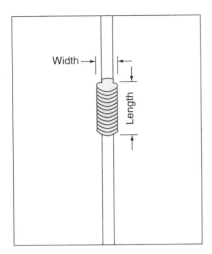

FIGURE 5–25 Butt and lap welds must be centered along the joint.

Plug Weld

- The weld diameter must be at least 3/8 inch (10 mm) but no larger than 1/2 inch (13 mm).
- No melt-through larger than 3/8 inch (10 mm) in diameter.
- The bead height must be at least flush with the plate but no higher than 1/8 inch (3 mm).

5.10 DESTRUCTIVE TESTING

Only after the weld has passed the visual inspection is it destructively tested. During the I-CAR qualification test, no welds will be destructively tested if they fail the visual inspection. However, during your practicing at school you should destructively test most or all of your welds. By destructively testing even bad welds, you should be able to better make changes in your welding techniques to improve your welding.

To destructively test your welds, you will need a solid vise with a jaw depth of at least 3 inches (76 mm), a 10- to 12-inch (250 mm to 300 mm) wedge-shaped cold chisel that has a tip that is approximately 7/8 of an inch (22 mm) wide, a 32-ounce (900 kg) shop hammer, gloves, safety glasses, and any other required protective gear (Figure 5-26). The destructive test procedures and acceptance criteria for welds are as follows:

BUTT WELD TEST

Coupon Disassembly

- Use all proper safety equipment and precautions during this part of the test.

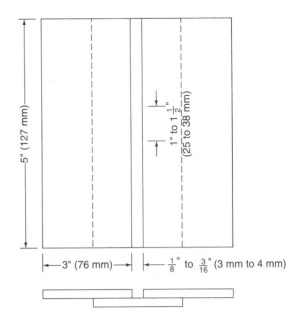

FIGURE 5–26 I-CAR butt weld test plate.

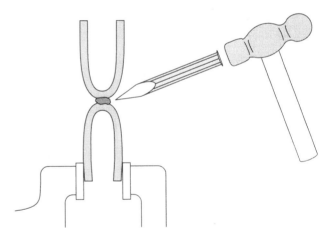

FIGURE 5–27 Clamp the test plate in a vice and use a chisel to break it apart.

- Use the hammer, chisel, and vise to bend the weld plates and backing plates as shown in Figure 5-27.
- Securely clamp the weld plates in the vise and chisel off the backing plate. Start chiseling at one end and work toward the other end.

Coupon Inspection

- There must be complete penetration into the backing plate along the entire length of the weld (Figure 5-28).
- There must be a clear, sharp fusion line along the edge of the weld.

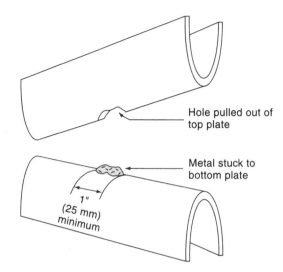

FIGURE 5–28   I-CAR minimum to pass butt weld.

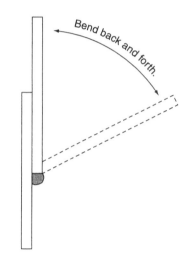

FIGURE 5–29   Testing a lap joint weld.

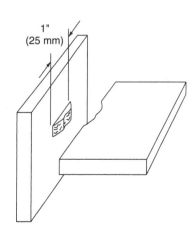

FIGURE 5–30   I-CAR minimum to pass lap weld.

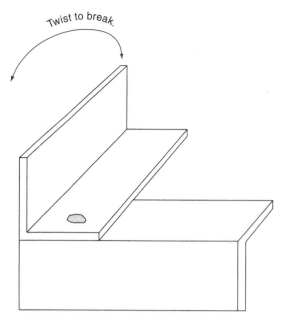

FIGURE 5–31   I-CAR plug weld specimen bent for testing.

LAP WELD TEST

Coupon Disassembly

- Use all proper safety equipment and precautions during this part of the test.
- Use the hammer, chisel, and vise to bend the weld plates back as shown in Figure 5-29.
- Securely clamp the bottom plate in the vise and bend the top plate back and forth until the weld breaks.

Coupon Inspection

- There must be at least 1 inch (25 mm) of base plate pulled off with the weld metal (Figure 5-30).

PLUG WELD TEST

Coupon Disassembly

- Use all proper safety equipment and precautions during this part of the test.
- Use the hammer and vise to bend the weld plates back as shown in Figure 5-31.
- Securely clamp the bottom plate in the vise and twist the top plate until the weld breaks free.

Coupons Inspection

There must be a nugget of the base metal removed with the weld, leaving a hole in the base plate at least 3/16 inch (5 mm) in diameter.

5.11   I-CAR WELDING

PRACTICE 5-38

Butt Weld with a Backing Plate in the Vertical Position

Using a properly set up and adjusted MIG welding machine (see Table 5-1 or manufacturer's operating manual for the welder for settings), welding gloves, safety glasses, appropriate clothing, approved welding respirator, three pairs of clamping pliers, marker or soapstone, rule, tape measure or weld gauge, 0.023-inch (0.6 mm) diameter wire, and three pieces of galvanized sheet metal 5 inches (127 mm) long by 3 inches (76 mm) wide and 18 gauge thick. You will make a square-grooved butt joint with backing in the vertical position.

Using the clamping pliers, clamp the joint together with a 1/16 to 3/23 inch (1.5 mm to 2 mm) root gap. Do not tack-weld the plates together.

Mark the weld length of 1 to 1 1/2 inches (25 mm to 38 mm) along the center of the joint. The marking must be bright enough for you to see it as you are welding.

Clamp the assembled test coupon in a welding stand so that it is in the vertical position. No portion of the back of the weld zone can be in contact with the weld stand. This could act as a heat sink and could cause weld penetration problems.

Hold the welding gun at the top of the plate with a 45-degree upward angle (Figure 5-32). Brace yourself; you might be able to rest one gloved hand against the plate to give yourself more control. Before starting the weld, practice moving along the joint to be sure that you have free movement along the length of the weld.

Lower your hood and start the weld at the top mark, and make a continuous weld down the joint to the bottom mark. Watch both the leading edge and sides of the molten weld pool for fusion. The leading edge should flow into the base metal, not curl over. The sides of the molten weld pool should also show fusion into the base metal and not be flashed (ragged) along the edges. Concentrate the heat and arc on the backing plate to insure good joint penetration (Figure 5-33).

The weld may be made with or without a weave pattern. If a weave pattern is used, it should be a very small C pattern. The C should follow the leading edge

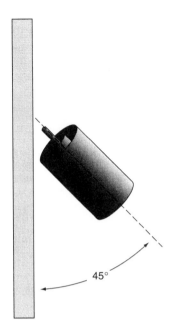

FIGURE 5-32 *Hold the welding gun at a 45-degree angle.*

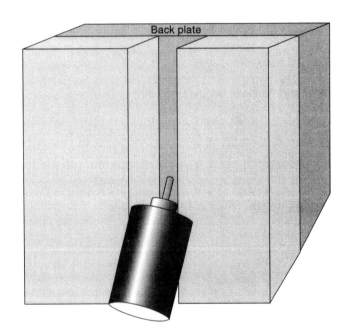

FIGURE 5-33 Concentrate arc on back plate.

of the weld. Small changes in the gun angle may help to increase or decrease penetration.

Repeat these welds until you can consistently pass the certification test. Turn off the welding machine, shielding gas supply, and clean up your work area when you are finished welding.

PRACTICE 5-39

Fillet Weld on a Lap Joint in the Vertical Position

Using the same equipment and setup as in Practice 5-38 but with just two 5- by 3-inch pieces of galvanized 18 gauge sheet metal, you will make a fillet weld on a lap joint.

Using the clamping pliers, clamp the joint together. Do not tack-weld the plates together.

Mark the weld length of 1 to 1 1/2 inches (25 mm to 38 mm) along the center of the joint. The marking must be bright enough for you to see it as you are welding.

Clamp the assembled test coupon in a welding stand so that it is in the vertical position. No portion of the back of the weld zone can be in contact with the weld stand.

Hold the welding gun at the top of the plate with a 45-degree upward angle. Brace yourself; you might be able to rest one gloved hand against the plate to give yourself more control. Before starting the weld, practice moving along the joint to be sure that you have free movement along the length of the weld.

Lower your hood and start the weld at the top mark, and make a continuous weld in a downward direction to the bottom mark. Watch both the leading edge and the sides of the molten weld pool for fusion. The leading edge should flow into the base metal, not curl over. The sides of the molten weld pool should also show fusion into the base metal and not be flashed (ragged) along the edges. Concentrate the heat and arc on the bottom plate very close to the edge of the top plate (Figure 5-34).

The weld may be made with or without a weave pattern. If a weave pattern is used, it should be a very small J pattern. The J should follow the leading edge of the weld. Small changes in the gun angle may help to increase or decrease penetration.

Repeat these welds until you can consistently pass the certification test. Turn off the welding machine, shielding gas supply, and clean up your work area when you are finished welding.

PRACTICE 5-40

Butt Weld with a Backing Plate in the Overhead Position

Using the same equipment, materials, and setup as listed in Practice 5-38, you will make a square-grooved butt joint with backing in the overhead position (Figure 5-35).

Using the clamping pliers clamp the joint together with a 1/16 to 3/23 inch (1.5 mm to 2 mm) root gap. Do not tack-weld the plates together.

Mark the weld length (1 to 1 1/2 inches) (25 mm to 38 mm) along the center of the joint. The marking must be bright enough for you to see it as you are welding.

Clamp the assembled test coupon in a welding stand so that it is in the overhead position. No portion of the back of the weld zone can be in contact with the weld stand.

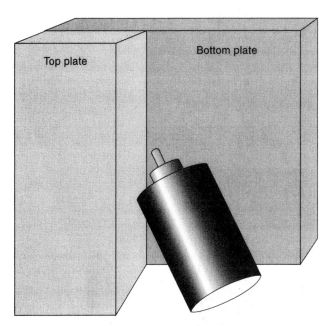

FIGURE 5-34 *Concentrate the arc on the bottom plate next to the edge of the top plate.*

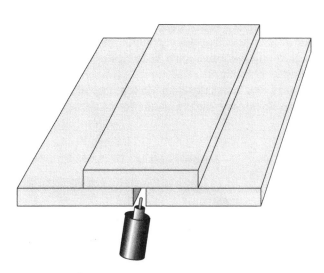

FIGURE 5-35 *Overhead butt joint.*

Hold the welding gun at the top of the plate with a 45-degree pulling angle. Brace yourself and, if possible, rest one gloved hand against the plate to give yourself more control. Before starting the weld, practice moving along the joint to be sure that you have free movement along the length of the weld. Make any adjustments in the height of the welding stand so that you are working in a comfortable position.

Lower your hood and start the weld at one mark and weld continuously to the other mark. You can stand so that you are welding directly toward or away from yourself or to one side, whichever position you prefer. Watch both the leading edge and the sides of the molten weld pool for fusion. The sides of the molten weld pool should also show fusion into the base metal and not be flashed (ragged) along the edges. Concentrate the heat and arc on the backing plate to insure good joint penetration.

The weld may be made with or without a weave pattern. If a weave pattern is used, it should be a very small C pattern. The C should follow the leading edge of the weld. Small changes in the gun angle may help to increase or decrease penetration.

Repeat these welds until you can consistently pass the certification test. Turn off the welding machine, shielding gas supply, and clean up your work area when you are finished welding.

PRACTICE 5–41

Fillet Weld on a Lap Joint in the Overhead Position

Using the same equipment, materials, and setup as listed in Practice 5-39, you will make a fillet weld on an overhead lap joint (Figure 5-36).

Using the clamping pliers, clamp the joint together. Do not tack-weld the plates together.

Mark the weld length as in Practice 5-39. Clamp the assembled test coupon in a welding stand so that it is in the overhead position. No portion of the back of the weld zone can be in contact with the weld stand.

Hold the welding gun at the top of the plate with a 45-degree pulling angle. Brace yourself and, if

possible, rest one gloved hand against the plate to give yourself more control. Before starting the weld, practice moving along the joint to be sure that you have free movement along the length of the weld.

Lower your hood and start the weld at one mark and weld continuously to the other mark. You can stand so that you are welding directly toward or away from yourself or to one side, whichever position you prefer. Watch both the leading edge and the sides of the molten weld pool for fusion. The leading edge should flow into the base metal, not curl over. The sides of the molten weld pool should also show fusion into the base metal and not be flashed (ragged) along the edges. Concentrate the heat and arc on the top plate very close to the edge of the bottom plate.

The weld may be made with or without a weave pattern. If a weave pattern is used, it should be a very small J pattern. The J should follow the leading edge of the weld. Small changes in the gun angle may help to increase or decrease penetration.

Repeat these welds until you can consistently pass the certification test. Turn off the welding machine, shielding gas supply, and clean up your work area when you are finished welding.

PRACTICE 5–42

Plug-Welded Lap Joint in the Vertical Position

Using the same equipment, materials and setup as listed in Practice 5-39, you will make a plug weld on a lap joint in the vertical position (Figure 5-37). one piece should be drilled or punched with a five-sixteenths-inch-diameter hole.

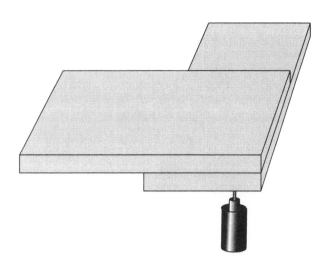

FIGURE 5–37 Overhead plug weld.

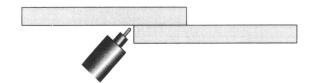

FIGURE 5–36 Overhead lap joint.

Using the clamping pliers, clamp the joint together so that the corners overlap. Do not tack-weld the plates together.

Clamp the assembled test coupons in a welding stand so that it is in the vertical position. No portion of the back of the weld zone can be in contact with the weld stand.

Hold the welding gun on the back plate at a 90-degree angle. Brace yourself and, if possible, rest one gloved hand against the plate to give yourself more control. Before starting the weld, practice moving around the inside of the hole to be sure that you have free movement around the weld.

Lower your hood and start the weld at the inside edge of the hole and continue all the way around the inside circumference. If the plug is not completely filled after the first pass, spiral inward toward the center of the hole. Do not overfill the plug. The maximum buildup is 1/8 inch (3 mm) or less.

Repeat these welds until you can consistently pass the certification test. Turn off the welding machine, shielding gas supply, and clean up your work area when you are finished welding.

PRACTICE 5-43

Plug-Welded Lap Joint in the Overhead Position

Using the same equipment, materials and setup as listed in Practice 5-42, you will make a plug weld on a lap joint in the overhead position.

Using the clamping pliers, clamp the joint together so that the corners overlap. Do not tack-weld the plates together.

Clamp the assembled test coupon in a welding stand so that it is in the overhead position. No portion of the back of the weld zone can be in contact with the weld stand.

Hold the welding gun on the top plate at a 90-degree angle. Brace yourself and, if possible, rest one gloved hand against the plate to give yourself more control. Before starting the weld, practice moving around the inside of the hole to be sure that you have free movement around the weld.

Lower your hood and start the weld at the inside edge of the hole and continue all the way around the inside circumference. If the plug is not completely filled after the first pass, spiral inward toward the center of the hole. Do not overfill the plug.

Repeat these welds until you can consistently pass the certification test. Turn off the welding machine, shielding gas supply, and clean up your work area when you are finished welding.

REVIEW QUESTIONS

1. Define mill scale and its purpose.
2. What is the most common problem with mill scale, and how can it be fixed?
3. Why is zinc applied to galvanized steel?
4. When welding on zinc-coated materials, why must you not breathe it?
5. If oil is removed from some cold- and hot-rolled steels, how much should be removed?
6. How should oil be removed from cold- and hot-rolled steels?
7. Which is easier to learn to weld on, thick or thin sheet metal?
8. What are MIG plug welds?
9. What positions are used on the I-CAR test?
10. Why are only these positions used?
11. Name the two types of I-CAR welding tests.
12. Whose responsibility is it to select the best welds for examination during the I-CAR testing?
13. List the visual inspection acceptance criteria for welds.
14. List the criteria for butt and lap welds.
15. List the criteria for visual inspections of plug welds.
16. Describe how to disassemble and inspect butt welds during the destructive phase of the I-CAR test.
17. Describe how to disassemble and inspect lap welds during the destructive phase of the I-CAR test.
18. Describe how to disassemble and inspect plug welds during the destructive phase of the I-CAR test.

Chapter 6

Resistance Spot Welding (RSW)

Objectives

After reading this chapter, you should be able to:

- Describe how a resistance spot welder works.
- List and explain the functions of the various resistance spot welding components.
- List the different spot welder adjustments.
- Describe the safety and setup procedures for resistance spot welding.
- List and describe the factors affecting a spot welding operation.
- Describe the three basic methods for determining a spot weld's integrity.
- Explain the causes of weak welds.
- Describe some other spot welding functions.

Resistance welding is defined as a process wherein coalescence (fusion between parts) is produced by the heat obtained from the resistance of the workpiece to the flow of low-voltage, high-density electric current in a circuit of which the workpiece is a part (Figure 6-1). Pressure is always applied to ensure a continuous electrical circuit and to forge the heated parts together. Heat is developed in the assembly to be welded, and pressure is applied by the welding machine through the electrodes. During the welding cycle, the mating surfaces of the parts do not have to melt in order for the weld to occur. The parts are usually joined as the result of heat and pressure and not their being melted together. Fluxes or filler metals are not needed for this welding process.

Resistance welding is one of the most useful and practical methods of joining metal. This process is ideally suited to production methods. It also requires workers who are less skilled. Resistance spot welding is the most widely used of all welding processes in the automotive industry.

6.1 RESISTANCE WELDING

The current for resistance welding is usually supplied by a transformer capacitor. The transformer is used to convert the high line voltage (low-amperage) power to the welding high-amperage current at a low voltage. The required pressure, or electrode force, is applied to the workpiece by pneumatic, hydraulic, or mechanical means (Figure 6-2).

Most resistance welding machines consist of the following three components:

- The mechanical system to hold the workpiece and to apply the electrode.
- The electrical circuit made up of a transformer, a current regulator, and a secondary circuit to conduct the welding current to the workpiece.
- The control system to regulate the time of the welding cycle.

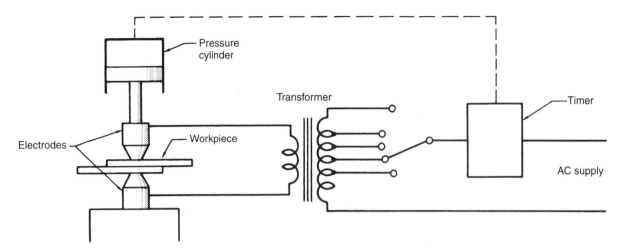

FIGURE 6–1   Fundamental resistance welding machine circuit.

FIGURE 6–2   Typical squeeze-type resistance spot welder. (Courtesy of Lors Machinery, Inc.)

There are several basic resistance welding processes. These processes include resistance spot welding (RSW), seam (RSEW), high-frequency seam (RSEW-HF), projection (PW), flash (FW), upset (UW), and percussion (PEW). More than one of these processes can be used during the original fabrication of the vehicle, but resistant spot welding is the only process used during repair.

The process is used to weld thousands of different industrial and consumer products, from nuclear piping heat shields and railway cars to major home appliances, automobiles, and trucks. In the United States, it is also widely used in the automotive aftermarket for sunroof installations and vehicle conversions, including recreational vehicles and stretch limousines. RSW is fast, easy to use, and produces strong welds.

Although resistance spot welding has been used in the European and Japanese unibody collision repair industry for more than 25 years, it is a relatively new process in the American repair industry. This is regrettable, because the resistance spot welder is ideal for welding many of the unibody's thin-gauge sections that call for good strength and no distortion. Since the early 1980s, a growing number of car manufacturers have published factory collision repair manuals that specify the use of resistance spot welding to repair-weld many areas of their unibody cars (Figure 6-3). These areas include rocker panels, radiator core supports, window and door opening flanges, roofs, many exterior panels, and pinch weld areas (Figure 6-4).

These same manuals also specify those areas for which the MIG welding process should be used. Typical MIG weld areas usually include the heavier rail sections and a number of other areas where the metal thickness or restricted accessibility to both sides of the repair joint would not be suitable for a hand-held squeeze-type resistance spot welder normally used for unibody work.

Never assume that one welding process is correct for all repairs or car models. Due to vehicle design differences, one automaker might recommend a different repair procedure for a given damaged area than would another automaker. The auto body technician should refer to the automaker's factory collision repair manual to determine which process and procedures are to be used.

While more than 90 percent of all factory welds in the unibody structure are resistance spot welds, the MIG welder is used for most repairs. The MIG welder and the resistance spot welder are two different machines designed for different applications. They are complementary in today's collision repair facilities.

MIG welders work best on inner structures, frameworks, gussets, and brackets. Conversely, resistance spot welders are designed for thin outer skins and

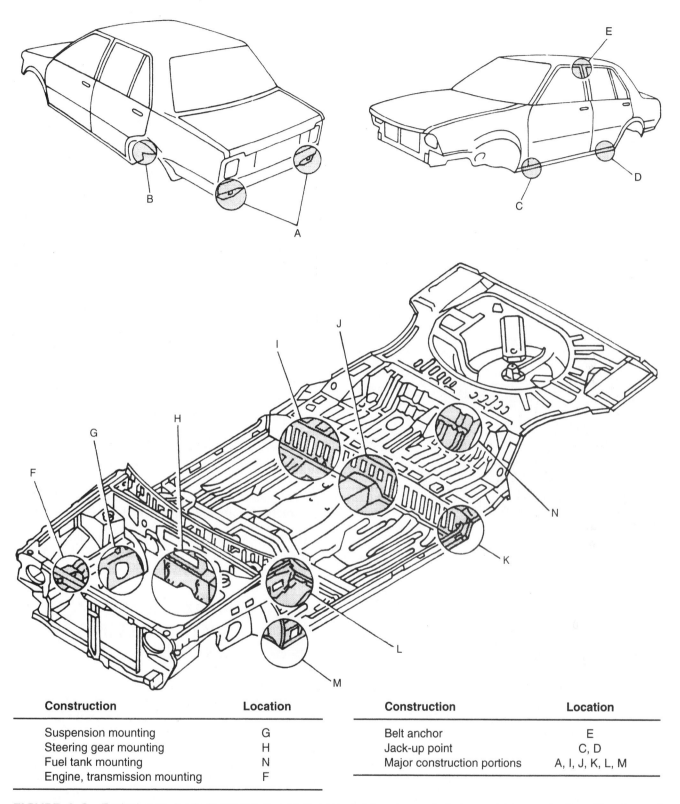

Construction	Location
Suspension mounting	G
Steering gear mounting	H
Fuel tank mounting	N
Engine, transmission mounting	F

Construction	Location
Belt anchor	E
Jack-up point	C, D
Major construction portions	A, I, J, K, L, M

FIGURE 6–3 Typical auto body locations in which spot welding is used during vehicle production (Courtesy of Nissan Motor Corp.)

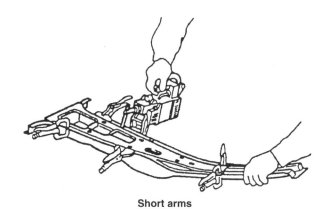

Short arms

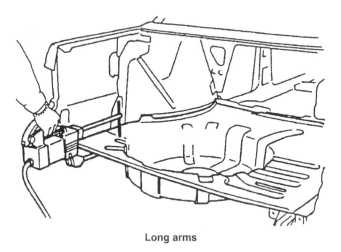

Long arms

FIGURE 6–4 *(A) Short and (B) long arms sqeeze-type resistance welders. (A) (Used with permission of Nissan Motor Corp.) (B) (Used with permission of Nissan Motor Corp.)*

cosmetic panels. On thin material like the 22- and 24-gauge sheet metal used on today's cars, resistance spot welding has the following advantages:

- Uses half as much heat as an MIG welder, which means less finishing time because of less heat distortion.
- Finishing time is further reduced because there is no weld metal to grind down.
- Duplicates OEM factory weld appearance.
- Clean; no smoke or fumes.
- Allows use of weld-through conductive zinc primers to restore corrosion protection to repair joints.
- Fast weld times of approximately 1 second, which produce strong welds on HSS steel, HSLA steel, and mild steels.
- Low initial cost and no consumables such as wire, rods, or gas are required, so it is economical.

Keep in mind that these two processes, resistance spot welding and MIG, should not be viewed as competitive, but as complementing each other. The properly equipped welding shop should use both spot welders and MIG welders.

The need for resistance spot welders in automotive repair is growing as many unibody automobiles are being manufactured and older unibody cars are being kept in service longer by their owners due to higher new-car costs. The resistance spot welder is becoming an increasingly important time and cost saver, as well as a necessary tool for repair welding today's unibody autos in compliance with the vehicle manufacturer's recommended repair procedures. For this reason, the competitive auto body technician must know how to use a resistance spot welder.

The sizes and shapes of the formed welds are controlled somewhat by the size and contour of the electrodes.

The welding time is controlled by a timer built into the machine. The timer controls four different steps (Figure 6-5). The steps are as follows:

- Squeeze time, or the time between the first application of electrode force and the first application of welding current.
- Weld time, or the actual time the current flows.
- Hold time, or the period during which the electrode force is applied and the welding current is shut off.
- Off period, or the time during which the electrodes are not contacting the workpieces.

Tables supplied by the machine manufacturer provide information for the exact time for each stage for different types and thicknesses of metal.

6.2 HOW RESISTANCE SPOT WELDING WORKS

Resistance spot welding relies on the resistance heat generated by a low-voltage electric current flowing through two pieces of metal held together, under pressure, by the squeeze force of the weld electrodes. One of the things that makes resistance spot welding unique is that the weld nugget is formed internally with relation to the surface of the base metal. Figure 6-6 shows a resistance spot weld nugget compared to an MIG spot weld. The MIG spot weld is made from one side only. The resistance spot weld is normally made with electrodes on each side of the workpiece. Resistance spot welds can be made with the workpiece in any position.

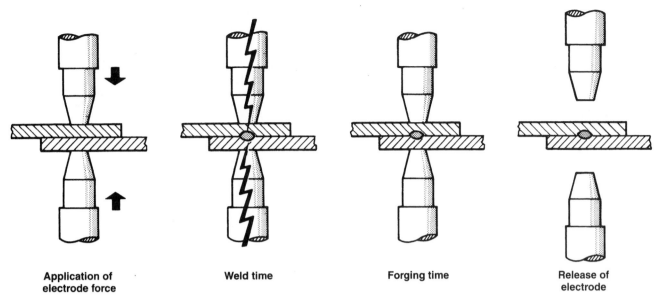

| Application of electrode force | Weld time | Forging time | Release of electrode |

FIGURE 6–5 Basic periods of spot welding.

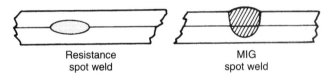

Resistance spot weld   MIG spot weld

FIGURE 6–6 Resistance and MIG spot weld comparison.

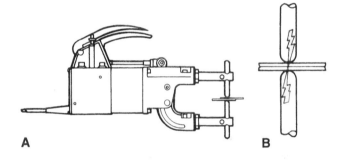

A   B

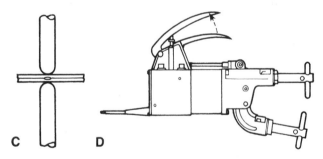

C   D

FIGURE 6–7 A resistance spot welding gun in the (A) squeeze position, (B) welding position, (C) hold position, and (D) off position.

The actual process of resistance spot welding is fairly simple. It has been compared to the forge or hammer welding done by blacksmiths. The blacksmith used a forge of hot coals to heat the metal parts to be welded. When the metal was white hot, the blacksmith would overlap the parts on the anvil and quickly forge them together under the pressure of the hammer blows. The smith depended on heat, pressure, and time to weld parts together.

A resistance spot welder also depends on heat, pressure, and time to make strong welds. The heat of the weld comes from the resistance of the metal to the passage of heavy current. In all cases, of course, the current must flow or the weld cannot be made. The pressure of the electrode tips on the workpiece holds the parts together in close and intimate contact. It is at this location where a pulse of current forms the weld. The force applied before, during, and after the current flow forges the heated parts together so that coalescence will occur. The welding time is

very short. Of course, the correct combination of welding current intensity, weld force, and weld time must be selected to make consistently strong welds. This is easy to do once it is understood how a resistance spot welding gun works (Figure 6-7).

- *Pressurization.* The mechanical welding bond between two pieces of sheet metal is directly related to the amount of force exerted on the sheet metal by the welding tips. As the tips squeeze the sheet metal together, an electrical current flows from the tips through the base metal, causing the two pieces to melt and fuse together. Weld spatter (internal or external) can result from too light a pressure on the tip or excessive current flow. Too heavy a tip pressure causes a small spot weld (Figure 6-8) and a reduced mechanical bond. In other words, as the tip pressure increases, the electrical current and subsequent heat are distributed over a wider area, thus reducing the diameter and the penetration of the weld.
- *Current Flow.* When pressure is applied to the metal, a large surge of electric current flows through the electrodes and through the two pieces of metal. The temperature rises rapidly at the joined portion of the metal where the resistance is greatest (Figure 6-9A). If the current continues to flow, the metal melts and fuses together (Figure 6-9B). If the current becomes too great or the pressure too low, internal spatter will result.
- *Holding.* If the current flow is stopped, the melted portion begins to cool and forms a round, flat bead of solidified metal known as a nugget. This structure becomes very dense due to the pressurization force, and its subsequent mechanical bonding is excellent (Figure 6-10). Pressurization time is very important; do not use less time than specified in the operator's manual.

HEAT

When current is passed through a conductor, the electrical resistance of the conductor to current flow will cause heat (or power) to be generated. The basic formula for heat generation is:

$H = I^2R$ where H = Heat
I^2 = Welding current squared
R = Resistance

The secondary portion of a resistance spot welding circuit, including the parts to be welded, is actually a series of resistances. The total additive value of this electrical resistance affects the current output of the resistance spot welding gun and the heat generation of the circuit.

Although current value is the same in all parts of the electrical circuit, the key fact is that the resistance values can vary considerably at different points in the circuit. The heat generated is directly proportional to the resistance at any point in the circuit. It is at the interface of the top and bottom workpieces where the

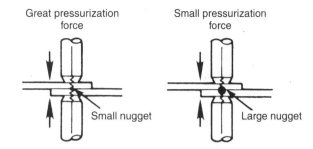

FIGURE 6–8 Electrode (tip) pressure.

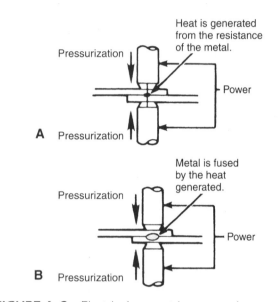

FIGURE 6–9 Electrical current (amperage).

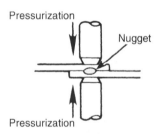

FIGURE 6–10 Electrical current (holding) flow time.

weld is to be made that the greatest relative resistance is required. (The term *relative* means with relation to the rest of the actual welding circuit.)

There are six major points of resistance in the work area.

1. Contact point between the electrode and top workpiece
2. Top workpiece
3. Interface of the top and bottom workpieces
4. Bottom workpiece
5. Contact point between the bottom workpiece and electrode
6. Resistance of electrode tips

The resistances are in series (which means the resistances are cumulative), and each point of resistance will retard current flow. The amount of resistance at the interface of the workpieces will depend on the heat transfer capabilities of the material, its electrical resistance, and the combined thickness of the materials at the weld joint. It is at this part of the circuit that the nugget of the weld is formed.

PRESSURE

The primary purpose of pressure is to hold the parts to be welded in intimate contact at the joint interface. This action assures consistent electrical resistance and conductivity at the point of the weld. The tong and electrode tips should not be used to pull the workpiece together for welding. The resistance spot welder is not designed to be an electrical C-clamp. The parts to be welded should be in intimate contact before pressure is applied.

Investigations have shown that high pressures exerted on the weld joint decrease the resistance at the point of contact between the electrode tip and the workpiece surface. The greater the pressure, the lower the resistance factor.

Where intimate contact of the electrode tip and the base metal exists, proper pressures will tend to conduct heat away from the weld. Higher currents are necessary with greater pressures and, conversely, lower pressures require less amperage from the resistance spot welding gun. This fact should be carefully noted, particularly when using a heat control with a resistance spot welder.

TIME

Resistance spot welding depends on the resistance of the base metal and the amount of current flowing to produce the heat necessary to make the spot weld. Another important factor is time. In most cases, several thousands of amperes are used in making the spot weld. Such amperage values, flowing through a relatively high resistance, will create a lot of heat in a short time. To make good resistance spot welds, it is necessary to have close control of the time the current is flowing.

Previously, the formula for heat generation was used. With the addition of the time element, the formula is completed as follows:

$$H = I^2RTK$$ where H = Heat
I^2 = Welding current squared
R = Resistance
T = Time
K = Heat losses

Welding heat is proportional to the square of the welding current. If the current is doubled, the heat generated is quadrupled. Welding heat is proportional

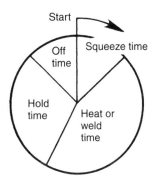

FIGURE 6–11 *Spot welding time cycle.*

to the total time of current flow; thus, if current is doubled the time can be reduced considerably. The welding heat generated is directly proportional to the resistance and is related to the material being welded and the pressure applied. The heat losses should be held to a minimum. It is therefore an advantage to shorten welding time.

Most resistance spot welds are made in very short time periods. Since alternating current is normally used for the welding process, procedures may be based on a 60-cycle time (60 cycles in 1 second). Figure 6-11 shows the resistance spot welding time cycle. The squeeze time is the time between pressure application and welding; the weld or heat time is measured in cycles; the hold time is the time that pressure is maintained after the weld is made; and the off time is the time wherein the electrodes separate to permit movement to the next spot.

Control of time is important. If the weld time element is too long, the base metal in the joint can exceed the melting point (and possibly the boiling point) of the material. This could cause faulty welds due to gas porosity. There is also the possibility of expulsion of molten metal from the weld joint (also called weld spatter), which could decrease the cross section of the joint and weaken the weld. Shorter weld times also decrease the possibility of excessive heat transfer in the base metal. Distortion of the welded parts is minimized, and the heat-affected zone around the weld nugget is substantially smaller. Not only is control of the welding time important, the holding time must be controlled also. As was mentioned earlier, the specified holding times given in the operator's manual should not be shortened.

6.3 RESISTANCE SPOT WELDING COMPONENTS

The components of a resistance spot welder (Figure 6-12) are the transformer, the welder control, and the gun (with interchangeable arm sets and electrode tips).

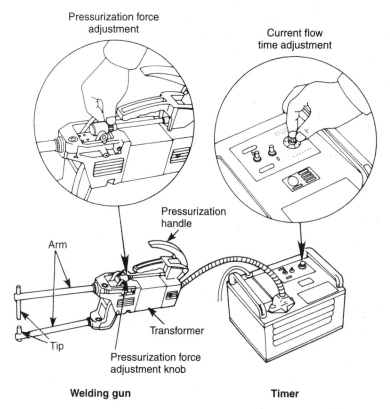

FIGURE 6–12 Components of a resistance spot welding system.

TRANSFORMER

The heart of a resistance spot welder is the transformer. It converts the shop line's high voltage (110 V to 220 V) and low amperage, (20 A to 30 A) current into a low secondary voltage (2 V to 5 V) and very high secondary amperage (8,000 A). It can be either built into the welding gun or mounted remotely and connected to the gun with cables. A built-in transformer is electrically more efficient, since there is little or no loss of welding current between the transformer and the gun. A remote transformer must be larger and draw more shop line current to compensate for power losses through the long cables connecting it to the gun. Remember that this high weld current will decrease when long-reach or wide-gap arm sets are used.

WELDER CONTROL

The weld current output can be adjusted to a lower intensity by the use of the welder control. The current is controlled by means of a solid-state weld timer and a phase shift weld current regulator. The control unit adjusts the transformer's weld current output and permits precise adjustment of the weld time.

The welder control must be capable of providing a full range of welding current adjustments. Weld current settings vary, depending upon the thickness of the steel to be welded and the length and gap of the arm sets needed to reach into the weld area. It might be necessary to decrease the weld current when welding with short-reach arm sets or to increase the weld current when using long-reach or wide-gap arm sets.

Weld timers are usually adjustable from 1/30 of a second to 1 second. The typical weld current adjustment range is from 30 to 100 percent of the maximum available weld current. A repeatable accuracy of at least 1/10 of a second is recommended for consistent weld quality.

Some manufacturers of resistance spot welders designed for unibody repair work offer additional control features that compensate for small amounts of surface scale or rust on the metal. Such features make it possible to determine when a poor weld condition exists.

Welders that do not have adjustable weld time and weld current capabilities or that use mechanical or manual timing are not satisfactory for consistent quality welding and thus are not recommended for unibody repair welding.

GUN

The muscle of a resistance spot welding operation is the gun that applies the squeeze force and delivers the welding current through the arms and electrodes to the weld area. It usually consists of a spring-assisted operating lever to generate forces in the range of 100 pounds to 250 pounds (45 kg to 113 kg) of actual clamping pressure. Most resistance spot welders are designed with a force-multiplying mechanism to produce the high electrode force required for consistent weld quality. These force-multiplying mechanisms can be spring or pneumatically assisted. Squeeze-type resistance welders that do not use a force-multiplying mechanism, and instead rely solely on the operator's manual grip for pressure, are not recommended for welding unibody structures.

This active clamping or squeeze force holds together the parts to be welded to maintain a good metal contact. This permits the welding current to flow easily through the repair joint to prevent expulsion or weld spatter from the molten weld nugget, which could weaken the weld. Active clamping force applies a follow-up pressure to force the weld as the welding current is turned off and the weld nugget contracts.

A toggle clamp or static force, like vise-grip pliers, cannot follow up; it loses clamping pressure as the weld nugget contracts, resulting in expulsion and weak welds. The squeeze pressure mechanism should be adjusted to develop sufficient clamping and follow-up force at the electrodes for the metal thickness being welded (Figure 6-13).

The majority of welding guns in collision repair facilities have a maximum capacity of up to two times 5/64-inch-thick (2 mm) steel when equipped with short-reach arm sets of 5 inches (125 mm) or less. Capacity with long-reach or wide-gap arm sets is at least two times 1/32-inch-thick (0.7 mm) steel. These capacities comply with the specifications listed in most factory body repair manuals.

Arms

Resistance spot welders used for unibody repair welding are available with a full range of interchangeable arm sets. Standard arm sets (Figure 6-14) are designed to reach difficult areas on most makes of cars, such as wheel well flanges, drip rails, taillight openings, and other tight pinch weld areas, as well as floor pan sections, rocker panels, and window and door openings. When working with new cars, be sure to check the factory repair manual, and look for special arm sets for the hard-to-reach areas.

Electrodes

Copper is the base metal normally used for resistance spot welding tongs and tips. The purpose of the electrode tips is to conduct the welding current to the

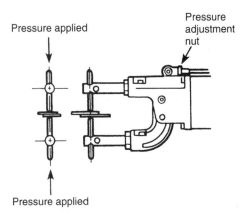

FIGURE 6–13 Adjustment of the squeeze pressure mechanism.

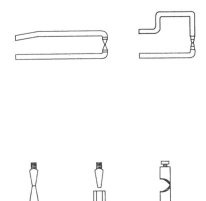

FIGURE 6–14 Standard tongs and tips for resistance spot welders. (Courtesy of Miller Electric)

workpiece, to be the focal point of the pressure applied to the weld joint, to conduct heat from the work surface, and to maintain their integrity of shape and characteristics of thermal and electrical conductivity under working conditions.

6.4 SPOT WELDER ADJUSTMENTS

To obtain sufficiently strong spot welds, perform the following checks and adjustments on the squeeze-type gun before starting the operation:

- **Arm set selection.** It is important to select the appropriate arm set for the area to be welded (Figure 6-15).
- **Arm adjustment.** Keep the gun arm as short as possible to obtain the maximum pressure for welding (Figure 6-16). Tighten the gun arm and tip securely so that they will not become loose during the operation.

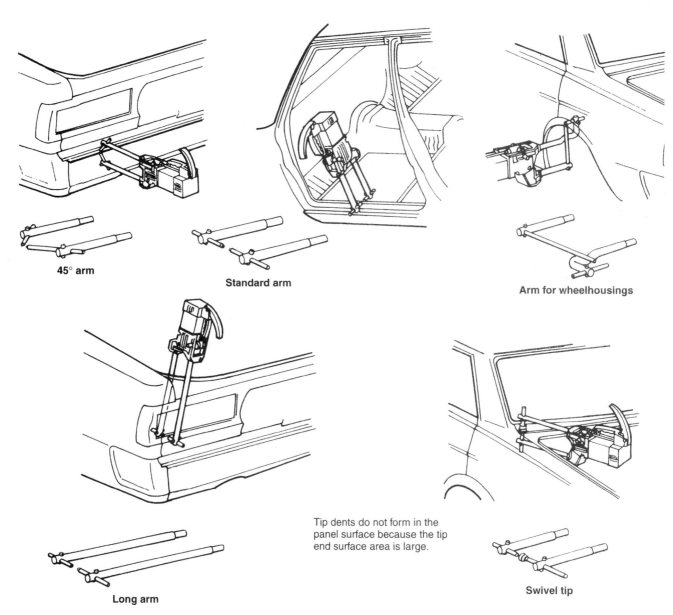

45° arm

Standard arm

Arm for wheelhousings

Long arm

Tip dents do not form in the panel surface because the tip end surface area is large.

Swivel tip

FIGURE 6–15   Select the proper arms for the job.

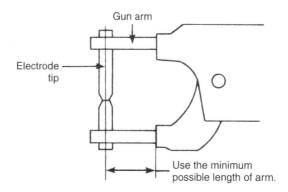

Gun arm

Electrode tip

Use the minimum possible length of arm.

FIGURE 6–16   Adjusting the gun arm.

- **Electrode tip alignment.** Align the upper and lower electrode tips on the same axis (Figure 6-17). Poor alignment of the tips causes insufficient pressure, which results in insufficient current density and reduced strength at the welded portions.
- **Electrode tip diameter.** The diameter of the spot weld decreases as the diameter of the electrode tip increases. However, if the electrode tip is too small, the spot weld will not increase in size. The tip diameter (Figure 6-18) must be properly controlled to obtain the desired weld strength. Before starting the operation, make sure that the tip diameter is kept the proper size, and file it cleanly to remove burnt or foreign matter from its surface. As the amount of dirt on the tip increases, the resistance at the tip also increases; this reduces both the current flow through the base metal and the weld penetration, resulting in an inferior weld. If the tips are used continuously over a long period of time, they will not dissipate heat properly and will become red hot. This will cause premature tip wear, increased resistance, and a drastic reduction in the welding current. If necessary, let the tips cool down after every five or six welds. If they are worn, use a tip dressing tool to reshape them (Figure 6-19).
- **Electrical current flow time.** When the electrical current flow time increases, the resulting heat increases the weld diameter and penetration. The amount of heat that is dissipated at the weld increases as the current flow time increases. Since the weld temperature will not rise after a certain amount of time, even if current flows longer than that, the spot weld size will not increase. However, tip pressure marks and heat warping might result.

The pressurization force and welding current cannot be adjusted on many spot welders, and the current value might be low. However, weld strength can be assured by lengthening the current flow time (in other words, letting the low current flow for a long time).

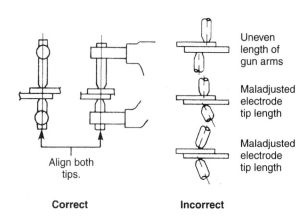

Correct   **Incorrect**

Uneven length of gun arms

Maladjusted electrode tip length

Maladjusted electrode tip length

Align both tips.

FIGURE 6–17 Correct the incorrect alignment of the electrode tip. (Courtesy of Nissan Motor Corp.)

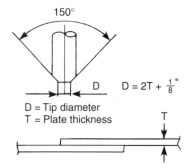

150°

$D = 2T + \frac{1}{8}"$

D = Tip diameter
T = Plate thickness

FIGURE 6–18 Determining tip diameter. (Courtesy of Nissan Motor Corp.)

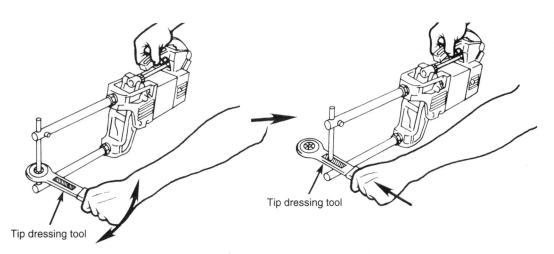

Tip dressing tool

Tip dressing tool

FIGURE 6–19 Reshaping the ends of the tips.

The best results can be obtained by adjusting the arm length or welding time according to the thickness of the panels. Although the welder instruction manual has these values listed inside, it is best to test the quality of the weld using the methods described later in this book as the adjustments are being done.

When spot-welding antirust steel panels, offset the drop in current density by raising the current value 10 to 20 percent above that for ordinary steel panels. Since the current value cannot be adjusted in spot welders ordinarily used for body repairs, lengthen the current flow time a little. It is important to differentiate between antirust sheet metal and ordinary sheet metal, since the protective zinc coating on the antirust panels must be removed along with the paint (when sanding prior to welding).

6.5 — SAFETY AND SETUP

Read the welder manufacturer's installation and operating instructions. Check the electrical service voltage in the shop to see if it matches the voltage requirements on the welder's nameplate. Make sure the welding equipment is connected to a positive ground, and have a qualified technician check for ground. Never connect a ground wire of single-phase electrical equipment to a leg of a three-phase electrical supply, because it could permanently damage the equipment and injure the operator. Check for the correct fuse and wire size recommended for the welder. As with all electrical equipment, be sure local electrical codes and legal requirements are met.

Unlike in other welding processes, a helmet and dark eye lens are not required with resistance spot welders. However, clear safety glasses should be worn at all times, and light shop gloves are usually recommended. Standard shop clothing in accordance with accepted shop safety practices is recommended.

If the welder has a separate control unit, connect the welding gun to the control following the manufacturer's instructions. If the welding gun is self-contained with a built-in control unit, this step will not be necessary.

Take two pieces of metal of the same thickness as the sheets to be repaired. Adjust the pressure mechanism to assure sufficient weld force on the two thicknesses of sample metal. Make sure that the welder arms are parallel and the tips make full contact when force is applied. If they are not, adjust the tips up or down until the arms are parallel. Secure the tips in place, and check to see that enough squeeze force can be applied to the metal piece being welded.

Now you are ready to connect the control unit or the self-contained gun, following the manufacturer's instructions, to a correct-voltage electrical supply that is fused with a long-delay type fuse. Remember, all electrical connections should comply with all applicable codes.

Adjust the weld current regulator to its lowest value. This is usually indicated as a percentage of current (30 to 100 percent). Adjust the weld timer to about 5 or 10 cycles (1/12 to 1/6 second).

Next, make sure that the metal used for sample welds is identical to the metal to be used on the vehicle. Scrap pieces cut from the repair area of the car are ideal. Be sure that the sample metal reaches as far into the welder throat (arm length) as the actual repair will.

Remember, the amount of steel placed in the throat of a welder will affect the welder's electrical efficiency and capacity. Squeeze the operating lever to apply pressure to the parts being welded. As pressure is being applied, a built-in switch turns on the welding current for the amount of time preset by the weld timer adjustment. The timer will automatically turn off the current at the end of the weld time. Since the weld time is less than 1 second, the entire process is very fast. Keep pressure on the weld for a moment to forge the weld before releasing the operating lever.

Test the sample weld for strength. The easiest and best method is the peel test. Simply peel the ends of the top sheet of metal upward, away from the bottom sheet. Then, peel the ends of the bottom sheet downward, away from the top sheet until the weld sample looks like two letter *U*s joined at the base (Figure 6-20).

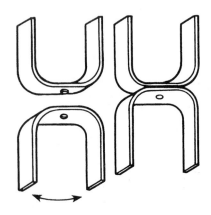

FIGURE 6–20   *Weld sample.*

Continue to rock the pieces apart until a hole is pulled in one of them. An alternate version of the peel test can be made using a vise, hammer, and chisel. Clamp the welded sample upright in a vise, gripping just below the weld. Place the chisel alongside the weld (never over the weld) and force the two welded pieces apart. A hole should be pulled in one of the two pieces of metal. The weld should always be stronger than the parent metal.

If the weld is not stronger, increase the weld current setting and make sample welds until expulsion or weld spatter is encountered. Expulsion may be visible as sparks shooting away from the weld. When expulsion is encountered, slightly reduce the weld current intensity and slightly increase the weld time. Make another sample weld and test it for strength. When a solid weld nugget can be peeled away, leaving a hole in one of the two sheets of metal, the weld is good. The actual repair work can now be started.

6.6 FACTORS AFFECTING OPERATION

Important operational procedures that should be considered when using a squeeze-type resistance spot welder are:

- **Clearance between welding surfaces.** Any clearance that exists between the surfaces to be welded causes poor current flow (Figure 6-21). Even if the weld can be made without removing the gap, the weld area will become smaller, resulting in insufficient strength. Flatten the two surfaces to remove the gap and clamp them tightly before welding.
- **Surfaces to be welded.** Paint film, rust, dust, or any other foreign matter on the metal surfaces to be welded can cause insufficient current flow and poor results and must be removed (Figure 6-22). Coat the surfaces to be welded with an anticorrosion agent that has higher conductivity. Apply the agent uniformly, even to the end face of the panel (Figure 6-23).
- **Performance of spot welding operations.** When performing spot welding operations, be sure to:
  —Use the direct welding method. For the portions to which direct welding cannot be applied, use MIG plug welding.
  —Apply electrodes at a right angle to the panel (Figure 6-24A). If the electrodes are not applied at right angles, the current density will be low, resulting in insufficient weld strength.
  —Spot-weld twice (Figure 6-24B).

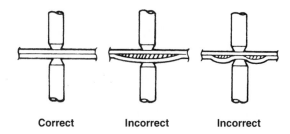

FIGURE 6-21 Correct and incorrect clearance between welding surfaces.

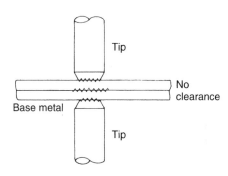

FIGURE 6-22 Condition of base metal surfaces.

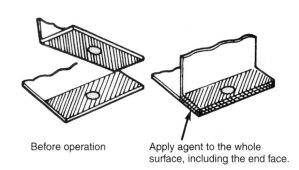

FIGURE 6-23 Areas to be protected with anticorrosion agent.

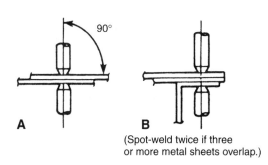

FIGURE 6-24 Precautions in performing spot welds.

- **Number of welding points.** Generally, the capacity of spot welding machines available in an auto repair shop is less than that of welding machines at the factory. Accordingly, the number of points of spot welding should be increased by 30 percent in a shop (Figure 6-25).
- **Minimum welding pitch.** The strength of individual spot welds is determined by the spot weld pitch (the distance between the welds) and edge distance (the distance from the spots to the panel edge). The bond between the panels becomes stronger as the weld pitch is shortened. However, past a certain point, the metal becomes saturated, and further shortening of the pitch will not increase the strength of the bond, because the current will flow to the spots that have previously been welded. This reactive current diversion increases as the number of spot welds increases, and the diverted current does not raise

the temperature at the welds (Figure 6-26). The distance of the weld pitch must be beyond the area influenced by the reactive current diversion. In general, the values presented in Table 6-1 should be observed.

- **Position of spot weld from the end of the panel.** The edge distance is also determined by the position of the welding tip. Even if the spot welds are normal, the welds will not have sufficient strength if the edge distance is insufficient. When welding near the end of a panel, observe the values presented in Table 6-2.
- **Spotting sequence.** Do not spot continuously in one direction only. This method promotes weak welds due to the shunt effect of the current (Figure 6-27). If the welding tips become hot and change their color, stop and allow the tips to cool.

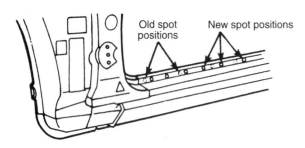

FIGURE 6–25 Number of points to spot weld.

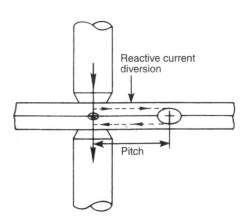

FIGURE 6–26 Minimum welding pitch (distance from the center of one weld to the center of the next weld).

	TABLE 6–1: SPOT WELDING POSITION		
Panel Thickness	Pitch (S)	Edge Distance (P)	
1/64″	7/16″ or more	13/64″ or more	
1/32″	9/16″ or more	13/64″ or more	
Less than 3/64″	11/16″ or more	1/4″ or more	
3/64″	7/8″ or more	9/32″ or more	
1/16″	1-9/64″ or more	5/16″ or more	

- **Welding corners.** Do not weld the corner radius portion, (Figure 6-28); this results in the concentration of stress that leads to cracks. The following locations require special consideration: the upper corner of the front and center pillars, the front upper portion of the rear fender, and the corner portion of the front and rear windows.

SPOT WELD INTEGRITY

The quality of spot welds can be determined largely through inspection and nondestructive and destructive testing. Inspection is used to judge the quality of the weld by its outward appearance and texture. The nondestructive and destructive tests are used to measure the strength of the weld.

INSPECTION

Check the finish of the weld visually and by touching. The items to check include:

- **Spot position.** The spot weld position should be in the center of the flange, with no tip holes and no welds overriding the edge. As a rule, the old spot position should be avoided.
- **Number of spots.** There should be at least 1.3 times the number made by the manufacturer. For example, 1.3 times four original factory welds equals roughly five new repair welds.
- **Pitch.** It should be a little shorter than that of the manufacturer, and the spots should be uniformly spaced. The minimum pitch should be at a distance where reactive current diversion will not occur.
- **Dents.** There should be no dents, or tip bruises, on the surfaces that exceed half the thickness of the panel.

TABLE 6–2: POSITION OF WELDING SPOT FROM THE END OF PANEL

Thickness (t)	Minimum pitch (★)	
1/64″	7/6″ or over	
1/32″	7/6″ or over	
Less than 3/64″	15/32″ or over	
3/64″	9/16″ or over	
1/16″	5/8″ or over	
5/64″	11/16″ or over	

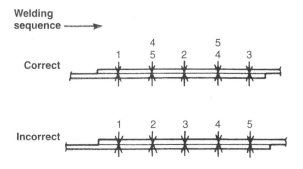

FIGURE 6-27 Proper welding of sequence. (Courtesy of Nissan Motor Corp.)

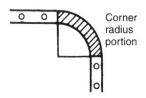

FIGURE 6-28 Proper method of welding corners. (Courtesy of Nissan Motor Corp.)

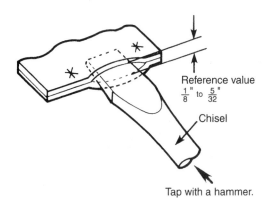

FIGURE 6–29 *Performing the nondestructive test.* (Courtesy of Nissan Motor Corp.)

- **Pinholes.** There should be no pinholes large enough to see.
- **Spatter.** A spatter is too large if a glove can catch on the surface when rubbed across it.

NONDESTRUCTIVE TESTING

To conduct a nondestructive test, use a chisel and hammer and proceed as follows:

1. Insert the tip of the chisel between the welded plates (Figure 6-29) and tap the end of the chisel until you have created a clearance of 1/8 to 5/32 inch (3 mm to 4 mm) between plates when the plate thickness is approximately 1/32 inch (0.7 mm). If the welded portions remain normal, the welding has been done properly. This clearance varies depending on the location of the welded spots, the length of the flange, plate thickness, welding pitch, and other factors. The values given here are only reference values.
2. If the thickness of the plates is not equal, the clearance between the plates must be limited to 1/16 to 5/64 inch (1.5 mm to 2 mm). Any further opening of the plates can become a destructive test.
3. Be sure to repair the deformed portion of the panel after inspection.

DESTRUCTIVE TESTING

Most destructive tests require the use of sophisticated equipment, a requirement that most collision repair facilities are unable to meet. For this reason, simpler methods have been developed.

One method of destructive testing was given earlier in this chapter. Here is another method: A test piece using the same metal as the welded piece is made; it should also be the same panel thickness and be welded in the positions shown in Figure 6-30. Next,

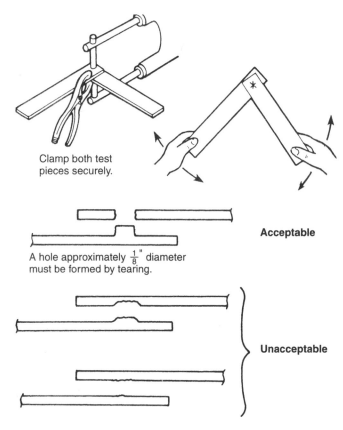

FIGURE 6–30 *Performing the destructive test.* (Courtesy of Nissan Motor Corp.)

force is applied in the direction of the arrows to separate the spots. If the weld pulls out cleanly (like a cork from a bottle), it is good. It should be noted that since the weld performance cannot be exactly duplicated by this test, these results should only serve as a reference.

CAUSES OF WEAK WELDS

The three most common causes of weak spot welds in auto body repair are failure to properly clean the areas to be welded, use of a nonconductive (insulating) primer at the repair joint, and mushrooming of the welder tips.

The first can be corrected by cleaning all sides of the repair joint with a wire brush (a wire cat-type brush in a power drill will clean two sides of a flange at the same time) or a recovery-type sandblaster. Use of only electrically conductive zinc-based weld-through primers on repair joint areas will eliminate the second cause.

Mushrooming of the welder tips is easily avoided by using the correct weld time and weld current for the job. Excessive weld times do not make better welds and will essentially "cook" the tips, causing them to mushroom.

TABLE 6–3: RESISTANCE SPOT WELDING DEFECTS AND THEIR CAUSES

Possible Cause of Weld Defect		Type of Weld Defect				
		Weak Weld	Excessive Expulsion	Electrode Mushrooming	Excessive Weld Marking	No Weld
Weld Current	Low	Primary			Secondary	Primary
	High		Primary		Secondary	
Weld Time	Short	Primary				Primary
	Long		Primary	Primary	Primary	
Low weld force		Secondary	Primary	Primary		
Electrode face diameter	Small	Primary				
	Large	Primary				Secondary
Poor metal fitup		Primary	Primary		Primary	Primary
Too close to edge		Secondary				
Dirty metal		Secondary	Primary			Primary
Weld spacing too close		Primary				Secondary

If a tip mushrooms to double its original size, its area has quadrupled. At that point, four times the weld current and four times the squeeze force previously needed with the correct size tip face are now required, because the weld current and force are spread out over a greater area. Mushroomed tips can be reshaped with a file or cutter tool, which can be fitted to an ordinary power drill.

The troubleshooting guide shown in Table 6-3 lists the common resistance welding problems encountered in auto body repair.

6.8 OTHER SPOT WELDING FUNCTIONS

Although the squeeze-type welding gun is used most often in the repair shop, there are other types of guns used with spot welding equipment. With the proper gun attachment, the spot welder can be used as a panel spotter, stud welder, and mold rivet welder.

PANEL SPOTTING

When one operates a panel spotter (Figure 6-31), the two electrode guns are placed on the nonstructural replacement panel. Figure 6-32 shows how both lap and flange joints can be made with a spliced or full panel installation. After the adjustments are made

FIGURE 6–31 *Typical panel spotter. (Courtesy of NLC, Inc. (LENCO))*

following the manufacturer's directions, push both electrodes against the panel and apply moderate pressure to close any gaps. Press the weld button on the switch handle and hold it down until the welding cycle stops automatically. When finger pressure is released from the weld button, the electrodes are moved to the next welding location.

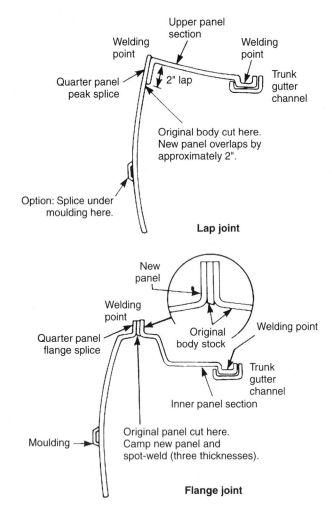

Lap joint

Flange joint

FIGURE 6–32 *Panel spotting lap and flange joints.*

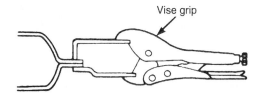

FIGURE 6–33 *Using vise grips to hold a flange joint together.*

Here are some other panel spotter operational tips:

- As in any spot welding operation, thoroughly clean the surfaces along the weld seam. If a replacement panel has been primed, strip off this coat on both sides and along the weld seam with a coarse abrasive paper. If the panel has a rust preventive film coating instead of a primer, it is sufficient to wipe off both sides of the seam with a clean rag and solvent.
- Use vise grips on all flange joint and drip rail applications (Figure 6-33). Weld near the vise grip jaws where the fitup is tight.
- A few sheet metal screws can be used on lap joints to position the panel for spot welding. Make sure that the paint has been removed from the joints.
- On long splice jobs, start in the middle of the panel and spot-weld in one direction only; for example, from the middle of the panel to the door

post. Start in the middle again and complete the panel welds to the taillight area. This helps to prevent distortion.
- Remove burrs on the newly cut panel to insure good metal-to-metal contact when pressure is applied to the electrodes. Burrs and dents cause air space between mating parts and prevent positive metal contact.

The twin electrodes of the panel spotter permit spot welding in many places where the squeeze-type spot welder has difficulty operating. In addition, the panel spotter can be converted into a squeeze-type spot welder with a gun attachment. However, this arrangement should be used only on nonstructural parts—never on structural parts.

STUD SPOT WELDING FOR DENT REMOVAL

Studs used in dent removal can be welded with a stud spot welder (Figure 6-34) or panel spotter (Figure 6-35). When you are using the latter, a stud-pulling kit (Figure 6-36) that contains a slide hammer and other necessary items must be used.

To remove a dent with either a stud or panel spot welder, you need a good-quality stud. The stud should offer the necessary combination of pull strength and tensile strength while remaining extremely flexible. Flexibility allows the stud to be bent out of the way when you are working on adjacent studs, then bent back when needed. The purpose of using studs is to minimize the heat required for repair and, therefore, to minimize the softening of the steel. This technique avoids the need for drilling or punching through the metal and undercoating, often an open invitation to corrosion. The procedure is as follows:

1. Fuse the stud to the dented area (Figure 6-37).
2. Use a dent puller or power jack to pull the stud.
3. When the dent is removed, snip off the stud with cutters and grind it flush with the panel (Figure 6-38). The entire procedure takes very little time, with no damage at all to the panel.

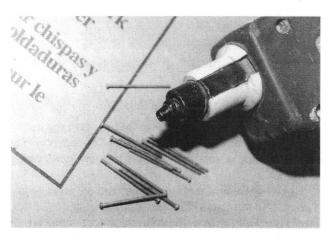

FIGURE 6–34 Typical stud welder for removing dents. (Courtesy of Larry Maupin)

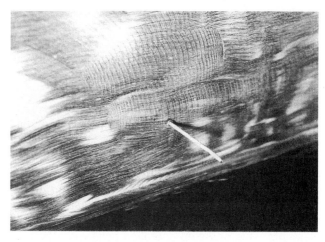

FIGURE 6–37 Stud welded to dent. (Courtesy of Larry Maupin)

FIGURE 6–35 Using a panel spotter to install spot studs. (Courtesy of Lenco Inc.)

FIGURE 6–38 Grind off the welded studs. (Courtesy of Larry Maupin)

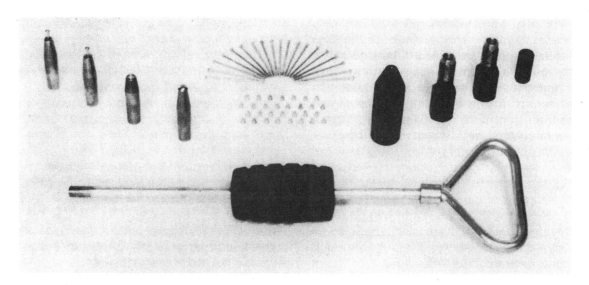

FIGURE 6–36 Typical panel spotter stud kit. (Courtesy of Lenco Inc.)

MOLD RIVET WELDING

Although some decorative strips are applied with adhesive, many mouldings are still applied with mold rivets and clips—for example, chrome strips on rocker panels and on window and vinyl roof mouldings.

When patching or refinishing areas that are susceptible to moisture, salt, or high humidity, a technician is usually apprehensive about drilling holes that expose inner panels. Mold rivet welding with a stud or spot welder is a logical alternative. One electrode has the mold rivet welding tip, the other has the ground tip. No holes are made; rivets can be relocated or replaced while not exposing vulnerable areas to outside elements. This one-step operation achieves a factory replica rivet and is ideal for placing rivets that need to be removed or relocated. They require very little grinding.

PRACTICE 6-1

Observing the Effect of Reducing Welding Current on Spot Welding of Thin-Gauge Mild Steel Sheets

Using an electrical resistance spot welder, the manufacturer's operating manual for the welder, welding gloves, safety glasses, appropriate clothing, pencil and paper, two or more pieces of 4-inch (100 mm) wide and 12-inch (300 mm) long, 18- to 24-gauge mild steel sheet metal, a vise, a chisel, and a hammer, you are going to observe the effect that reducing the current has on spot welds produced.

Clamp the sheet metal together with approximately 1 inch (25 mm) overlapping along the 12-inch (300 mm) side. Referring to the manufacturer's literature or Table 6-1, set the amperage, time, and other parameters to the recommended levels for the thickness of metal to be welded. Record the settings on the paper. Make a spot weld approximately 1 inch (25 mm) from the end of the plate. Lower the amperage and record the new setting on the paper. Make another spot weld approximately 2 inches (50 mm) from the first weld. Repeat this process until it is obvious that no fusion is occurring between the plates.

Turn off the welder and allow the plates to cool. Place the test plate in the vise and, using the chisel, separate the spot welds. Evaluate the welds to see what effect the lowering of the current had on the weld quality and acceptability. Check to see which weld currents produced acceptable welds.

Clean up your work station.

PRACTICE 6-2

Observing the Effect of Increasing Welding Current on Spot Welding of Thin-Gauge Mild Steel Sheets

Using the same equipment, setup, and materials as in Practice 6-1, you are going to observe the effect that increasing the current has on spot welds produced.

Clamp the sheet metal together with approximately 1 inch (25 mm) overlapping along the 12-inch (300 mm) side. Referring to the manufacturer's literature or Table 6-1, set the amperage, time, and other parameters to the recommended levels for the thickness of metal to be welded. Record the settings on the paper. Make a spot weld approximately 1 inch (25 mm) from the end of the plate. Raise the amperage and record the new setting on the paper. Make another spot weld approximately 2 inches (50 mm) from the first weld. Repeat this process until you see sparks being thrown out from under the electrode.

Turn off the welder and allow the plates to cool. Place the test plate in the vise and, using the chisel, separate the spot welds. Evaluate the welds to see what effect the raising of the current had on the weld quality and acceptability. Check to see which weld currents produced acceptable welds.

Clean up your work station.

PRACTICE 6-3

Observing the Effect of Reducing Welding Time on Spot Welding of Thin-Gauge Mild Steel Sheets

Using the same equipment, setup, and materials as in Practice 6-1, you are going to observe the effect that reducing the current has on spot welds produced.

Clamp the sheet metal together and set up the welder as in Practice 6-1. Record the settings on the paper. Make a spot weld approximately 1 inch (25 mm) from the end of the plate. Decrease the time and record the new setting on the paper. Make another spot weld approximately 2 inches (50 mm) from the first weld. Repeat this process until it is obvious that no fusion is occurring between the plates.

Turn off the welder and allow the plates to cool. Place the test plate in the vise and, using the chisel, separate the spot welds. Evaluate the welds to see what effect the lowering of the time had on the weld quality and acceptability. Check to see which weld times produced acceptable welds.

Clean up your work station.

PRACTICE 6–4

Observing the Effect of Increasing Welding Time on Spot Welding of Thin-Gauge Mild Steel Sheets

Using the same equipment, setup, and materials as in Practice 6–1, you are going to observe the effect that reducing the current has on spot welds produced.

Clamp the sheet metal together and set up the welder as in Practice 6–1. Record the settings on the paper. Make a spot weld approximately 1 inch (25 mm) from the end of the plate. Increase the time and record the new setting on the paper. Make another spot weld approximately 2 inches from the first weld. Repeat this process until you see sparks being thrown out from under the electrode or the plate turns bright red at the weld.

Turn off the welder and allow the plates to cool. Place the test plate in the vise and, using the chisel, separate the spot welds. Evaluate the welds to see what effect the raising of the time had on the weld quality and acceptability. Check to see which weld times produced acceptable welds.

Clean up your work station.

REVIEW QUESTIONS

1. Name the three components in most resistance welding machines.
2. What are the types of resistance welding processes?
3. What resistance process is used during repair?
4. What parts of a unibody car should be repaired with resistance spot welding?
5. What document should a welder check before beginning a repair?
6. List the advantages of spot welding over MIG welding.
7. What four steps does the welding timer control?
8. What are the three things that make resistance spot welding work?
9. Where is the greatest relative resistance encountered?
10. List the six major points of resistance in the work area.
11. List the components of a resistance spot welder.
12. What function does a transformer for a resistance spot welder perform?
13. What checks and adjustments should be performed on the squeeze-type gun before starting a spot welding operation?
14. How can the welder compensate for antirust steel panels during spot welding?
15. List some of the safety procedures for resistance spot welding.
16. When one is performing a sample weld, where should the weld current regulator and weld timer be set?
17. Describe the peel test and its alternate.
18. What are the eight factors that affect operation?
19. What do you look for when inspecting a weld visually and by touching?
20. Describe the procedures for nondestructive testing of welds.

Chapter
7

Shielded Metal Arc Welding (Stick Welding)

Objectives

After completing this chapter, you should be able to:
- Set up a shielded metal arc welding machine.
- Explain how to control arc blow.
- Make shielded metal arc welds in mild steel.
- Demonstrate safe working practices.

Shielded metal arc welding (SMAW), commonly referred to as "stick welding," is a welding process that uses a flux-covered metal electrode to carry an electrical current (Figure 7-1). The current forms an arc across the gap between the end of the electrode and the work. The electric arc creates sufficient heat to melt both the electrode and the work. Molten metal from the electrode travels across the arc to the molten pool on the base metal, where they mix together. The end of the electrode and the molten pool of metal are surrounded, purified, and protected by a gaseous cloud and a covering of slag produced as the flux coating of the electrode burns or vaporizes. As the arc moves away, the mixture of molten electrode and base metal solidifies and becomes one piece.

This welding method provides a high temperature and concentration of heat that allow a small molten weld pool to be built up quickly. The addition of filler metal from the electrode adds reinforcement and increases the strength of the weld. SMAW can be performed on steel metal 1/16 inch (1.5 mm) thick or thicker. A minimum of equipment is required, and it can be portable.

SMAW has limited use in most collision repair facilities. It is more widely used for making repairs on heavier sections like those found on older vehicles, some trucks, and recreational vehicles. Although stick

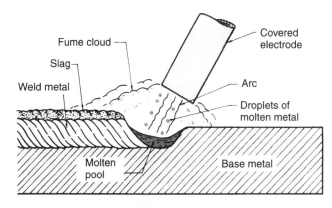

FIGURE 7-1 *Shielded metal arc welding.*

welding has limited use in most high-production shops, it is a skill that is needed in most smaller shops. In addition to its use on some vehicles, stick welding is often used as part of maintenance work commonly performed in or around smaller shops.

SMAW is the most widely used welding process because of its low cost, flexibility, portability, and versatility. The machine and the electrodes are low cost. The machine itself can be as simple as a 110-volt, step-down transformer. The electrodes are available from a large number of manufacturers in packages from 1 pound (0.5 kg) to 50 pounds (22 kg).

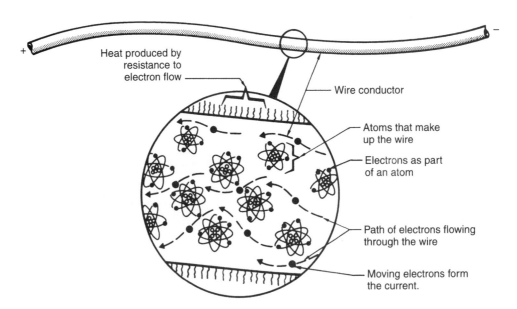

Heat produced by
resistance to
electron flow

Wire conductor

Atoms that make
up the wire

Electrons as part
of an atom

Path of electrons flowing
through the wire

Moving electrons form
the current.

FIGURE 7–2 *Electrons traveling along a conductor.*

The SMAW process is very flexible in terms of the metal thicknesses that can be welded and the variety of positions it can be used in. Metal with thickness varying from 16 gauge to several feet can be welded using the same machine with different settings.

7.1 — WELDING CURRENT

The welding current is an electric current. An electric current is a flow of electrons. Electrons flow through a conductor from negative (−) to positive (+) (Figure 7-2). Resistance to the flow of electrons produces heat. The greater the resistance, the greater the heat. Air has a high resistance to current flow. As the electrons jump the air gap between the end of the electrode and the work, a great deal of heat is produced. Electrons flowing across an air gap produce an arc.

TEMPERATURE

The temperature of a welding arc exceeds 11,000° F (6,000° C). The exact temperature depends on the resistance to the current flow. The resistance is affected by the arc length and the chemical composition of the gases formed as the electrode covering burns and vaporizes. As the arc lengthens, the resistance increases, thus causing a rise in the arc voltage and temperature. The shorter the arc, the lower the arc voltage and temperature produced.

Most shielded metal arc welding electrodes have chemicals added to their coverings to stabilize the arc. These arc stabilizers reduce the arc resistance,

making it easier to hold an arc. By lowering the resistance, the arc stabilizers also lower the arc temperature. Other chemicals within the gaseous cloud around the arc may raise or lower the resistance.

CURRENTS

The three different types of current used for welding are alternating current (AC), direct-current electrode negative (DCEN), and direct-current electrode positive (DCEP). The terms DCEN and DCEP have replaced the former terms direct-current straight polarity (DCSP) and direct-current reverse polarity (DCRP). DCEN and DCSP are the same currents, and DCEP and DCRP are the same currents. Some electrodes can be used with only one type of current. Others can be used with two or more types of current. Each welding current has a different effect on the weld.

- **DCEN.** In direct-current electrode negative, the electrode is negative, and the work is positive (Figure 7-3). DCEN welding current produces a high electrode melting rate.
- **DCEP.** In direct-current electrode positive, the electrode is positive, and the work is negative (Figure 7-4). DCEP current produces the best welding arc characteristics.
- **AC.** In alternating current, the electrons change direction every 1/120 of a second so that the electrode and work alternate from anode to cathode (Figure 7-5). The rapid reversal of the current flow causes the welding heat to be evenly distributed on both the work and the electrode—that is, half on the work and half on the electrode. The even heating gives the weld bead a balance between penetration and buildup.

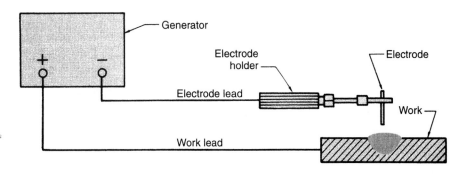

FIGURE 7–3   Straight polarity (DCSP), electrode negative (dcen).

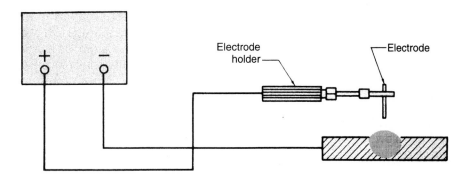

FIGURE 7–4   Reverse polarity (dcrp), electrode positive (dcep).

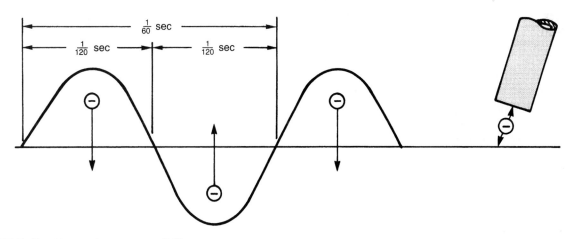

FIGURE 7–5   Alternating current (AC).

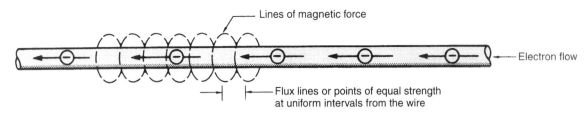

FIGURE 7–6 Magnetic force around a wire.

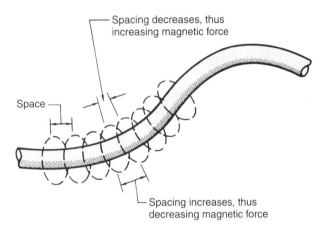

FIGURE 7–7 Flux lines along a bent wire.

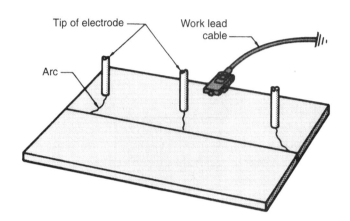

FIGURE 7–8 Arc blow.

7.2 ARC BLOW

When electrons flow, they create lines of magnetic force that circle around the line of flow (Figure 7-6). Lines of magnetic force are referred to as magnetic flux lines. These lines space themselves evenly along a current-carrying wire. If the wire is bent, the flux lines on one side are compressed, and those on the other side are stretched out (Figure 7-7). The unevenly spaced flux lines try to straighten the wire so that the lines can be evenly spaced once again. The force that they place on the wire is usually small.

The welding current flowing through a plate or any residual magnetic fields in the plate will result in uneven flux lines. These uneven flux lines can, in turn, cause an arc to move during a weld. This movement of the arc is called *arc blow*. Arc blow makes the arc drift like a string would drift in the wind. Arc blow is most noticeable in corners, at the ends of plates, and when the work lead is connected to only one side of a plate (Figure 7-8). If arc blow is a problem, it can be controlled by connecting the work lead to the end of the weld joint, and the weld then should be made in the direction away from the work lead (Figure 7-9). Another way of controlling arc blow is to use two work leads, one on each side of the weld. The best way to

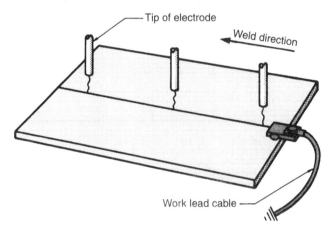

FIGURE 7–9 Correct current connections to control arc blow.

eliminate arc blow is to use alternating current. AC usually does not allow the flux lines to build long enough to bend the arc before the current changes direction. If it is impossible to move the work connection or to change to AC, a very short arc length can help control arc blow. A large tack weld or a change in the electrode angle can also help control arc blow.

Arc blow may not be a problem as you are learning to weld in the shop, because most welding tables are all steel. However, if you are using a pipe stand to hold your welding practice plates, arc blow can become a problem. Try reclamping your practice plates.

7.3 TYPES OF POWER SOURCES

Two types of electrical devices can be used to produce the low-voltage, high-amperage current combination that arc welding requires. One type uses electric motors or internal combustion engines to drive alternators or generators. The other type uses step-down transformers. Because transformer-type welding machines are quieter, more energy efficient, require less maintenance, and are less expensive, they are now the industry standards. However, engine-powered generators are still widely used for portable welding.

7.4 DUTY CYCLE

Welding machines produce internal heat at the same time they produce the welding current. Except for automatic welding machines, welders are rarely used every minute for long periods of time. The auto body technician must take time to change electrodes, change positions, or change parts. Shielded metal arc welding never continues for long periods of time.

The duty cycle is the percentage of time a welding machine can be used continuously. A 60 percent duty cycle means that out of every ten minutes, the machine can be used for six minutes at the maximum rated current. When providing power at this level, it must be cooled off for four minutes out of every ten minutes. The duty cycle increases as the amperage is lowered and decreases as amperage is raised (Figure 7-10). Most welding machines weld at a 60 percent rate or less. Therefore, most manufacturers list the amperage rating for a 60 percent duty cycle on the nameplate that is attached to the machine. Other duty cycles are given on a graph in the owner's manual.

7.5 WELDING CABLES

The terms _welding cables or welding leads_ are used to mean the same thing. Cables to be used for welding must be flexible, well-insulated, and the correct

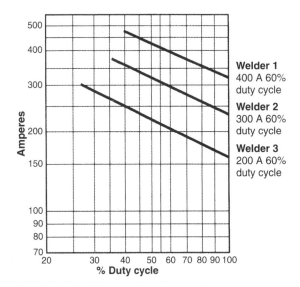

FIGURE 7-10 _Duty cycle of a typical shielded metal arc welding machine._

size for the job. Most welding cables are made from stranded copper wire. Some manufacturers sell a newer type of cable made from aluminum wires. The aluminum wires are lighter and less expensive than copper. Because aluminum as a conductor is not as good as copper for a given wire size, the aluminum wire should be one size larger than would be required for copper.

The insulation on welding cables will be exposed to hot sparks, flames, grease, oils, sharp edges, impact, and other types of wear. To withstand such wear, only specially manufactured insulation should be used for welding cable. Several new types of insulation are available that will give longer service against these adverse conditions.

As electricity flows through a cable, the resistance to the flow causes the cable to heat up and increase the voltage drop. To minimize the loss of power and prevent overheating, the electrode cable and work cable must be the correct size. Table 7-1 lists the minimum size cable that is required for each amperage and length. Large welding lead sizes make electrode manipulation difficult. Smaller cable can be spliced to the electrode end of a large cable to make it more flexible. This so-called whip-end cable must not be over 10 feet (3 m) long.

CAUTION: A splice in a cable should not be within 10 feet (3 m) of the electrode because of the possibility of electrical shock.

TABLE 7–1:   COPPER AND ALUMINUM WELDING LEAD SIZES

		Copper Welding Lead Sizes								
Amperes		100	150	200	250	300	350	400	450	500
ft	m									
50	15	2	2	2	2	1	1/0	1/0	2/0	2/0
75	23	2	2	1	1/0	2/0	2/0	3/0	3/0	4/0
100	30	2	1	1/0	2/0	3/0	4/0	4/0		
125	38	2	1/0	2/0	3/0	4/0				
150	46	1	2/0	3/0	4/0					
175	53	1/0	3/0	4/0						
200	61	1/0	3/0	4/0						
250	76	2/0	4/0							
300	91	3/0								
350	107	3/0								
400	122	4/0								

		Aluminum Welding Lead Sizes								
Amperes		100	150	200	250	300	350	400	450	500
ft	m									
50	15	2	2	1/0	2/0	2/0	3/0	4/0		
75	23	2	1/0	2/0	3/0	4/0				
100	30	1/0	2/0	4/0						
125	38	2/0	3/0							
150	46	2/0	3/0							
175	53	3/0								
200	61	4/0								
225	69	4/0								

(Length of Cable)

7.6 — ELECTRODE HOLDERS

The electrode holder should be of the proper amperage rating and in good repair for safe welding. Electrode holders are designed to be used at their maximum amperage rating or less. Higher amperage values will cause the holder to overheat and burn up. If the holder is too large for the amperage range being used, manipulation is hard, and operator fatigue increases. Make sure that the correct size holder is chosen.

CAUTION: Never dip a hot electrode holder in water to cool it off. The problem causing the holder to overheat should be repaired.

7.7 — SETUP

Arc welding machines should be located near the welding site, but far enough away so that they are not covered with spark showers. The machines may be stacked to save space, but there must be room enough between the machines for air to circulate to keep the machines from overheating. The air that is circulated through the machine should be as free as possible of dust, oil, and metal filings. Even in a good location, the power should be turned off periodically and the machine blown out with compressed air.

The welding machine should be located away from cleaning tanks and any other sources of corrosive fumes that could be blown through it. Water leaks must be fixed and puddles cleaned up before a

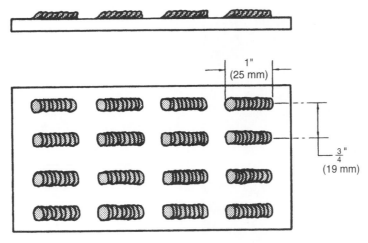

FIGURE 7–11 Practice plate for striking an arc.

machine is used. The workstation must be free of combustible materials. Screens should be provided to protect other workers from the arc light.

Check the surroundings before starting to weld. If heavy materials or vehicles are being moved in the area around you, there should be a safety watch. A safety watch can warn a person of danger while that person is welding.

PRACTICE 7–1

Shielded Metal Arc Welding Safety

Using a welding workstation, welding machine, welding electrodes, welding helmet, eye and ear protection, welding gloves, proper work clothing, and any special protective clothing that may be required, demonstrate the safe way to prepare yourself and the welding workstation for welding. Include in your demonstration appropriate references to burn protection, eye and ear protection, material specification data sheets, ventilation, electrical safety, general work clothing, special protective clothing, and area cleanup.

PRACTICE 7–2

Striking the Arc

Using a properly set up and adjusted arc welding machine, the proper safety protection as demonstrated in Practice 7-1, E6011 welding electrodes having a 1/8-inch (3 mm) diameter, and one piece of mild steel plate, 1/4-inch (6 mm) thick, you will practice striking an arc (Figure 7-11).

With the electrode held over the plate, lower your helmet. Scratch the electrode across the plate (like

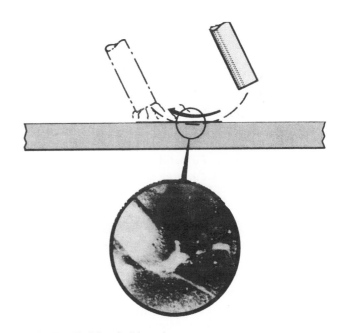

FIGURE 7–12 Striking the arc.

striking a large match) (Figure 7-12). As the arc is established, slightly raise the electrode to the desired arc length. Hold the arc in one place until the molten weld pool builds to the desired size. Slowly lower the electrode as it burns off and move it forward to start the bead.

If the electrode sticks to the plate, quickly squeeze the electrode holder lever to release the electrode. Break the electrode free by bending it back and forth a few times. Do not touch the electrode without gloves, because it will still be hot. If the flux breaks away from the end of the electrode, throw out the electrode, because restarting the arc will be very difficult (Figure 7-13).

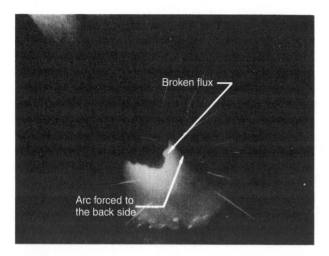

FIGURE 7-13 If the flux is broken off the end completely or on one side, the arc can be erratic or forced to the side.

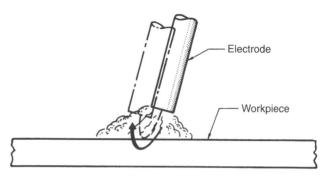

FIGURE 7-14 Striking the arc on a spot.

Break the arc by rapidly raising the electrode after completing a 1-inch (25-mm) weld bead. Restart the arc as you did before, and make another short weld. Repeat this process until you can easily start the arc each time. Turn off the welding machine and clean up your work area when you are finished welding.

PRACTICE 7-3

Striking the Arc Accurately

Using the same materials and setup as described in Practice 7-2, you will start the arc at a specific spot in order to prevent damage to the surrounding plate.

Hold the electrode over the desired starting point. After lowering your helmet, swiftly bounce the electrode against the plate (Figure 7-14). A lot of practice is required to develop the speed and skill needed to prevent the electrode from sticking to the plate.

A more accurate method of starting the arc involves holding the electrode steady by resting it on your free hand like a pool cue. The electrode is rapidly pushed forward so that it strikes the metal exactly where it should. This is an excellent method of striking an arc. Striking an arc in an incorrect spot may cause damage to the base metal.

Practice starting the arc until you can start it within 1/4 inch (6 mm) of the desired location. Turn off the welding machine and clean up your work area when you are finished welding.

7.8 ELECTRODE SIZE AND HEAT

The selection of the correct size electrode for a weld is determined by the skill of the auto body technician, the thickness of the metal to be welded, and the size of the metal. Using small-diameter electrodes requires less skill than using large-diameter electrodes. The deposition rate—the rate that weld metal is added to the weld—is slower when small-diameter electrodes are used. Small-diameter electrodes will make acceptable welds on thick plate, but more time is required to make the weld.

Large-diameter electrodes may overheat the metal if they are used with thin or small pieces of metal. To determine if a weld is too hot, watch the shape of the trailing edge of the molten weld pool (Figure 7-15). Rounded ripples indicate the weld is cooling uniformly and that the heat is not excessive. If the ripples are pointed, the weld is cooling too slowly because of excessive heat. Extreme overheating can cause a burn-through, which is hard to repair.

To correct an overheating problem, you can turn down the amperage, use a shorter arc, travel at a faster rate, use a chill plate (a large piece of metal used to absorb excessive heat), or use a smaller electrode at a lower current setting.

7.9 ARC LENGTH

The arc length is the distance the arc must jump from the end of the electrode to the plate. As the weld progresses, the electrode becomes shorter as it is consumed. To maintain a constant arc length, the electrode must be lowered continuously. Maintaining

Amount of heat directed at weld	Weld pool
Too low	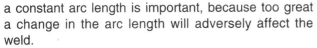
Correct	
Too hot	

FIGURE 7–15 _The effect on the shape of the molten weld pool caused by the heat input._

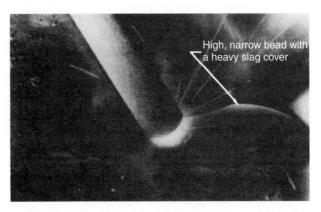

High, narrow bead with a heavy slag cover

FIGURE 7–16 _Welding with too short an arc length._

FIGURE 7–17 _Welding with too long an arc length._

a constant arc length is important, because too great a change in the arc length will adversely affect the weld.

As the arc length is shortened, metal transferring across the gap may short out the electrode, causing it to stick to the plate. The weld that results is narrow and has a high buildup (Figure 7-16).

Long arc lengths produce more spatter, because the metal being transferred may drop outside of the molten weld pool. The weld is wide and has little buildup (Figure 7-17).

There is a narrow range for the arc length in which it is stable, metal transfer is smooth, and the bead shape is controlled. Factors affecting the length are the type of electrode, joint design, metal thickness, and current setting.

7.10 — ELECTRODE ANGLE

The electrode angle is the angle that the electrode departs from the surface of the metal. The angle is generally described as either leading or trailing, based on its relation to the direction of travel (Figure 7-18). The relative angle is important, because there is a jetting force that blows the metal and flux from the end of the electrode to the plate.

LEADING ANGLE

A leading electrode angle pushes molten metal and slag ahead of the weld (Figure 7-19). When this angle

is used in the flat position, caution must be taken to prevent cold lap and slag inclusions. The solid metal ahead of the weld cools and solidifies the molten filler metal and slag before they can melt the solid metal. This rapid cooling prevents the metals from fusing together (Figure 7-20). As the weld passes over this area, heat from the arc may not melt it. As a result, some cold lap and slag inclusions are left. The following are suggestions for preventing cold lap and slag inclusions:

- Use as little leading angle as possible.
- Ensure that the arc melts the base metal completely (Figure 7-21).
- Use a penetrating-type electrode that causes little buildup.
- Move the arc back and forth across the molten weld pool to fuse both edges.

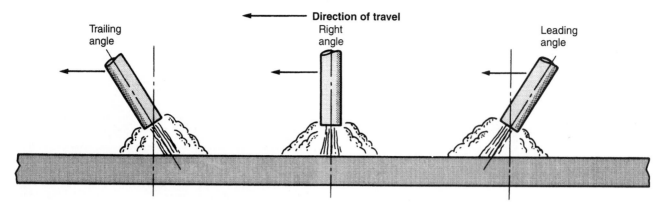

FIGURE 7-18 Direction of travel and electrode angle.

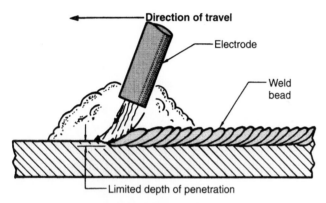

FIGURE 7-19 Leading electrode angle.

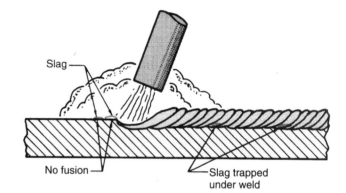

FIGURE 7-20 Some electrodes, such as E7018, may not remove the deposits ahead of the molten weld pool, resulting in discontinuities within the weld.

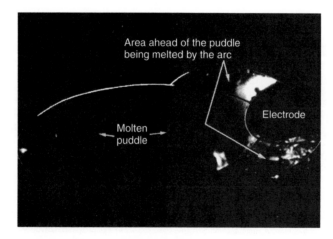

FIGURE 7-21 Metal being melted ahead of the molten weld pool helps to ensure good weld fusion.

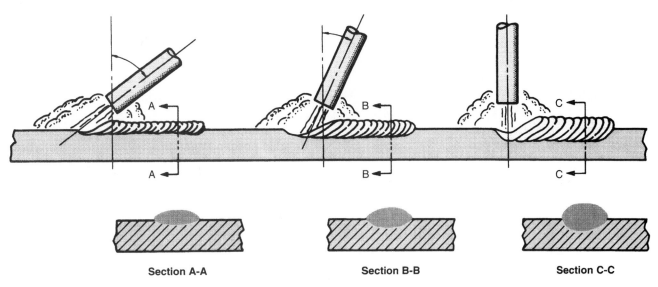

FIGURE 7–22 Effect of a leading angle on weld bead buildup, width, and penetration. As the angle decreases toward the vertical position (C), penetration increases.

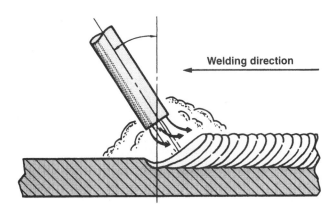

Welding direction

FIGURE 7–23 Trailing electrode angle.

A leading angle can be used to minimize penetration or to help hold metal in place for vertical welds (Figure 7-22).

TRAILING ANGLE

A trailing electrode angle pushes the molten metal away from the leading edge of the molten weld pool toward the back, where it solidifies (Figure 7-23). As the molten metal is forced away from the bottom of the weld, the arc melts more of the base metal, which results in deeper penetration. The molten metal pushed to the back of the weld solidifies and forms reinforcement for the weld (Figure 7-24).

7.11 ELECTRODE MANIPULATION

The movement or weaving of the welding electrode can control the following characteristics of the weld bead: penetration, buildup, width, porosity, undercut, overlap, and slag inclusions. The exact weave pattern for each weld is often the personal choice of the auto body technician. However, some patterns are especially helpful for specific welding situations. The pattern selected for a flat (1G) butt joint is not as critical as is the pattern selection for other joints and other positions.

Many weave patterns are available for the auto body technician to use. Figure 7-25 shows ten different patterns that can be used for most welding conditions.

The circular pattern is often used for flat position welds on butt, tee, and outside corner joints, and for buildup or surfacing applications. The circle can be made wider or longer to change the bead width or penetration (Figure 7-26).

The C and square patterns are both good for most 1G (flat) welds, but they can also be used for vertical (3G) positions. These patterns can also be used if there is a large gap to be filled when both pieces of metal are nearly the same size and thickness.

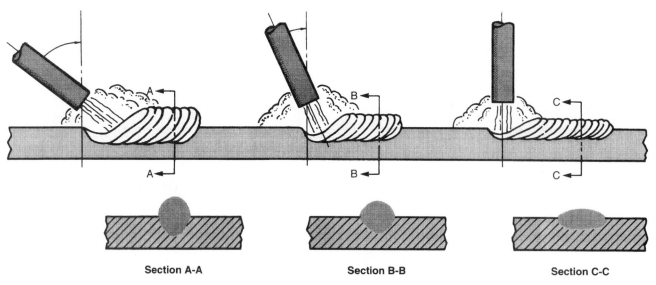

Section A-A   Section B-B   Section C-C

FIGURE 7–24 Effect of a trailing angle on weld bead buildup, width, and penetration. Section A-A shows a greater weld buildup due to the large angle of the electrode from a vertical position.

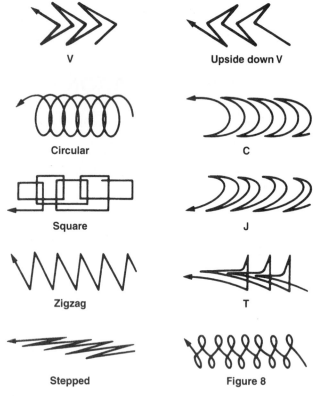

V

Upside down V

Circular

C

Square

J

Zigzag

T

Stepped

Figure 8

FIGURE 7–25 Weave patterns.

This weave pattern results in a narrow bead with deep penetration.

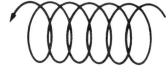

This weave pattern results in a wide bead with shallow penetration.

FIGURE 7–26 Changing the weave pattern size to change the weld bead characteristics.

a uniform bead contour is maintained during out-of-position welds.

The T pattern works well with fillet welds in the vertical (3F) and overhead (4F) positions (Figure 7-28). It also can be used for deep groove welds for the hot pass. The top of the T can be used to fill in the toe of the weld to prevent undercutting.

The straight step pattern can be used for stringer beads, root-pass welds, and multiple-pass welds in all positions. For this pattern, the smallest quantity of metal is molten at one time as compared to other patterns. Therefore, the weld is more easily controlled. At the same time that the electrode is stepped forward, the arc length is increased so that no metal is deposited ahead of the molten weld pool (Figures 7-29 and 7-30). This action allows the molten weld pool to cool to a controllable size. In addition, the arc burns

The J pattern works well on flat (1F) lap joints, all vertical (3G) joints, and horizontal (2G) butt and lap (2F) welds. This pattern allows the heat to be concentrated on the thicker plate (Figure 7-27). It also allows the reinforcement to be built up on the metal deposited during the first part of the pattern. As a result,

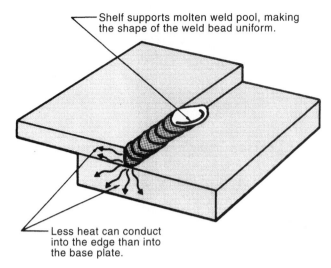

Shelf supports molten weld pool, making the shape of the weld bead uniform.

Less heat can conduct into the edge than into the base plate.

FIGURE 7–27 The J pattern allows the heat to be concentrated on the thicker plate.

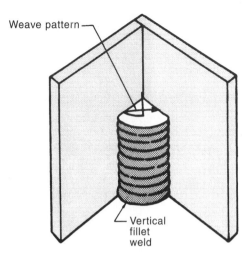

Weave pattern

Vertical fillet weld

FIGURE 7–28 T pattern.

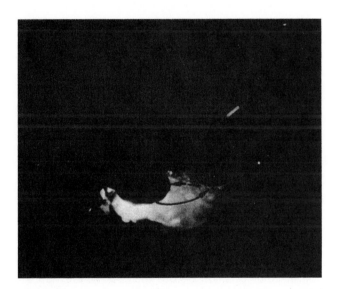

FIGURE 7–29 The electrode is moved slightly forward and then returned to the weld pool.

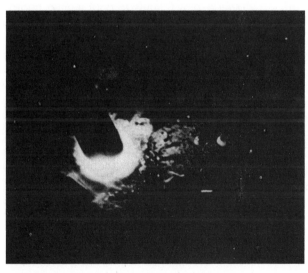

FIGURE 7–30 The electrode does not deposit metal or melt the base metal.

off any paint, oil, or dirt from the metal before it can contaminate the weld.

The figure-8 pattern and the zigzag pattern are used as cover passes in the flat and vertical positions. Do not weave more than 2 1/2 times the width of the electrode. These patterns deposit a large quantity of metal at one time. A shelf can be used to support the molten weld pool when making vertical welds using either of these patterns (Figure 7-31).

POSITIONING OF THE WELDER AND THE PLATE

7.12

The auto body technician should be in a relaxed, comfortable position before starting to weld. A good position is important both for your comfort and for the quality of the welds. Welding in an awkward position

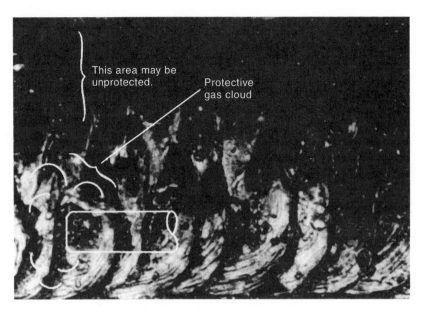

FIGURE 7–31 *Using the shelf to support the molten pool for vertical welds.*

can cause fatigue, which leads to poor welder coordination and poor-quality welds. Auto body technicians must have enough freedom of movement so that they do not need to change position during a weld. Body position changes should be made only during electrode changes.

When the welding helmet is down, the auto body technician is blind to the surroundings. Due to the arc, the technician's field of vision is also very limited. These factors often cause the technician to sway. To stop this swaying, the technician should lean against or hold onto a stable object. Even if one is seated when welding, touching a stable object will make welding more relaxing.

Welding can be made easier if the auto body technician first finds the most comfortable angle of movement. To do this, you should be in either a seated or a standing position in front of the welding table. The welding machine should be turned off. With an electrode in place in the electrode holder, draw a straight line along the plate to be welded. Turn the plate to several different angles, and determine by moving the electrode along the line which angle is most comfortable for welding (Figure 7-32).

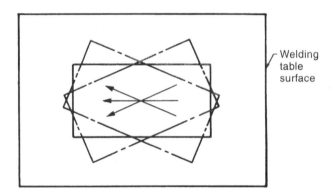

FIGURE 7–32 *Change the plate angle to find the most comfortable welding position.*

are made. The stringer beads should be practiced first in each position before the technician tries the different joints in each position. Some time can be saved by starting with the stringer beads. If this is done, it is not necessary to cut or tack the plate together, and a number of beads can be made on the same plate.

Students will find it easier to start with butt joints. The lap, tee, and outside corner joints are all about the same level of difficulty.

Starting with the flat position allows the technician to build skills slowly, so that out-of-position welds become easier to do. The horizontal tee and lap welds are almost as easy to make as the flat welds. Overhead welds are as simple to make as vertical welds, but they are harder to position. Horizontal butt welds are more difficult to perform than most other welds.

7.13 PRACTICE WELDS

Practice welds are grouped by type of joint and type of welding electrode. The auto body technician or instructor should select the order in which the welds

ELECTRODES

Arc welding electrodes used for practice welds are grouped into three filler metal classes according to their major welding characteristics. The groups are E6010 and E6011, E6012 and E6013, and E7016 and E7018.

- **F3 E6010 and E6011 Electrodes.** Both of these electrodes have cellulose-based fluxes. As a result, these electrodes have a forceful arc with little slag left on the weld bead.
- **F2 E6012 and E6013 Electrodes.** These electrodes have rutile-based fluxes, giving a smooth, easy arc with a thick slag left on the weld bead.
- **F4 E7016 and E7018 Electrodes.** Both of these electrodes have a mineral-based flux. The resulting arc is smooth and easy, with a very heavy slag left on the weld bead.

The cellulose- and rutile-based groups of electrodes have characteristics that make them the best electrodes for starting specific welds. The electrodes with the cellulose-based fluxes do not have heavy slags that may interfere with the welder's view of the weld. This feature is an advantage for flat tee and lap joints. Electrodes with the rutile-based fluxes (giving an easy arc with low spatter) are easier to control and are used for flat stringer beads and butt joints.

STRINGER BEADS

A straight weld bead on the surface of a plate, with little or no side-to-side electrode movement, is known as a stringer bead. Stringer beads are used by students to practice maintaining arc length, weave patterns, and electrode angle so that their welds will be straight, uniform, and free from defects. Stringer beads (Figure 7-33) are also used to set the machine amperage and for buildup or surfacing applications.

The stringer bead should be straight. A beginning auto body technician needs time to develop the skill of viewing the entire welding area. At first, the technician sees only the arc (Figure 7-34). With practice, one begins to see parts of the molten weld pool. After much practice, one will see the molten weld pool (front, back, and both sides), slag, buildup, and the surrounding plate (Figure 7-35). Often, at this skill level, the auto body technician may not even notice the arc.

A straight weld is easily made once the technician develops the ability to view the entire welding zone, because the technician will occasionally glance around to ensure that the weld is straight. The technician can also note if the weld is uniform and free from defects. One's ability to view the entire weld area is demonstrated by making consistently straight and uniform stringer beads.

FIGURE 7–33   Stringer bead.

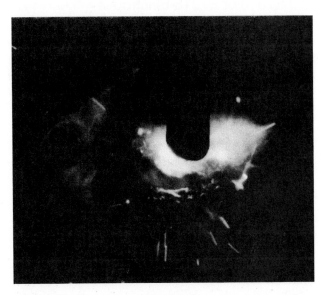

FIGURE 7–34   Beginning welders frequently see only the arc and sparks from the electrode.

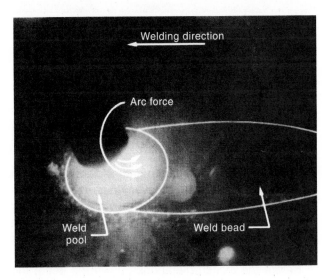

FIGURE 7–35   More experienced welders can see the molten pool, metal being transferred across the arc, and penetration into the base metal.

After the practice stringer beads, a variety of weave bead patterns should be practiced to improve your ability to control the molten weld pool when welding out of position.

PRACTICE 7-4

Straight Stringer Beads in the Flat Position Using E6010 or E6011 Electrodes, E6012 or E6013 Electrodes, and E7016 or E7018 Electrodes

Using a properly set up and adjusted arc welding machine, proper safety protection as demonstrated in Practice 7-1, arc welding electrodes with a 1/8-inch (3 mm) diameter, and one piece of mild steel plate 6 inches (150 mm) long × 1/4 inch (6 mm) thick, you will make straight stringer beads.

- Starting at one end of the plate, make a straight weld the full length of the plate.
- Watch the molten weld pool at the points, not at the end of the electrode. As you become more skillful, it is easier to watch the molten weld pool.
- Repeat the beads with all three F groups of electrodes until you have consistently good beads.
- Cool, chip, and inspect each bead for defects after completing it.

Turn off the welding machine and clean up your work area when you are finished welding.

PRACTICE 7-5

Stringer Beads in the Vertical Up Position Using E6010 or E6011 Electrodes, E6012 or E6013 Electrodes, and E7016 or E7018 Electrodes

Using the same setup, materials, and electrodes as listed in Practice 7-4, you will make vertical up stringer beads. Start with the plate at a 45-degree angle.

This technique is the same as that used to make a vertical weld. However, a lower level of skill is required at 45 degrees, and it is easier to develop your skill. After you master the 45-degree angle, the angle is increased successively until a vertical position is reached (Figure 7-36).

Before the molten metal drips down the bead, the back of the molten weld pool will start to bulge (Figure 7-37). When this happens, increase the speed of travel and the weave pattern.

Cool, chip, and inspect each completed weld for defects. Repeat the beads as necessary with all three F groups of electrodes until consistently good beads are obtained in this position. Turn off the welding machine and clean up your work area when you are finished welding.

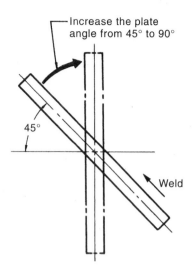

FIGURE 7-36 *Once the 45-degree angle is mastered, the plate angle is increased successively until a vertical position is reached.*

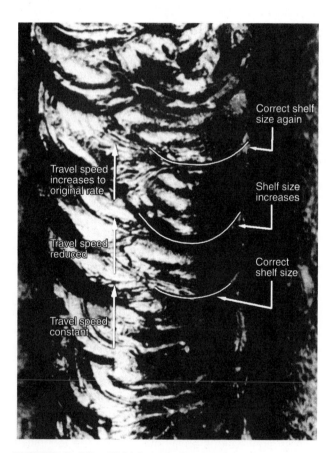

FIGURE 7-37 *E7018 vertical up weld.*

PRACTICE 7–6

Horizontal Stringer Beads Using E6010 or E6011 Electrodes, E6012 or E6013 Electrodes, and E7016 or E7018 Electrodes

Using the same setup, materials, and electrodes as listed in Practice 7-4, you will make horizontal stringer beads on a plate. When you begin to practice the horizontal stringer bead, the plate may be reclined slightly (Figure 7-38). This placement allows you to build the required skill by practicing the correct techniques successfully. The J weave pattern is suggested for this practice. As the electrode is drawn along the straight back of the J, metal is deposited. This metal supports the molten weld pool, resulting in a bead with a uniform contour (Figure 7-39).

Angling the electrode up and back toward the weld causes more metal to be deposited along the top edge of the weld. Keeping the bead small allows the surface tension to hold the molten weld pool in place.

Gradually increase the angle of the plate until it is vertical and the stringer bead is horizontal. Repeat the beads as needed with all three F groups of electrodes until consistently good beads are obtained in this position. Turn off the welding machine and clean up your work area when you are finished welding.

SQUARE BUTT JOINT

The square butt joint is made by first tack-welding two flat pieces of plate together (Figure 7-40). The space

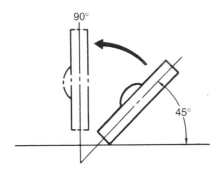

FIGURE 7–38 Change the plate angle as welding skill improves.

FIGURE 7–40 The tack weld should be small and uniform to minimize its effect on the final weld.

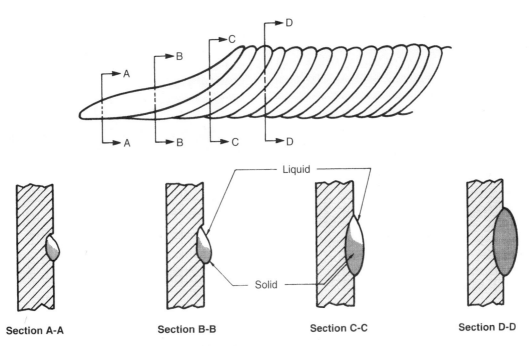

| Section A-A | Section B-B | Section C-C | Section D-D |

FIGURE 7–39 The progression of a horizontal bead.

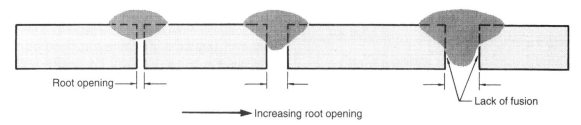

FIGURE 7–41 *Effect of root opening on weld penetration.*

between the plates is called the root opening or root gap. Changes in the root opening will affect penetration. As the space increases, the weld penetration also increases. The root opening for most butt welds will vary from 0 to about 1/8 of an inch (3 mm). Excessively large openings can cause burn-through or a cold lap at the weld root (Figure 7-41).

After a butt weld is completed, the plate can be cut apart so it can be used for rewelding. The strips for butt welding should be no smaller than 1 inch (25 mm) wide. If they are too narrow, there will be a problem with heat buildup.

If the plate strips are no longer flat after the weld has been cut out, they can be tack-welded together and flattened with a hammer (Figure 7-42).

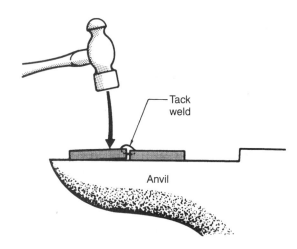

FIGURE 7–42 *After the plates are tack welded together, then can be forced into alignment by striking them with a hammer.*

PRACTICE 7–7

Welded Square Butt Joint in the Flat Position (1G) Using E6010 or E6011 Electrodes, E6012 or E6013 Electrodes, and E7016 or E7018 Electrodes

Using a properly set up and adjusted arc welding machine, proper safety protection, arc welding electrodes having a 1/8-inch (3 mm) diameter, and two or more pieces of mild steel plate 6 inches (150 mm) long by 1/4 inch (6 mm) thick, you will make a welded square butt joint in the flat position (Figure 7-43).

Tack-weld the plates together along the 6-inch side and place them flat on the welding table. Starting at one end, establish a molten weld pool on both plates. Hold the electrode in the molten weld pool until it flows together (Figure 7-44). After the gap is bridged by the molten weld pool, start weaving the electrode slowly back and forth across the joint. Moving the electrode too quickly from side to side may result in slag being trapped in the joint (Figure 7-45).

Continue the weld along the length of the joint. Normally, deep penetration is not required for this type of weld. If full plate penetration is required, the edges of the butt joint should be beveled or a larger-than-normal root gap should be used. Cool, chip, and inspect the weld for uniformity and soundness. Repeat the welds as needed to master all three F groups of electrodes in this position. Turn off the welding machine and clean up your work area when you are finished welding.

PRACTICE 7–8

Vertical (3G) Up-Welded Square Butt Weld Using E6010 or E6011 Electrodes, E6012 or E6013 Electrodes, and E7016 or E7018 Electrodes

Using the same setup, materials, and electrodes as listed in Practice 7-7, you will make vertical up-welded square butt joints.

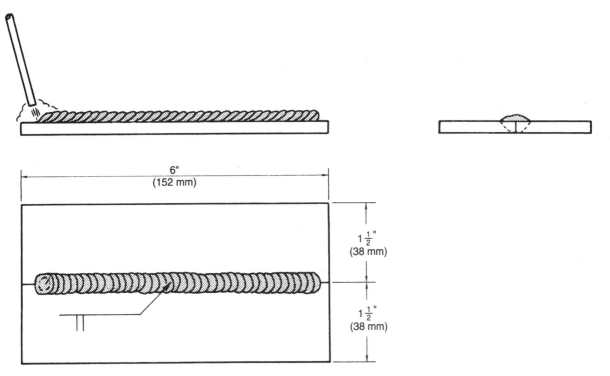

FIGURE 7–43 Square butt joint in the flat position.

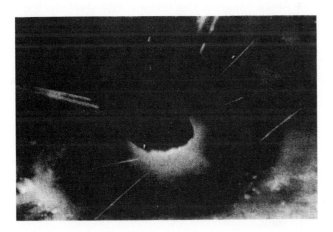

FIGURE 7–44 Hold the arc in one area long enough to establish the size of the molten weld pool desired.

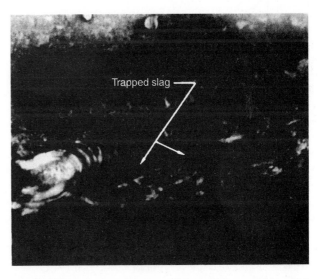

FIGURE 7–45 Moving the electrode from side to side too quickly can result in slag being trapped between the plates.

With the plates at a 45-degree angle, start at the bottom and make the molten weld pool bridge the gap between the plates (Figure 7-46). Build the bead size slowly so that the molten weld pool has a shelf for support. The C, J, or square weave pattern works well for this joint.

As the electrode is moved up the weld, the arc is lengthened slightly so that little or no metal is deposited ahead of the molten weld pool. When the electrode is brought back into the molten weld pool, it should be lowered to deposit metal (Figure 7-47).

As skill is developed, increase the plate angle until it is vertical. Cool, chip, and inspect the weld for uniformity and defects. Repeat the welds with all three F groups of electrodes until you can consistently make welds free of defects. Turn off the welding machine and clean up your work area when you are finished welding.

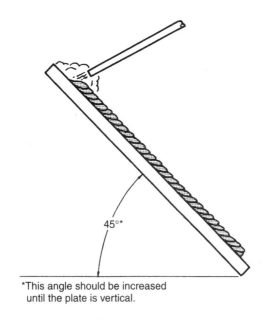

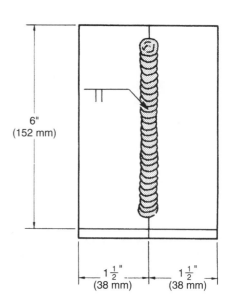

FIGURE 7–46   Square butt joint in the vertical up position.

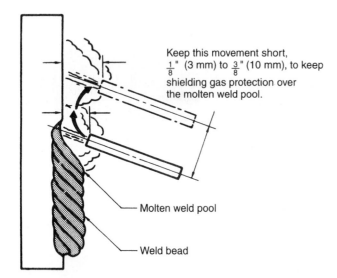

FIGURE 7–47   Electrode movement for vertical up weld.

PRACTICE 7–9

Welded Horizontal (2G) Square Butt
Weld Using E6010 or E6011 Electrodes,
E6012 or E6013 Electrodes,
and E7016 or E7018 Electrodes

Using the same setup, materials, and electrodes as described in Practice 7-7, you will make a welded horizontal square butt joint.

- Start practicing these welds with the plate at a slight angle.
- Strike the arc on the bottom plate and build the molten weld pool until it bridges the gap.

If the weld is started on the top plate, slag will be trapped in the root at the beginning of the weld because of poor initial penetration. The slag may cause the weld to crack when it is placed in service.

The J weave pattern is recommended in order to deposit metal on the lower plate so that it can support the bead. By pushing the electrode inward as you cross the gap between the plates, you achieve deeper penetration.

As you acquire more skill, gradually increase the plate angle until it is vertical and the weld is horizontal.

- Cool, chip, and inspect the weld for uniformity and defects.
- Repeat the welds with all three F groups of electrodes until you can consistently make welds free of defects.

Turn off the welding machine and clean up your work area when you are finished welding.

LAP JOINT

A lap joint is made by overlapping the edges of the two plates (Figure 7-48). The joint can be welded on one side or both sides with a fillet weld. In Practice 7-10, both sides should be welded unless otherwise noted.

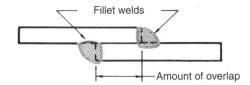

Fillet welds

Amount of overlap

FIGURE 7–48   Lap joint.

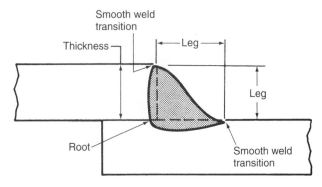

Smooth weld transition

Thickness

Leg

Leg

Root

Smooth weld transition

FIGURE 7–49   The legs of a fillet weld generally should be equal to the thickness of the base metal.

As the fillet weld is made on the lap joint, the buildup should equal the thickness of the plate (Figure 7-49). A good weld will have a smooth transition from the plate surface to the weld. If this transition is abrupt, it can cause stresses that will weaken the joint.

Penetration for lap joints does not improve their strength; complete fusion is required. The roots of fillet welds must be melted to ensure a completely fused joint. If the molten weld pool shows a notch during the weld (Figure 7-50), this is an indication that the root is not being fused together. The weave pattern will help prevent this problem (Figure 7-51).

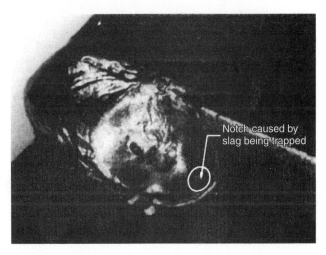

Notch caused by slag being trapped

FIGURE 7–50   Watch the root of the weld bead to be sure there is complete fusion.

PRACTICE 7-10

Welded Lap Joint in the Flat Position (1F) Using E6010 or E6011 Electrodes, E6012 or E6013 Electrodes, and E7016 or E7018 Electrodes

Using a properly set up and adjusted arc welding machine, proper safety protection, arc welding electrodes having a 1/8-inch (3 mm) diameter, and two or more pieces of mild steel plate 6 inches (150 mm) long by 1/4 inch (6 mm) thick, you will make a welded lap joint in the flat position (Figure 7-52).

Hold the plates together tightly with an overlap of no more than 1/4 inch (6 mm). Tack-weld the plates together. A small tack weld may be added in the center to prevent distortion during welding (Figure 7-53). Chip the tacks before you start to weld.

The J, C, or zigzag weave pattern works well on this joint. Strike the arc and establish a molten pool directly in the joint. Move the electrode out on the bottom plate and then onto the weld to the top edge of the top plate (Figure 7-54). Follow the surface of the plates with the arc. Do not follow the trailing edge of the weld bead. Following the molten weld pool will not allow for good root fusion and will also cause slag to collect in the root. If slag does collect, a good weld is not possible. Stop the weld and chip the slag to remove it before the weld is completed. Cool, chip, and

inspect the weld for uniformity and defects. Repeat the welds with all three F groups of electrodes until you can consistently make welds free of defects. Turn off the welding machine and clean up your work area when you are finished welding.

PRACTICE 7-11

Welded Lap Joint in the Horizontal Position (2F) Using E6010 or E6011 Electrodes, E6012 or E6013 Electrodes, and E7016 or E7018 Electrodes

Using the same setup, materials, and electrodes as listed in Practice 7-10, you will make a welded horizontal lap joint.

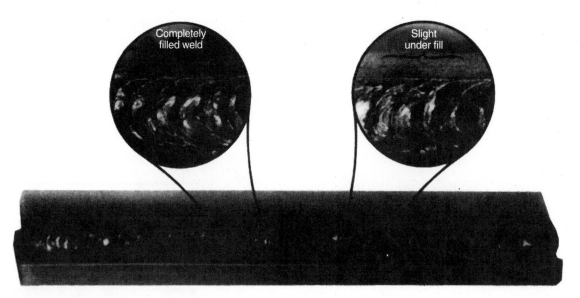

FIGURE 7–51 Lap joint.

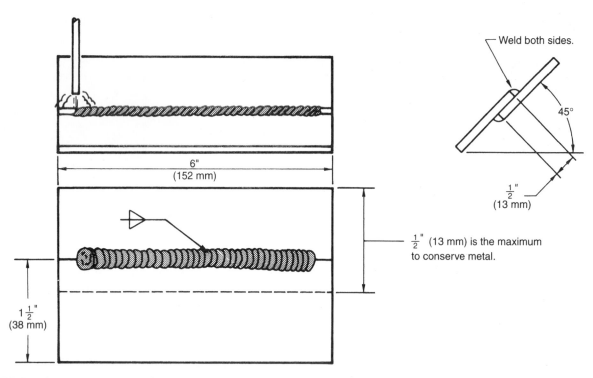

FIGURE 7–52 Lap joint in the flat position.

The horizontal lap joint and the flat lap joint require nearly the same technique and skill to achieve a proper weld (Figure 7-55). Use the J, C, or zig-zag weave pattern to make the weld. Do not allow slag to collect in the root. The fillet must be equally divided between both plates for good strength. After completing the weld, cool, chip, and inspect the weld for uniformity and defects. Repeat the welds using all three F groups of electrodes until you can consistently make welds free of defects. Turn off the welding machine and clean up your work area when you are finished welding.

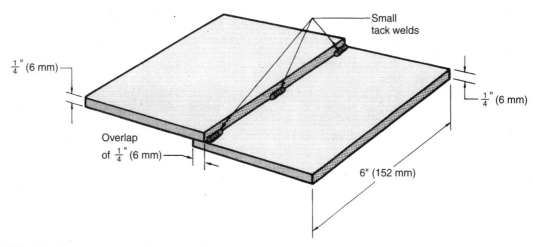

FIGURE 7–53 Tack welding the plates together.

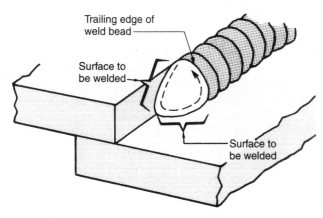

FIGURE 7–54 Follow the surface of the plate to ensure good fusion.

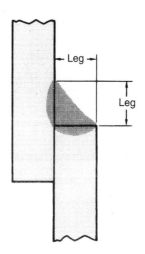

FIGURE 7–55 The horizontal lap joint should have a fillet weld that is equal on both plates.

TEE JOINT

The tee joint is made by tack welding one piece of metal on another piece of metal at a right angle (Figure 7-56). After the joint is tack-welded together, the slag is chipped from the tack-welds. If the slag is not removed, it will cause a slag inclusion in the final weld.

The heat is not distributed uniformly between both plates during a tee weld. Because the plate that forms the stem of the tee can conduct heat away from the arc in only one direction, it will heat up faster than the base plate. Heat escapes into the base plate in two directions. When using a weave pattern, most of the heat should be directed to the base plate to keep the weld size more uniform and to help prevent undercut.

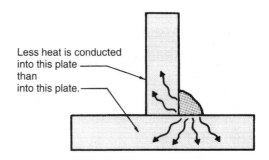

FIGURE 7–56 Tee joint.

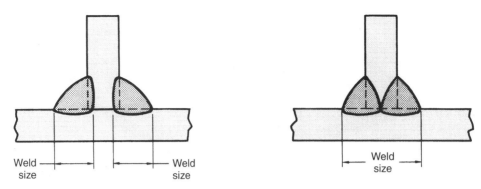

FIGURE 7–57 If the total weld sizes are equal, then both tee joints will have equal strength.

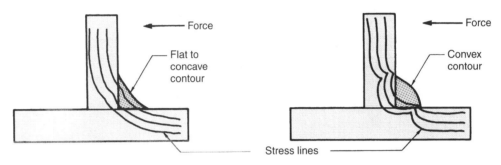

FIGURE 7–58 The stresses are distributed more uniformly through a flat or concave fillet weld.

A welded tee joint can be strong if it is welded on both sides, even without having deep penetration (Figure 7-57). The weld will be as strong as the base plate if the size of the two welds equals the total thickness of the base plate. The weld bead should have a flat or slightly concave appearance to ensure the greatest strength and efficiency (Figure 7-58).

PRACTICE 7–12

Tee Joint in the Flat Position (1F) Using E6010 or 6011 Electrodes, E6012 or E6013 Electrodes, and E7016 or E7018 Electrodes

Using a properly set up and adjusted arc welding machine, proper safety protection, arc welding electrodes having a 1/8-inch (3 mm) diameter, and two or more pieces of mild steel plate 6 inches (150 mm) long by 1/4 inch (6 mm) thick, you will make a welded tee joint in the flat position (Figure 7-59).

After the plates are tack-welded together, place them on the welding table so the weld will be flat. Start at one end and establish a molten weld pool on both plates. Allow the molten weld pool to flow together before starting the weave pattern. Any of the weave patterns will work well on this joint. To prevent slag inclusions, use a slightly higher-than-normal amperage setting.

When the weld is completed, cool, chip, and inspect it for uniformity and soundness. Repeat the welds as needed for all of these groups of electrodes until you can consistently make welds free of defects. Turn off the welding machine and clean up your work area when you are finished welding.

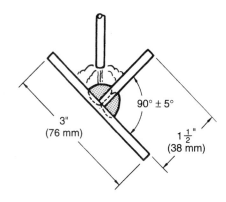

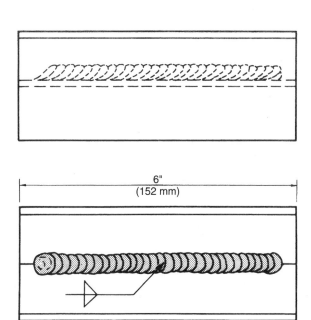

FIGURE 7–59   Tee joint in the flat position.

REVIEW QUESTIONS

1. What is shielded metal arc welding (SMAW)?
2. Why is SMAW the most widely used welding process?
3. What are the three different types of current used for welding?
4. What are two types of electrical devices that can be used to produce the low-voltage, high-amperage current combination that arc welding requires?
5. Where should an arc welding machine be located?
6. What determines the selection of the size of welding electrode for a weld?
7. What is the arc length?
8. What is the straight step pattern used for?
9. What are the three filler metal classes of electrodes used for practice welds?
10. How is a square butt joint made?
11. How is a lap joint made?
12. How is a tee joint made?

Chapter

8

Oxyacetylene Welding

Objectives

After completing this chapter, you should be able to:

- Make a variety of welded joints in any position on thin-gauge, mild steel sheet.
- Describe how to maintain the major components of oxyfuel welding equipment.
- Explain the method of testing an oxyfuel system for leaks.
- Demonstrate how to set up, light, adjust, extinguish, and disassemble oxyfuel welding equipment safely.
- Explain how to set up and weld mild steel.

Oxyacetylene is a form of fusion welding in which oxygen and acetylene are used in combination. With the advances in recent years in other welding processes, particularly MIG, the use of oxy acetylene has declined greatly. MIG welding is faster, cleaner, and causes less distortion than oxyacetylene. In addition, because it is difficult to concentrate the flame's heat in one place, the heat of oxyacetylene welding flame affects the surrounding areas and reduces the strength of the steel panels. For this reason, automakers do not recommend the use of oxyacetylene for the repair of damaged vehicles. But although oxyacetylene is in disfavor with most new car manufacturers, it has some use in the collision repair facility. The oxyacetylene flame is used in some heat-shrinking operations, brazing and soldering, surface cleaning, and the cutting of non-structural parts.

Oxyfuel welding, cutting, brazing, heating, and other processes use the same basic equipment. When storing, handling, assembling, testing, adjusting, lighting, shutting off, and disassembling this basic equipment, the same safety procedures must be followed for each process. Improper or careless work habits can cause serious safety hazards. Proper attention to all details makes these processes safe.

Certain basic equipment is common to all gas welding and cutting. Cylinders, regulators, hoses, hose fittings, safety valves, torches, and tips are some of the basic equipment used. Although there are many manufacturers producing a variety of gas equipment, it all works on the same principle. When you are not sure how a new piece of equipment is operated, seek professional assistance. Never experiment with any equipment.

All oxyfuel processes use a high-heat, high-temperature flame produced by burning a fuel gas mixed with pure oxygen. The gases are supplied in pressurized cylinders. The gas pressure from the cylinders must be reduced by a regulator. The gas then flows through flexible hoses to the torch. The torch controls this flow and mixes the gases in the proper proportion for good combustion at the end of the tip.

Acetylene is the most widely used fuel gas, but about twenty-five other gases are available. The tip is usually the only equipment change required in order to use another fuel gas. The adjustment and skill required are often different, but the storage, handling, assembling, and testing are the same. When changing gases, make sure the tip can be used safely with the new gas.

8.1 PRESSURE REGULATORS

Regulators all work on the same principle: they reduce a high pressure to a lower, working pressure. The lower pressure must be held constant over a range of flow rates.

CAUTION: Although all regulators work the same way, they cannot be safely used interchangeably on different types of gas or for different pressure ranges without the possibility of a fire or an explosion.

REGULATOR GAUGES

There may be one or two gauges on a regulator. One gauge shows the working pressure, and the other indicates the cylinder pressure (Figure 8-1). The working pressure gauge shows the pressure going to the torch.

The high-pressure gauge on a regulator shows cylinder pressure only. This gauge may be used to indicate the amount of gas that remains in a cylinder. The pressure shown on a gauge is read as pounds per square inch gauge (psig) or kilopascals (kPag).

SAFETY RELEASE DEVICE

Regulators may be equipped with either a safety release valve or a safety disc to prevent excessively high pressures from damaging the regulator. A safety release valve features a small ball held tightly against a seat by a spring. The release valve will reseat itself after excessive pressure has been released. A safety disc is a thin piece of metal held between two seals (Figure 8-2). The disc will burst when subjected to excessive pressure. The disc must

then be replaced before the regulator can be used again.

FITTINGS

A variety of inlet or cylinder fittings are available to ensure that the regulator cannot be connected to the wrong gas or pressure (Figures 8-3). A few adapters are available that will allow some regulators to be attached to different types of fittings. The two most common types of adapters are those that (1) adapt a left-hand male acetylene cylinder fitting to a right-hand female regulator fitting, or vice versa, and (2) adapt an argon or mixed-gas male fitting to a female flat-washer-type CO_2 fitting (Figure 8-4).

FIGURE 8–1   *Safety release valve on an oxygen regulator. (Courtesy of Victor Equipment Company)*

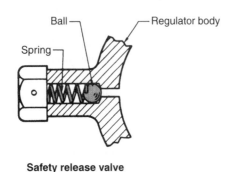

Safety release valve

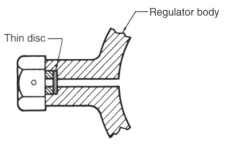

Safety disc valve

FIGURE 8–2   *Pressure release valve.*

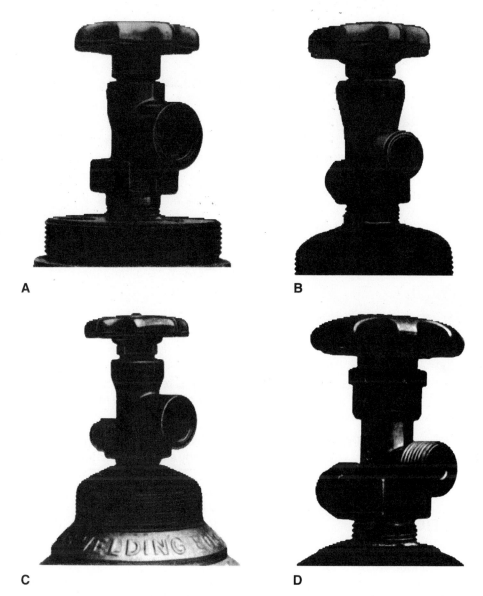

FIGURE 8–3 (A) Acetylene cylinder valve (left-hand thread). (B) Oxygen cylinder valve. (C) Argon cylinder valve. (D) Carbon dioxide (CO_2) cylinder valve.

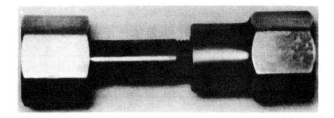

FIGURE 8–4   Carbon dioxide-to-argon adapter.

The connections to the cylinder and to the hose must be kept free of dirt and oil. Fittings should screw together freely by hand and should only require light wrench pressure to be leak-tight. If the fitting does not tighten freely on the connection, both parts should be cleaned. If the joint leaks after it has been tightened with a wrench, the seat should be checked. Examine the seat and threads for damage. If the seat is damaged, it can be repaired with a reamer specially made for this purpose (Figure 8-5). The threads can be repaired by using a die or thread file (Figure 8-6). Care should be taken not to gouge or score the soft brass when repairing a connector. Severely damaged connections must be replaced.

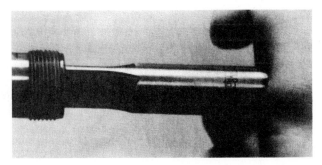

FIGURE 8–5 *A reamer is used to repair damaged torch seats.*

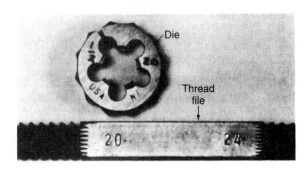

FIGURE 8–6 *The tools used to repair damaged threads.*

FIGURE 8–7 Left-hand threaded fittings are identified with a notch.

SAFETY PRECAUTIONS

If the adjusting screw is not loosened when the work is completed, and the cylinder valve is turned off, long-term damage can occur to the diaphragm and springs. In addition, when the cylinder valve is re-opened, some high-pressure gas can pass by the open high-pressure valve before the diaphragm can close it. This condition may cause the diaphragm to rupture or the low-pressure gauge to explode, or both. High-pressure valve seats that leak will result in a creeping or rising pressure on the working side of the regulator. If the leakage at the seat is severe, the maximum safe pressure can be exceeded on the working side, resulting in damage to the diaphragm, gauge, hoses, or other equipment.

CAUTION: Regulators that creep should not be used.

A gauge that gives a faulty reading or that is damaged can result in dangerous pressure settings. Gauges that do not read 0 (zero) pressure when the

pressure is released, or those that have a damaged glass or case, must be repaired or replaced.

CAUTION: All work on regulators must be done by properly trained repair technicians.

CAUTION: Regulators should be located far enough from the actual work that flames or sparks cannot reach them.

The outlet connection on a regulator is either a right-hand fitting for oxygen or a left-hand fitting for fuel gases. A left-hand threaded fitting has a notched nut (Figure 8-7).

REGULATOR CARE AND USE

There are no internal or external moving parts on a regulator or a gauge that require oiling (Figure 8-8).

CAUTION: Oiling a regulator is unsafe and may cause a fire or an explosion.

If the adjusting screw becomes tight and difficult to turn, it can be removed and cleaned with a dry, oil-free rag. When replacing the adjusting screw, be sure it does not become cross-threaded. Many regulators use a nylon nut in the regulator body, and the nylon is easily cross-threaded.

FIGURE 8-8   Never oil a regulator. (Courtesy of Air Products and Chemicals, Inc.)

FIGURE 8-9   A torch body or handle used for welding or cutting. (Courtesy of Victor Equipment Company)

FIGURE 8-10   A torch used only for cutting.

When welding is finished and the cylinders are turned off, the gas pressure must be released and the adjusting screw backed out. This procedure is done so that the diaphragm, gauges, and adjusting spring are not under load. A regulator that is left pressurized causes the diaphragm to stretch, the Bourdon tube to straighten, and the adjusting spring to compress. These changes result in a less accurate regulator with a shorter life expectancy.

8.2   WELDING AND CUTTING TORCH DESIGN AND SERVICE

The oxyacetylene hand torch is the most common type of oxyfuel gas torch used in industry. The hand torch may be either a combination welding and cutting torch or a cutting torch only (Figures 8-9 and 8-10).

The combination welding and cutting torch offers more flexibility because a cutting head, welding tip, or heating tip can be attached quickly to the same torch body (Figure 8-11). Combination torch sets are often used in schools, automotive repair shops, collision repair facilities, small welding shops, or in any other situation where flexibility is needed. The combination torch sets usually are more practical for portable welding, since the single unit can be used for both cutting and welding.

FIGURE 8-11   A combination welding and cutting torch kit. (Courtesy of ESAB Welding and Cutting Products)

Straight cutting torches are usually longer than combination torches. The longer length helps keep the operator farther away from heat and sparks. In addition, thicker material can be cut with greater comfort.

Most manufacturers make torches in a variety of sizes for different types of work. There are small torches for jewelry work (Figure 8-12) and large torches for heavy plates. Specialty torches for heating, brazing, or soldering are also available. Some of these use a fuel-air mixture (Figure 8-13). Fuel-air torches are often used by plumbers and air-conditioning technicians for brazing and soldering copper pipe or tubing. There are no industrial standards for tip size identification, tip threads, or seats. Therefore, each style, size, and type of torch can be used only with the tips made by the same manufacturer to fit the specific torch.

TORCH CARE AND USE

The torch body contains threaded connections for the hoses and tips. These connections must be protected from any damage. Most torch connections are external and made of soft brass that is easily damaged. Some connections, however, are more protected because they have either internal threads or stainless steel threads for the tips. The best protection against damage and dirt is to leave the tip and hoses connected when the torch is not in use.

Because the hose connections are close to each other, a wrench should never be used on one nut

FIGURE 8–12 Lightweight torch for small, delicate jobs. (Courtesy of National Torch Tip Co.)

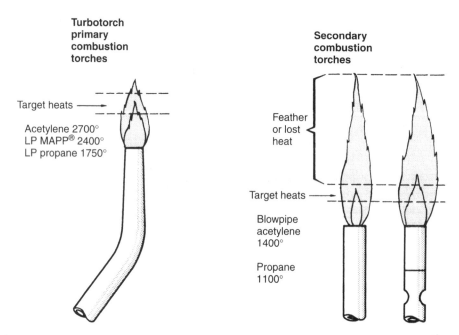

FIGURE 8–13 Some air/gas torches use a special tip that improves the combustion for a hotter, more effective flame. (Courtesy of Victor Equipment Company)

unless the other connection is protected with a hose-fitting nut (Figure 8-14).

The hose connections should not leak after they are tightened with a wrench. If leaks are present, the seat should be repaired or replaced. Some torches have removable hose connection fittings so that replacement is possible.

The valves should be easily turned on and off and should stop all gas flowing with minimum finger pressure. To find leaking valve seats, set the regulators to a working pressure. With the torch valves off, spray the tip with a leak-detecting solution. The presence of bubbles indicates a leaking valve seat (Figure 8-15). The gas should not leak past the valve stem packing when the valve is open or when it is closed. To test

leaks around the valve stem, set the regulator to a working pressure. With the valves off, spray the valve stem with a leak-detecting solution and watch for bubbles, indicating a leaking valve packing. The valve stem packing can now be tested with the valve open. Place a finger over the hole in the tip and open the valve. Spray the stem and watch for bubbles, which would indicate a leaking valve packing (Figure 8-16). If either test indicates a leak, the valve stem packing nut can be tightened until the leak stops. After the leak stops, turn the valve knob. It should still turn freely. If it does not, or if the leak cannot be stopped, replace the valve packing.

The valve packing and valve seat can be easily repaired on most torches by following the instructions given in the repair kit. On some torches, the entire valve assembly can be replaced, if necessary.

TIP CARE AND USE

Because no industrial standard tip size identification system exists, the student must become familiar with the size of the orifice (hole) in the tip and the thickness range for which it can be used. Comparing the overall size of the tip can be done only for tips made by the same manufacturer for the same type and style of torch (Figure 8-17). Learning a specific manufacturer's system is not always the answer, because on older, worn tips the orifice may have been enlarged by repeated cleaning. Finding the correct tip size is often determined by trial and error.

Torch tips may have metal-to-metal seals, or they may have an O ring or gasket between the tip and the torch seat. Metal-to-metal seal tips must be tightened

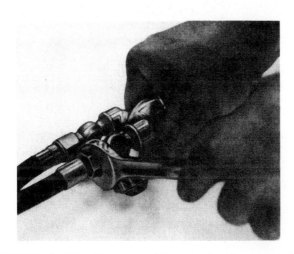

FIGURE 8-14   One hose-fitting nut will protect the threads when the other nut is loosened or tightened.

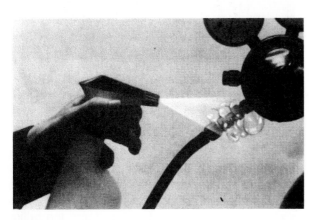

FIGURE 8-15   Check all connections for possible leaks and tighten if necessary.

FIGURE 8-16   The torch valves should be cleared for leaks, and the valve packing nut should be tightened if necessary.

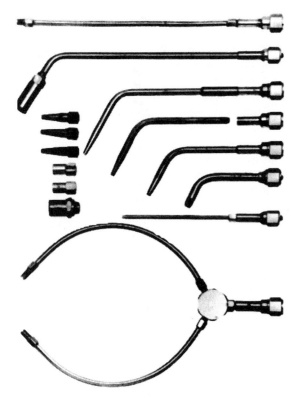

FIGURE 8–17 A variety of tip styles and sizes for one torch body. (Courtesy of Victor Equipment Company)

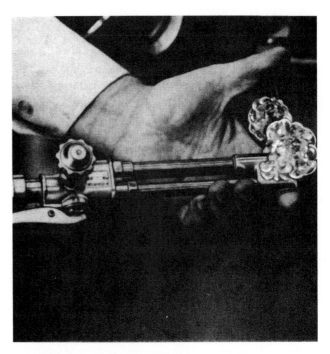

FIGURE 8–18 Checking a torch tip for a leaking seat. (Courtesy of D. Rhodes)

FIGURE 8–19 Standard set of tip cleaners. (Courtesy of Uniweld Products, Inc.)

with a wrench. Tips with an O ring or gasket may be tightened with a wrench or by hand, depending on the manufacturer's recommendations. Using the wrong method of tightening the tip fitting may result in damage to the torch body or the tip.

After a tip is installed, it should be checked for leaks. To test for leaks, set the regulator to a working pressure, turn the torch valve on, and put a finger over the holes in the end of the tip. Spray a leak-detecting solution around the base of the tip at the fitting and watch for bubbles that indicate a leaking seat (Figure 8-18).

If the leaking tip seat has an O ring or gasket, then the O ring or gasket is replaced with a new one. If the leaking tip seat is the metal-to-metal type, then a special reamer is used to smooth the seat.

Used, dirty tips can be cleaned with a set of tip cleaners. A little oxygen flow should be turned on so that dirt loosened during cleaning will be blown away. Using the file provided in the tip-cleaning set (Figure 8-19), file the end of the tip smooth and square. Next, select the size of tip cleaner that fits easily into the orifice. The tip cleaner is a small, round file and should only be moved in and out of the orifice

a few times (Figure 8-20). Be sure the tip cleaner is straight and that it is held in a steady position to prevent it from bending or breaking off in the tip. Excessive use of the tip cleaner tends to ream the orifice, making it too large. Therefore, use the tip cleaner only as required.

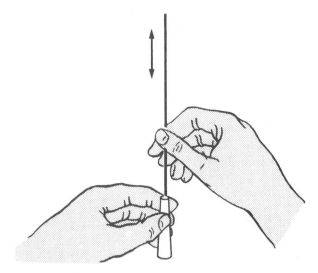

FIGURE 8-20 *Cleaning a tip with a standard tip cleaner.*

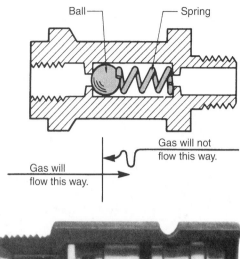

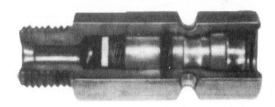

FIGURE 8-21 *Reverse flow valve only. (Courtesy of Airco Welding Products)*

REVERSE FLOW AND FLASHBACK VALVES

The purpose of the reverse flow valve is to prevent gases from accidently flowing through the torch and into the wrong hose. If the gases being used are allowed to mix in the hose or regulator, they might explode. The reverse flow valve is a spring-loaded check valve that closes when gas pressure from a backflow tries to occur through the torch valves (Figure 8-21). Some torches have reverse flow valves built into the torch body, but most torches must have these safety devices added. If the torch does not come with a reverse flow valve, one must be added to either the torch end or regulator end of the hose.

A reverse flow of gas will occur if the torch is not turned off or bled properly. The torch valves must be opened one at a time so that the gas pressure in that hose will be vented into the atmosphere and not into the other hose (Figure 8-22).

CAUTION: If both valves are opened at the same time, one gas may be pushed back up the hose of the other gas.

A reverse flow valve will not stop the flame from a flashback from continuing through the hoses. A flashback arrestor will do the job of a reverse flow valve, and it will stop the flame of a flashback (Figure 8-23). The flashback arrestor is designed to quickly stop the flow of gas during a flashback. These valves work on a similar principle to the gas valve at a service station. They are very sensitive to any

backpressure in the hose and stop the flow if any backpressure is detected.

Both devices must be checked on a regular basis to see that they are working correctly. The internal valves may become plugged with dirt or they may become sticky and not operate correctly. To test the check valve, you can try to blow air backwards through the valve. To test the flashback arrestor, follow the manufacturer's recommended procedure. If the safety device does not function correctly, it should be replaced.

8.3 HOSES AND FITTINGS

Most welding hoses used today are molded together as one piece and are referred to as Siamese hose. Hoses that are not of the Siamese type, or hose ends that have separated, may be taped together. When you tape the hoses, you must not wrap them solidly.

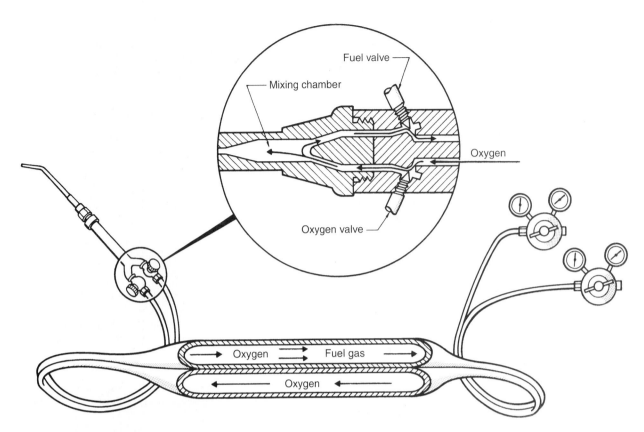

FIGURE 8-22 Gas may flow back up the hose if both valves are opened at the same time when the system is being bled down after use. Installing reverse flow valves on the torch can prevent this.

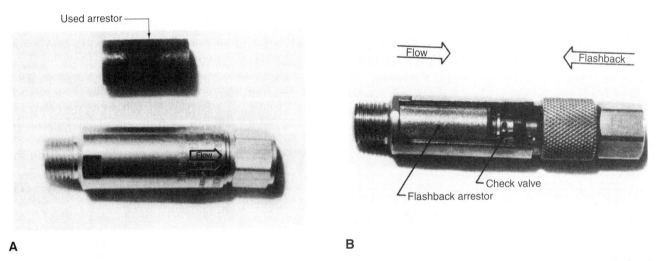

A   **B**

FIGURE 8-23 Flashback arrestor and reverse low valve combination. (A) Arrestor cartridge can be replaced after it has done its job. (B) Cutaway.

The hoses should be wrapped for about two inches (51 mm) every foot (300 mm), allowing the colors of the hose to be seen.

Fuel gas hoses must be red and have left-hand threaded fittings. Oxygen hoses must be green and have right-hand threaded fittings.

When hoses are not in use, the gas should be turned off and the pressure bled off. Turning off the equipment and releasing the pressure prevents any undetected leaks from causing a fire or explosion. This action also eliminates a dangerous situation that would be created if a hose were cut by equipment or materials being handled by workers who were unfamiliar with welding equipment. If the auto body technician is not certain that the hoses have been bled, they should be purged before the torch is lit.

Hoses are resistant to burns, but they are not burn-proof. They should be kept out of direct flame, sparks, and hot metal. You must be especially cautious when using a cutting torch. If a hose becomes damaged, the damaged section should be removed and the hose repaired with a splice. Damaged hoses should never be taped to stop leaks.

Hoses should be checked periodically for leaks. To test a hose for leaks, adjust the regulator to a working pressure with the torch valves closed. Wet the hose with a leak-detecting solution by rubbing it with a wet rag, spraying it, or dipping it in a bucket. Then watch for bubbles, which indicate that the hose leaks.

8.4 BACKFIRES AND FLASHBACKS

A backfire occurs when a flame goes out with a loud snap or pop. A backfire may be caused by: (1) touching the tip against the workpiece, (2) overheating the tip, (3) operating the torch when the flame settings are too low, (4) a loose tip, (5) damaged seats, or (6) dirt in the tip. The problem that caused the backfire must be corrected before relighting the torch. A backfire may cause a flashback.

A flashback occurs when the flame burns back inside the tip, torch, hose, or regulator. If the torch does flashback, close the oxygen valve at once and then close the fuel valve. The order in which the valves are closed is not as important as the speed at which they are closed. A flashback that reaches the cylinder may cause a fire or an explosion.

A flashback usually makes a high-pitched squealing or hissing sound. Closing the torch oxygen valve stops the flame inside at once. Then the fuel gas valve should be closed and the torch allowed to cool off before repairing the problem. When a flashback occurs, there is usually a serious problem with the

equipment, and a qualified technician should be called. After locating and repairing the problem, gas should be blown through the tip for a few seconds to clear out any soot that may have accumulated in the passages. A flashback that burns in the hose leaves a carbon char inside that may explode and burn in a pressurized oxygen system. A fuel gas is not required to kindle a hot, severe fire inside such hose sections. Discard hose sections in which a flashback has occurred and obtain new hose.

8.5 TYPES OF FLAMES

There are three distinctly different oxyacetylene flame settings. A *carburizing* flame has an excess of fuel gas. This flame has the lowest temperature and may put extra carbon in the weld metal. A *neutral* flame has a balance of fuel gas and oxygen. It is the most commonly used flame because it adds nothing to the weld metal. An *oxidizing* flame has an excess of oxygen. This flame has the highest temperature and may put oxides in the weld metal.

PRACTICE 8–1

Setting up an Oxyfuel Torch Set

This practice requires a disassembled oxyfuel torch set, consisting of two regulators, two reverse flow valves, one set of hoses, a torch body, a welding tip, two cylinders, a portable cart or supporting wall, and a wrench. The student is to assemble the equipment in a safe manner.

1. Safety chain the cylinders in the cart or to a wall (Figure 8-24). Then remove the valve protection caps (Figure 8-25).

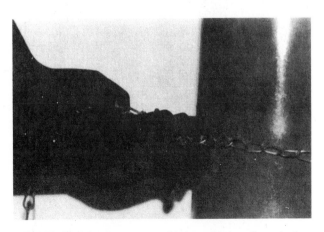

FIGURE 8–24 *Securing cylinder with a safety chain.*

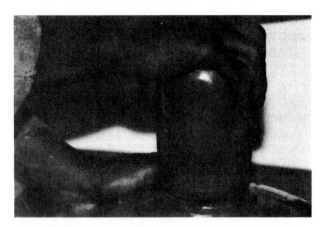

FIGURE 8–25 *Unscrew the valve protector cap. Put the cap in a safe place; they must be replaced on empty cylinders before they are returned.*

2. Crack the cylinder valve on each cylinder for a second to blow away dirt that may be in the valve.

CAUTION: If a fuel gas cylinder does not have a valve hand wheel permanently attached, then you must use a nonadjustable wrench to open the cylinder valve. The wrench must stay with the cylinder as long as the cylinder is on (Figure 8-26).

CAUTION: Always stand to one side. Point the valve away from anyone in the area and be sure there are no sources of ignition when cracking the valve.

---

3. Attach the regulators to the cylinder valves (Figure 8-27A). The nuts can be started by hand and then tightened with a wrench (Figure 8-27B).
4. Attach a reverse flow valve or flashback arrestor, if the torch does not have them built in, to the hose connection on the regulator or to the hose connection on the torch body, depending on the type of reverse flow valve in the set (Figure 8-28). Occasionally test each reverse flow valve by blowing through it to make sure it works properly.
5. Connect the hoses. The red hose has a left-hand nut and attaches to the fuel gas regulator. The green hose has a right-hand nut and attaches to the oxygen regulator.

Small combination wrench

Large combination wrench

T-wrench

FIGURE 8–26 *Nonadjustable wrenches for acetylene cylinders. (Courtesy of ESAB Welding and Cutting Products)*

6. Attach the torch to the hoses (Figure 8-29). Connect both hose nuts finger-tight before using a wrench to tighten either one.
7. Check the tip seals for nicks or O rings, if used, for damage. Check the owner's manual, or a supplier, to determine if the torch tip should be tightened by hand only or should be tightened with a wrench (Figure 8-30).

CAUTION: Tightening a tip the incorrect way may be dangerous and might damage the equipment.

A

B

FIGURE 8–27 Attach the oxygen regulator (A) to the oxygen cylinder valve. Using a wrench (B), tighten the nut.

FIGURE 8–28 Attach reverse flow valves.

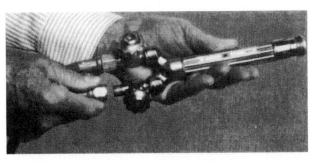

FIGURE 8–29 Connect the free ends of the oxygen and the acetylene hoses to the welding torch. (Courtesy of Albany Calcium Light Co., Inc.)

FIGURE 8–30 Select the proper tip or nozzle and install it on the torch body. (Courtesy of Albany Calcium Light Co., Inc.)

Check all connections to be sure they are tight. The oxyfuel equipment is now assembled and ready for use.

PRACTICE 8–2

Turning On and Testing a Torch

Using the oxyfuel equipment that was properly assembled in Practice 8-1, a nonadjustable tank wrench, and a leak-detecting solution, you will pressurize the system and test for leaks.

1. Back out the regulator pressure adjusting screws until they are loose (Figure 8-31).
2. Standing to one side of the regulator, open the cylinder valve *slowly* so that the pressure rises on the gauge slowly (Figure 8-32).

CAUTION: If the valve is opened quickly, the regulator or gauge may be damaged, or the gauge may explode.

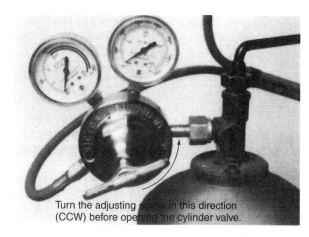

FIGURE 8–31 Back out both regulator adjusting screws before opening the cylinder valve. (Courtesy of Albany Calcium Light Co., Inc.)

FIGURE 8–32 Stand to one side when opening the cylinder valve. (Courtesy of Albany Calcium Light Co., Inc.)

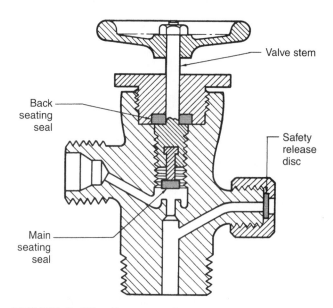

FIGURE 8–33 Cutaway of an oxygen cylinder valve showing the two separate seals. The back seating seal prevents leakage around the valve stem when the valve is open.

FIGURE 8–34 Open the cylinder valve slowly.

3. Open the oxygen valve all the way until it is sealed at the top (Figure 8-33).
4. Open the acetylene or other fuel gas valve 1/4 turn, or just enough to get gas pressure (Figure 8-34). If the cylinder valve does not have a handwheel, use a nonadjustable wrench and leave it in place on the valve stem while the gas is on.

CAUTION: The acetylene valve should never be opened more than 1 1/2 turns, so that in an emergency it can be turned off quickly.

---

5. Open one torch valve, and point the tip away from any source of ignition. Slowly turn in the pressure adjusting screw until gas can be heard escaping from the torch. The gas should flow long enough to allow the hose to be completely purged (emptied) of air and replaced by the gas before the torch valve is closed. Repeat this process with the other gas.

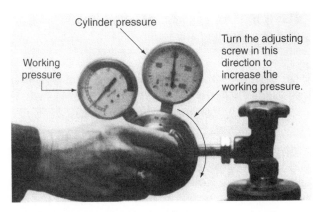

FIGURE 8–35 Adjust the regulator to read 5 psig (0.35 kg/cm₂g) working pressure.

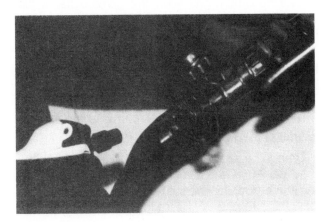

FIGURE 8–36 Spray fittings with a leak-detecting solution.

6. After purging is completed, and with both torch valves off, adjust both regulators to read 5 psig (35 kPag) (Figure 8-35).
7. Spray a leak-detecting solution on each hose and regulator connection and on each valve stem on the torch and cylinders. Watch for bubbles that indicate a leak. Turn off the cylinder valve before tightening any leaking connections (Figure 8-36).

CAUTION: Connections should not be overtightened. If they do not seal properly, repair or replace them.

FIGURE 8–37 Identify any cylinder that has a problem by marking it.

CAUTION: Leaking cylinder valve stems should not be repaired. Turn off the valve, disconnect the cylinder, mark the cylinder, and notify the supplier to come and pick up the bad cylinder (Figure 8-37).

The assembled oxyfuel welding equipment is now tested and ready to be ignited and adjusted.

PRACTICE 8–3

Lighting and Adjusting an Oxyacetylene Flame

Using the assembled and tested oxyfuel welding equipment from Practice 8-2, a sparklighter, gas welding goggles, gloves, and proper protective clothing, you will light and adjust an oxyacetylene torch for welding.

1. Wearing proper clothing, gloves, and gas welding goggles, turn both regulator adjusting screws in until the working pressure gauges read 5 psig (35 kPag). If you mistakenly turn on more than 5 psig (35 kPag), open the torch valve to allow the pressure to drop as the adjusting screw is turned outward.
2. Turn on the torch fuel-gas valve just enough so that some gas escapes.

CAUTION: Be sure the torch is pointed away from any sources of ignition or any object or person that might be damaged or harmed by the flame when it is lit.

3. Using a sparklighter, light the torch. Hold the lighter near the end of the tip (Figure 8-38) but not covering the end (Figure 8-39).

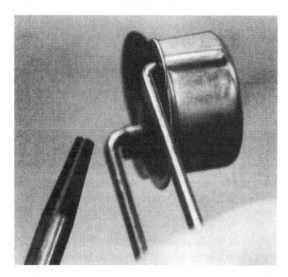

FIGURE 8–38 *Correct position to hold a sparklighter.*

FIGURE 8–39 *Sparklighter held too close over the end of the tip.*

CAUTION: A sparklighter is the only safe device to use when lighting any torch.

4. With the torch lit, increase the flow of acetylene until the flame stops smoking.
5. Slowly turn on the oxygen and adjust the torch to a neutral flame.

This flame setting uses the minimum gas flow rate for this specific tip. The fuel flow should never be adjusted to a rate below the point where the smoke stops. This is the minimum flow rate at which the cool gases will pull the flame heat out of the tip. If excessive heat is allowed to build up in a tip, it can cause a backfire or flashback.

The maximum gas flow rate gives a flame that, when adjusted to the neutral setting, does not settle back on the tip.

PRACTICE 8–4

Shutting Off and Disassembling Oxyfuel Welding Equipment

Using the properly lit and adjusted torch from Practice 8-3 and a wrench, you will extinguish the flame and disassemble the torch set.

1. First, quickly turn off the torch fuel-gas valve. This action blows the flame out and away from the tip, ensuring that the fire is out. In addition, it prevents the flame from burning back inside the torch. On large tips or hot tips, turning the fuel off first may cause the tip to pop. The pop is caused by a lean fuel mixture in the tip.

   If you find that the tip pops each time you turn the fuel off first, turn the oxygen off first to prevent the pop. Be sure that the flame is out before putting the torch down.
2. After the flame is out, turn off the oxygen valve.
3. Turn off the cylinder valves.
4. Open one torch valve at a time to bleed off the pressure.
5. When all of the pressure is released from the system, back both regulator adjusting screws out until they are loose.
6. Loosen both ends of both hoses and unscrew them.
7. Loosen both regulators and unscrew them from the cylinder valves.
8. Replace the valve protection caps.

8.6 — FUEL GASES

ACETYLENE (C₂H₂)

Acetylene is the most frequently used fuel gas. The mixture of oxygen and acetylene produces a high heat and high temperature flame that is widely used for welding, cutting, brazing, and heating.

Acetylene is colorless, lighter than air, and has a strong garlic smell. Acetylene is also unstable at pressures above 30 psig (200 kPag), or at temperatures above 1,435° F (780° C). Above the critical pressure or temperature, an explosion may occur if acetylene rapidly decomposes. This decomposition can occur without the presence of oxygen and may occur as a result of electrical shock or extreme physical shock.

CAUTION: Because of the instability of acetylene, it must never be used at pressures above 15 psig (100 kPag) or subjected to any possible electrical shock, excessive heat, or rough handling. Acetylene must not be manifolded or distributed through copper lines because it forms copper acetylide, which is explosive.

Acetone is used inside acetylene cylinders to absorb the gas and make it more stable. The cylinder is filled with a porous material, and then acetone is added to the cylinder, where it absorbs about twenty-four times its own weight in acetylene. Acetone absorbs acetylene in the same manner as water absorbs carbon dioxide in a carbonated drink. As the acetylene is drawn off for use, it evaporates out of the acetone.

The neutral oxyacetylene flame burns at a temperature of approximately 5,589° F (3,087° C). The flame burns in two parts, the inner cone and the outer envelope; refer to Figure 8-40. The high temperature produced by the oxyacetylene flame is concentrated around the inner cone (Figure 8-41). As the flame is moved back away from the work, the localized heating is reduced quickly. For welding, this highly concentrated temperature is the greatest advantage of the oxyacetylene flame over other oxyfuel gases. The molten weld pool is easily controlled by the torch angle and the position of the inner cone in relation to the work.

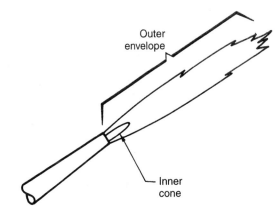

FIGURE 8–40   The two parts of a flame.

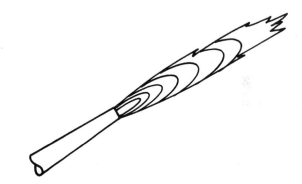

FIGURE 8–41   Thermal gradients of a flame.

METHYLACETYLENE-PROPADIENE (MPS)

Many different MPS gases are in use today as fuel gases for oxyfuel cutting, heating, brazing, metallizing, and to a limited extent for welding. MPS gases are mixtures of two or more of the following gases: propane (C₃H₄), butane (C₄H₁₀), butadiene (C₄H₆), methyl acetylene (C₃H₄), and propadiene (C₃H₆). The mixtures vary in composition and characteristics from one manufacturer to another. However, all manufacturers provide MPS gases as liquefied gases in pressurized cylinders.

MPS gases have a neutral oxyfuel flame temperature of about 5,301° F (2,927° C). The exact temperature varies with the specific mixture. The flame temperature and heat are high enough to be used for welding, but MPS gases are seldom used for this purpose.

MAPP® gas is the trade name for the stabilized liquefied mixture of methylacetylene (CH_3:C:CH) and propadiene (CH_2:C:CH_2) gases. Oxy MAPP® combusts with a high-heat, high-temperature flame that works well for cutting, heating, brazing, and metallizing (Figure 8-42). Oxy MAPP® produces a neutral flame temperature of 5,301° F (2,927° C).

8.7   FILLER METALS

Filler metals specifically designed to be used with an oxyfuel torch are generally divided into three groups. One group of rods used for welding is designated with the prefix letter R. Another group of rods used for brazing is designated with the prefix letter B. A third group is used for buildup, wear resistance surfacing, or both. Welding rods in this group may also be classified in one of the other groups, may be patented and use a trade name, or may be tubular with a granular material in the center. The tubular welding rods are designated with an RWC prefix. Some filler metals, for example BRCuZn, are classified both as a braze welding rod (R) and a brazing rod (B) because they can be used either way.

Mild steel filler metals are welding rods that are mainly iron. They may have other elements added to change their strength, corrosion resistance, weldability, or another physical property. There are three major AWS specifications for ferrous filler metals. These three specifications are A5.2 low-carbon, low-alloy steel, A5.15 cast iron, and A5.9 stainless steel. Within each specification, there are classes; for example, in group A5.2 the classes are RG45, RG60, and RG65.

Small shops sometimes use other types of wire for weld filler metals. The most popular substitution is often coat-hanger wire. Using such substitutes can cause weld failure. Coat-hanger wires were not manufactured for welding purposes, and their chemistry varies greatly.

8.8   OXYFUEL WELDING

Oxyacetylene welding is limited to thin metal sections or to times when portability is important. Welding practices will concentrate on the use of gas welding on metal having a thickness of 16 gauge (approximately 1/6 inch [2 mm]) or thinner.

Mild steel is the easiest metal to gas weld. With this metal, it is possible to make welds with 100 percent integrity (lack of porosity, oxides, or other defects) that have excellent strength, ductility, and other positive characteristics. The secondary flame shields the molten weld pool from the air, which would cause oxidation.

FACTORS AFFECTING THE WELD
Torch Tip Size

The torch tip size should be used to control the weld bead width, penetration, and speed. Penetration is the depth into the base metal that the weld fusion or melting extends from the surface, excluding any reinforcement. Because each tip size has a limited operating range, tip sizes must be changed to suit the thickness and size of the metal being welded. Never lower the size of the torch flame when the correct tip size is unavailable. Other factors that can be changed to control the weld size are the torch angle, the flame-to-metal distance, the welding rod size, or the way the torch is manipulated.

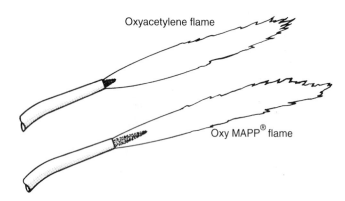

Oxyacetylene flame

Oxy MAPP® flame

FIGURE 8–42 *The inner cone and outer envelope of a MAPP flame are longer than an equal-volume flame of acetylene because of the slower burn rate of MAPP.*

Torch Angle

The torch angle and the angle between the inner cone and the metal have a great effect on the speed of melting and size of the molten weld pool. The ideal angle for the welding torch is 45 degrees. As this angle increases toward 90 degrees, the rate of heating increases. As the angle decreases toward 0 degrees, the rate of heating decreases, as illustrated in Figure 8-43. The distance between the inner cone and the metal ideally should be 1/8 inch to 1/4 inch (3 mm to 6 mm). As this distance increases, the rate of heating decreases; as it decreases, the heating rate increases (Figure 8-44).

Welding Rod Size

Welding rod size and torch manipulation can be used to control the weld bead characteristics. A larger-size welding rod can be used to cool the molten weld pool, increase buildup, and reduce penetration (Figure 8-45). The torch can be manipulated so

that the direct heat from the flame is flashed off the molten weld pool for a moment to allow it to cool (Figure 8-46).

CHARACTERISTICS OF THE WELD

The molten weld pool must be protected by the secondary flame to prevent the atmosphere from contaminating the metal. If the secondary flame is suddenly moved away from a molten weld pool, the pool will throw off a large number of sparks. These sparks are caused by the rapid burning of the metal and its alloys as they come into contact with oxygen in the air. This is a particular problem when a weld is stopped. The weld crater is especially susceptible to cracking. This tendency is greatly increased if the molten weld pool is allowed to burn out (Figure 8-47). To prevent burnout, the torch should be raised or tilted, keeping the outer flame envelope over the molten weld pool until it solidifies. As the molten weld pool is being cooled it should also be filled with welding rod so that

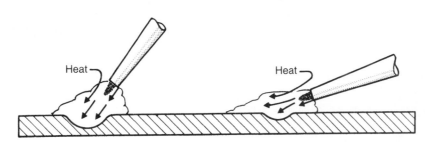

FIGURE 8–43 Changing the torch angle changes the percentage of heat that is transferred into the metal.

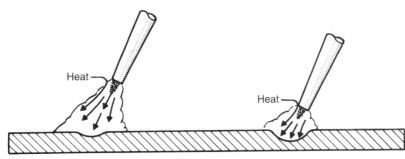

FIGURE 8–44 Changing the distance between the torch tip and the metal changes the percentage of heat input into the metal.

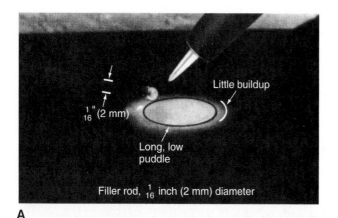

A

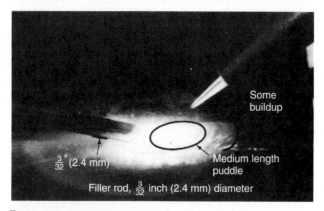

B

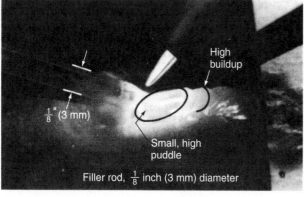

C

FIGURE 8–45 *If all other conditions remain the same, changing the size of the filler rod will affect the weld as shown in (A), (B), and (C).*

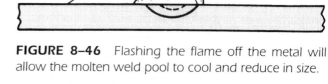

FIGURE 8–46 *Flashing the flame off the metal will allow the molten weld pool to cool and reduce in size.*

FIGURE 8–47 *Building up the molten weld pool before it is ended will help prevent crater cracking.*

it is a uniform height compared to the surrounding weld bead.

An increase in sparks on clean metal means an increase in weld temperature. A decrease in sparks indicates a decrease in weld temperature. Often the number of sparks in the air increases just before a burn-through takes place; that is, burning out the

molten metal that appears on the back side of the plate. This burnout does not happen to molten metal until it reaches the kindling temperature (the temperature that must be attained before something begins to burn). Small amounts of total penetration usually will not cause a burnout. When the sparks increase quickly, the torch should be pulled back to allow the metal to cool and prevent a burn-through.

PRACTICE 8–5

Flame Effect on Metal

The first practice examines how the flame affects mild steel. Use a piece of 16 gauge mild steel and the proper size torch tip. Light and adjust the flame by turning down the oxygen so that the flame has excessive acetylene. Hold the flame on the metal until it melts, and observe what happens (Figure 8-48). Now adjust the flame by turning up the oxygen so that the flame is neutral. Hold this flame on the metal until it melts, and observe what takes place (Figure 8-49). Next, adjust the flame by turning up the oxygen so

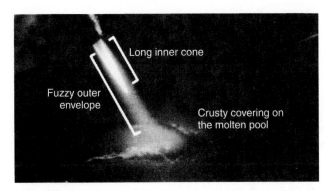

FIGURE 8-48 *Carbonizing flame (excessive acetylene).*

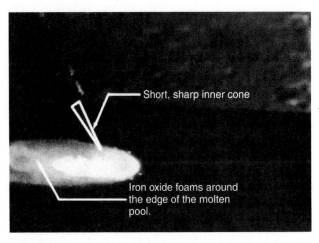

FIGURE 8-50 *Oxidizing flame (excessive oxygen).*

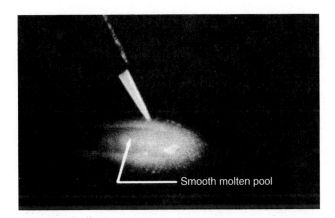

FIGURE 8-49 *Neutral flame (balanced oxygen and acetylene).*

that the flame has excessive oxygen (Figure 8-50). Hold this flame on the metal until it melts, and observe what happens. Repeat this experiment until you can easily identify each of the three flame settings by the flame, molten weld pool, sound, and sparks.

Before welding begins, it is a good idea to find a comfortable position. The more comfortable you are, the easier it will be for you to make uniform welds. The angle of the plate to you and the direction of travel are important.

Place a plate on the table in front of you and, with the torch off, practice moving the torch in one of the suggested patterns along a straight line, as illustrated in Figure 8-51. Turn the plate and repeat this step until you determine the most comfortable direction of travel. Later, when you have mastered several joints, you should change this angle and try to weld in a less comfortable position. Welding in the field or shop

must often be done in positions that are less than comfortable, so the auto body technician needs to be somewhat versatile.

It is important to feed the welding wire into the molten weld pool at a uniform rate. Figure 8-52 shows some suggested methods of feeding the wire by hand. It is also suggested that you not cut the welding wire in two pieces for welding. Short lengths are easy to use, but this practice is not widely accepted in industry. The end of the welding wire may be rested on your shoulder so that it is easy to handle.

CAUTION: The end of the filler rod should have a hook bent in it so that you can readily tell which end may be hot, and so that the sharp end will not be a hazard to a technician who may be working next to you (Figure 8-53).

The torch hoses may be stiff and may therefore cause the torch to twist. Before you start welding, move the hoses so that there is no twisting of the torch. This will make the torch easy to manipulate and will be more relaxing for you.

PRACTICE 8-6

Pushing a Molten Weld Pool

Using a clean piece of 16 gauge mild steel sheet about six inches long, and a torch that is lit and adjusted to a neutral flame, push a molten weld pool in a

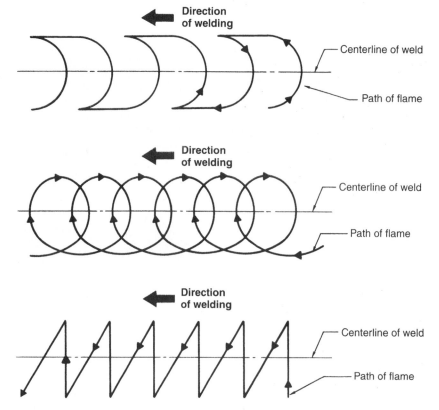

FIGURE 8–51 A few torch patterns.

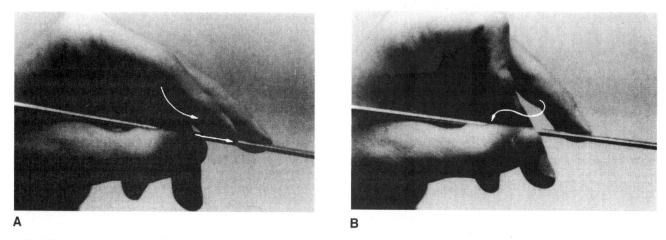

A

B

FIGURE 8–52 Feed the filler rod by using your index finger.

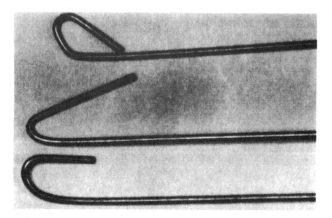

FIGURE 8–53 The end of the filler rod should be bent for safety and easy identification.

straight line down the sheet. Start at one end and hold the torch at a 45-degree angle in the direction of the weld (Figure 8-54). When the metal starts to melt, move the torch in a circular pattern down the sheet toward the other end. If the size of the molten weld pool changes, speed up or slow down to keep it the same size all the way down the sheet (Figure 8-55). Repeat this practice until you can keep the width of the molten weld pool uniform and the direction of travel in a straight line.

Uniformity in width shows that you have control of the molten weld pool. A straight line indicates that

you can see more than the molten weld pool itself. Students just learning to weld usually see only the molten weld pool. As you master the technique of welding, your visual range will increase. A broad visual range is important so that later you will be able to follow the joint, watch for distortion, see if other adjustments are needed, and relax as you weld. Turn off the cylinders, bleed the hoses, back out the regulator adjusting screw, and clean up your work area when you are finished.

PRACTICE 8-7

Stringer Bead, Flat Position

Repeat Practice 8-7 until you have mastered a straight and uniform weld with any of the three sizes of filler rod.

Joining two or more clean pieces of metal to form a welded joint is the next step in learning to weld. The joints must be uniform in width and reinforcement so that they will have maximum strength. For each type of joint, the amount of penetration required to give

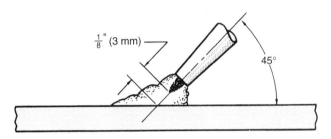

FIGURE 8–54 Hold the torch at a 45-degree angle in the direction of the weld.

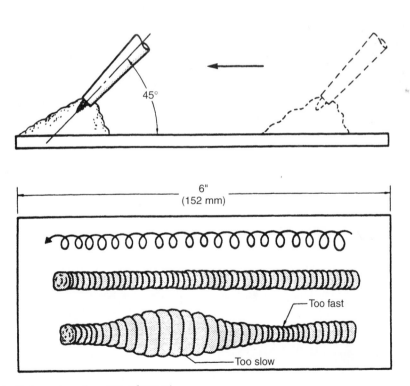

FIGURE 8–55 Effect of changing the rate of travel.

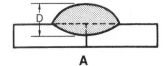 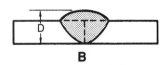

FIGURE 8–56 Welds (A) and (B) both have approximately the same strength, but only if the reinforcement does not have to be removed.

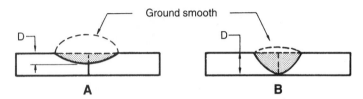

FIGURE 8–57 If the reinforcement on both welds is removed, weld (B) would be the stronger weld.

maximum strength will vary and may not be 100 percent (Figure 8-56). For example, in thin sheet metal there is usually enough reinforcement on the weld to give the weld adequate strength. But if that reinforcement has to be removed, then 100 percent penetration is important (Figure 8-57). Some joints, such as the lap joint, never need 100 percent penetration. However, they do need 100 percent fusion. Turn off the cylinders, bleed the hoses, back out the regulator adjusting screw, and clean up your work area when you are finished.

8.9 FLANGE JOINT

The flat flange joint can be made with or without the addition of filler metal. This joint is one of the easiest welded joints to make. If the sheets are tacked together properly, the addition of filler metal is not needed. However, if filler metal is added, it should be added uniformly, as in the stringer beads in Practice 8-7.

PRACTICE 8–8

Flange Joint Flat Position

Using a properly lit and adjusted torch, two clean pieces of 16 gauge mild steel 6 inches long, and filler metal, you will make a flat flange edge-welded joint (Figure 8-58).

Place one of the pieces of metal in a jig or on a firebrick and hold or brace the other piece of metal

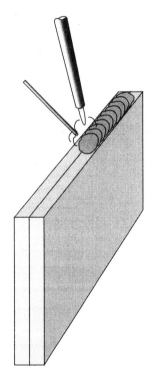

FIGURE 8–58 Flange edge weld.

vertically on it. Tack the ends of the two sheets together. Then set it upright and put two or three more tacks on the joint. Holding the torch as shown in Figure 8-58, make a uniform weld along the joint. Repeat this until the weld can be made without defects. Turn off the cylinders, bleed the hoses, back out the regulator adjusting screw, and clean up your work area when you are finished.

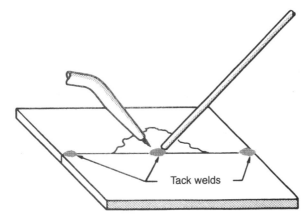

FIGURE 8–59   Making a tack weld.

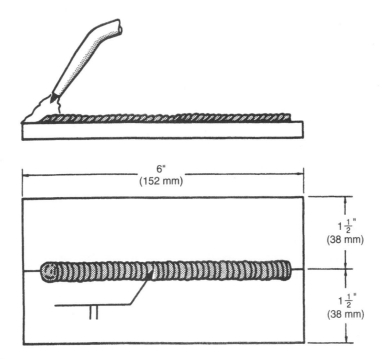

6"
(152 mm)

1 1/2 "
(38 mm)

1 1/2 "
(38 mm)

FIGURE 8–60   Butt joint.

BUTT JOINT

The flat butt joint is a welded joint and one of the easiest to make. To make the butt joint, place two clean pieces of metal flat on the table and tack-weld both ends together as illustrated in Figure 8-59. Tack welds may also be placed along the joint before welding begins. Point the torch so that the flame is distributed equally on both sheets. The flame is to be in the direction that the weld is to progress. If the sheets to be welded are of different sizes or thicknesses, the torch should be pointed so that both pieces melt at the same time.

When both sheet edges have melted, add the filler rod in the same manner as in Practice 8-7. Turn off the cylinders, bleed the hoses, back out the regulator adjusting screw, and clean up your work area when you are finished.

PRACTICE 8–9

Butt Joint Flat Position

Using a properly lit and adjusted torch, two clean pieces of 16-gauge mild steel, 6 inch (152 mm) long, and filler metal, you will make a welded butt joint (Figure 8-60).

Place the two pieces of metal in a jig or on a firebrick and tack-weld both ends together. The tack on the ends can be made by simply heating the ends and allowing them to fuse together or by placing a small drop of filler metal on the sheet and heating the filler metal until it fuses to the sheet. The latter method is especially convenient if you have to use one hand to hold the sheets together and the other to hold the torch. After both ends are tacked together, place one or two small tacks along the joint to prevent warping during welding.

With the sheets tacked together, start welding from one end to the other using the technique learned in Practice 8-7. Repeat this weld until you can make a welded butt joint that is uniform in width and reinforcement and has no visual defects. The penetration of this practice weld may vary. Turn off the cylinders, bleed the hoses, back out the regulator adjusting screw, and clean up your work area when you are finished.

8.11 — LAP JOINT

The flat lap joint can be easily welded if some basic manipulations are used. When one is heating the two clean sheets, caution must be exercised to ensure that both sheets start melting at the same time. Heat is not distributed uniformly in the lap joint (Figure 8-61). Because of this difference in heating rate, the flame must be directed on the bottom sheet and away from the top sheet (Figure 8-62). The filler rod should be added to the top sheet. Gravity will pull the molten weld pool down to the bottom sheet, so it is therefore not necessary to put metal on the bottom sheet. If the filler metal is not added to the top sheet or if it is not added fast enough, surface tension will pull the molten weld pool back from the joint (Figure 8-63). When this happens, the rod should be added directly into this notch, and it will close. The weld appearance and strength will not be affected.

8.12 — TEE JOINT

The flat tee joint is more difficult to make than the butt or lap joints. The tee joint has the same problem with uneven heating as the lap joint does. It is important to hold the flame so that both sheets melt at the same time (Figure 8-64). Another problem that is unique to the tee joint is that a large percentage of the welding

heat is reflected back on the torch. This reflected heat can cause even a properly cleaned and adjusted torch to backfire or pop. To help prevent this from happening, angle the torch more in the direction of the weld travel.

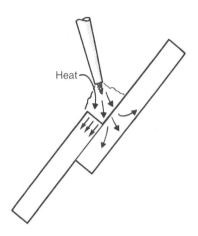

FIGURE 8–61   *Heat is conducted away more quickly in the bottom plate, resulting in the top plate's melting more quickly.*

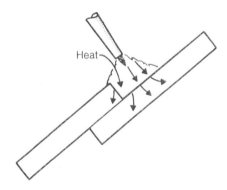

FIGURE 8–62   *Flame heat should be directed at the bottom plate to compensate for thermal conductivity.*

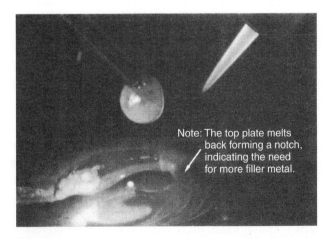

Note: The top plate melts back forming a notch, indicating the need for more filler metal.

FIGURE 8–63   *Add filler metal fast enough to prevent a notch from forming.*

8.13 OUT-OF-POSITION WELDING

A part to be welded cannot always be welded in the flat position. Whenever a weld is performed in a position other than flat, it is said to be out-of-position. Welds made in the vertical, horizontal, or overhead positions are out-of-position and somewhat more difficult than flat welds.

VERTICAL WELDS

A vertical weld is the most common out-of-position weld that an auto body technician is required to perform. When you make a vertical weld, it is important to control the size of the molten weld pool. If the molten weld pool size increases beyond that which the shelf will support, it will overflow and drip down the weld (Figure 8-65). These drops, when cooled, look like the drips of wax on a candle. To prevent the molten weld pool from dripping, you must watch the trailing edge of the pool. The trailing edge will

constantly be solidifying, forming a new shelf to support the molten weld pool as the weld progresses upward (Figure 8-66). Small molten weld pools are less likely than large ones to drip.

The less vertical the sheet, the easier the weld is to make, but the type of manipulation required is the same. Welding on a sheet at a 45-degree angle requires the same manipulation and skill as welding on a vertical sheet. However, the speed of manipulation is slower, and the skill is less critical than at vertical. Your welding techniques should be mastered at a 45-degree angle. Then you should increase the angle of the sheet until you can make totally vertical welds. Each practice weld in this section should be started on an incline, and, as skill is gained, the angle should be increased until the sheets are vertical.

PRACTICE 8-10

Stringer Bead at a 45-Degree Angle

Using a properly lit and adjusted torch, two clean pieces of 16 gauge mild steel 6 inches (152 mm) long, and filler metal, you will make a bead at a 45-degree angle.

The filler metal should be added as you did in Practice 8-7. It may be necessary to flash the torch off the molten weld pool to allow it to cool (Figure 8-67). Flashing the torch off allows the molten weld pool to cool by moving the hotter inner cone away from the molten weld pool itself. While still protecting the molten metal with the outer flame envelope, a rhythm of moving the torch and adding the rod should be established. This rhythm helps make the bead uniform. Repeat this practice until the weld can be made without defects. Turn off the cylinders, bleed the hoses, back out the regulator adjusting screw, and clean up your work area when you are finished.

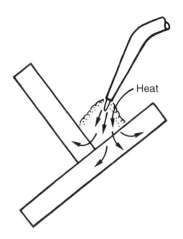

FIGURE 8–64 Direct the heat on the bottom plate to equalize the heating rates.

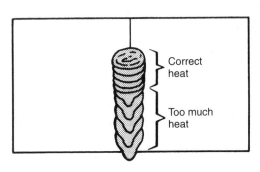

FIGURE 8–65 Vertical weld showing effect of too much heat.

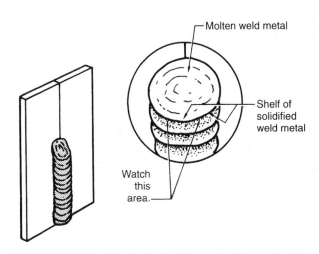

FIGURE 8–66 Watch the trailing edge to see that the molten pool stays properly supported on the shelf.

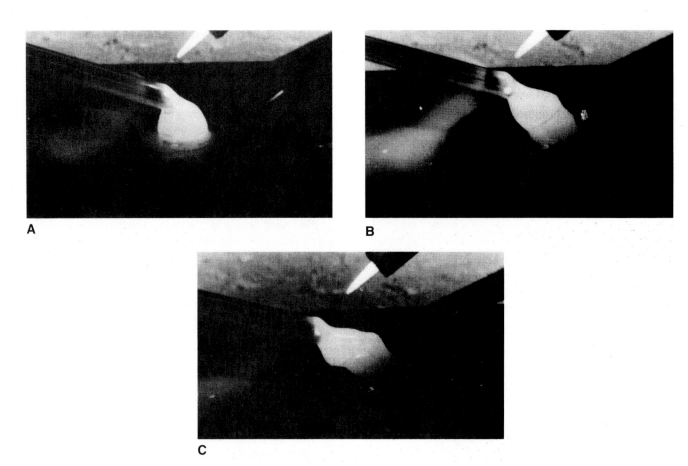

A   B

C

FIGURE 8–67   By flashing the flame off and controlling the pool size, a weld can be built up (A), and up, (B) and over (C).

PRACTICE 8-11

Stringer Bead, Vertical Position

Repeat Practice 8-10 until you have mastered a straight and uniform weld bead in a vertical position. Turn off the cylinders, bleed the hoses, back out the regulator adjusting screw, and clean up your work area when you are finished.

PRACTICE 8-12

Butt Joint at a 45-Degree Angle

Using the same equipment and materials as in Practice 8-10, you will make a welded butt joint at a 45-degree angle (Figure 8-68).

Tack the sheets together and support them at a 45-degree angle. Weld using the method of flashing the torch off the molten weld pool to control penetration and weld contour. Make a weld that has uniform width and reinforcement. Repeat this practice until the weld can be made without defects. Turn off the cylinders, bleed the hoses, back out the regulator adjusting screw, and clean up your work area when you are finished.

PRACTICE 8-13

Butt Joint, Vertical Position

Using the same equipment, materials, and setup as described in Practice 8-12, make a welded butt joint in the vertical position. Make a weld that is uniform in width and reinforcement and has no visual defects. The penetration of this practice weld may vary. Turn off the cylinders, bleed the hoses, back out the regulator adjusting screw, and clean up your work area when you are finished.

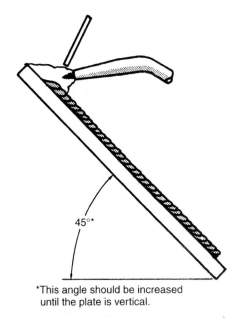

*This angle should be increased until the plate is vertical.

FIGURE 8–68   Butt joint at a 45-degree angle.

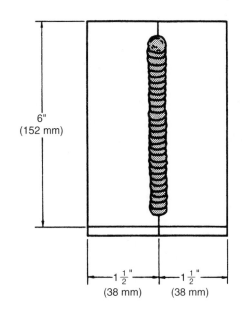

PRACTICE 8-14

Lap Joint at a 45-Degree Angle

Using the same equipment, materials, and setup as in Practice 8-12, you will weld a lap joint at a 45-degree angle.

After tacking the sheets together and supporting them at a 45 degree angle, use the same method of adding rod as you did for the flat lap joint. Again, flash off the torch as needed to control the molten weld pool.

Repeat this weld until you can make a weld that is uniform in width and reinforcement and has no visual defects. Both sides of the joint can be welded. Turn off the cylinders, bleed the hoses, back out the regulator adjusting screw, and clean up your work area when you are finished.

PRACTICE 8-15

Lap Joint Vertical Position

Using the same equipment, materials, and setup as listed in Practice 8-14, weld a lap joint in the vertical position. Make a weld that is uniform in width and reinforcement and has no visual defects. Both sides of the joint can be welded. Repeat this practice until the weld can be made without defects. Turn off the cylinders, bleed the hoses, back out the regulator

FIGURE 8–69   A J weave pattern for horizontal welds.

adjusting screw, and clean up your work area when you are finished.

HORIZONTAL WELDS

Horizontal welds, like vertical welds, must rely on some part of the weld bead to support the molten weld pool as the weld is made. The shelf that supports a horizontal weld must be built up under the molten weld pool and at the same time keep the weld bead uniform. The weave pattern required for a horizontal weld is completely different from that of any of the other positions. The pattern (Figure 8-69) builds a shelf on the bottom side of the bead to support the molten weld pool, which is elongated across the top. The sheet may be tipped back at a 45-degree angle for the stringer bead. Doing this allows you to acquire the needed skills before proceeding to the more difficult, fully horizontal position. As with the vertically inclined sheet, the skills required are the same.

When starting a horizontal bead, it is important to start with a small bead and build it to the desired size. If too large a molten weld pool is started, the shelf

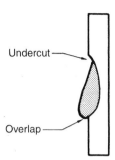

FIGURE 8–70 Too large a molten weld pool.

FIGURE 8–72 Horizontal lap joint.

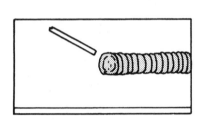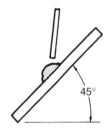

FIGURE 8–71 Horizontal stringer bead at a reclining angle.

does not have time to form properly. The weld bead will tend to sag downward and not be uniform. As a result, there may be an undercut of the top edge and an overlap on the bottom edge (Figure 8-70).

PRACTICE 8–16

Horizontal Stringer Bead at a 45-Degree Sheet Angle

Using a properly lit and adjusted torch, one clean piece of 16 gauge mild steel 6 inches long, and filler metal, you will make a horizontal bead at a 45-degree reclining angle (Figure 8-71).

Add the filler metal along the top leading edge of the molten weld pool. Surface tension will help to hold it on the top. The weld should be uniform in width and reinforcement and have no visual defects. Repeat this practice until the weld can be made without defects. Turn off the cylinders, bleed the hoses, back out the

regulator adjusting screw, and clean up your work area when you are finished.

PRACTICE 8–17

Stringer Bead, Horizontal Position

Using the same equipment, materials, and setup as listed in Practice 8-16, make a stringer bead in the horizontal position. The stringer bead should be uniform in width and reinforcement and have no visual defects. Turn off the cylinders, bleed the hoses, back out the regulator adjusting screw, and clean up your work area when you are finished.

PRACTICE 8–18

Lap Joint, Horizontal Position

Using a properly lit and adjusted torch, two clean pieces of 16 gauge mild steel, 6 inches long, and filler metal, you will weld a lap joint in the horizontal position.

After tacking the sheets together, support the assembly as illustrated in Figure 8-72. The weld must be uniform in width and reinforcement and have no visual defects. The sheet can be turned over, and the other side can be welded. Repeat this practice until the weld can be made without defects. Turn off the cylinders, bleed the hoses, back out the regulator adjusting screw, and clean up your work area when you are finished.

REVIEW QUESTIONS

1. How can it be determined how much liquified gas is remaining in a cylinder?
2. List the steps to stop a joint leak between the cylinder and the hose.
3. Describe the direction of the threads on the outlet connection on a regulator and how to tell the types of connections apart.
4. List the steps to determine where and if a valve seat is leaking.
5. Determine how to tighten a torch tip by its type.
6. List the steps to clean a tip.
7. How do you prevent a reverse gas flow?
8. Describe fuel gas and oxygen hoses.
9. What causes a backfire?
10. What do you do if the torch flashes back?
11. After a flashback has occurred, what should be done to prevent it from happening again?
12. List the three distinctly different oxyacetylene flame settings.
13. What are some of the problems with acetylene?
14. What gases can be mixed to make up MPS gases?
15. What should be done to a cylinder containing an MPS gas before use?
16. What are some of the safety features of a MAPP® gas?
17. Describe the three groups of filler metal classifications designed to be used with an oxyfuel torch.
18. What are the AWS specifications for mild steel filler metals?
19. Name a common popular substitute for filler metal and explain why it should not be used.
20. What are three basic factors that affect the weld?

Chapter 9

Brazing and Soldering

Objectives

After reading this chapter, you should be able to:

- Define the terms **soldering, brazing,** and **braze welding.**
- List the advantages of brazing.
- Describe the three functions of flux in the brazing process.
- Perform the general brazing procedure.
- Identify the unique characteristics of torch, induction, resistance, and diffusion brazing.
- Perform the general soldering procedure.

Soldering and brazing are both classified by the American Welding Society as liquid-solid phase-bonding processes. Liquid means that the filler metal is melted; solid means that the base material or materials are not melted. The phase is the temperature at which bonding takes place between the solid base material and the liquid filler metal. The bond between the base material and filler metal is a metallurgical bond, because no melting or alloying of the base metal occurs. If done correctly, this bond results in a joint having four or five times the tensile strength of that of the filler metal itself. Brazing is a popular method of joining metal in the automotive trades.

Soldering and brazing differ only in that soldering takes place at a temperature below 840° F (450° C), and brazing occurs at a temperature above 840° F (450° C). Because only the temperature separates the two processes, it is possible to do both soldering and brazing using different mixtures of the same metals, depending upon the alloys used and their melting temperatures.

Brazing is divided into two major categories: brazing and braze welding. In brazing, the parts being joined must be fitted so that the joint spacing is very small, approximately .025 to .002 inches (0.6 to 0.06 mm) (Figure 9-1). This small spacing allows capillary action to draw the filler metal into the joint when the parts reach the proper phase temperature.

Capillary action is the force that pulls water up into a paper towel or pulls a liquid into a very fine straw, Figure 9-2. Braze welding does not need capillary action to pull filler metal into the joint. Examples of brazing and braze-welded joint designs are shown in Figure 9-3.

9.1 ADVANTAGES OF SOLDERING AND BRAZING

There are several reasons for the widespread use of soldering or brazing in collision repair:

- Because the pieces of base metal are joined at a relatively low temperature at which the base metal does not melt, there is less risk of distortion and stress in the base metal.
- Because of their excellent flow characteristics, both soldering and brazing penetrate well into narrow gaps and are therefore very convenient for filling gaps in body seams.

FIGURE 9–1   A brazed lap joint (A) and a braze-welded lap joint (B).

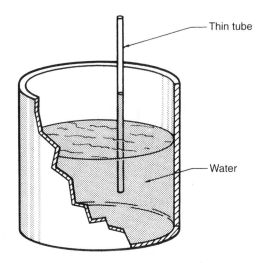

FIGURE 9–2   Capillary action pulls water into a thin tube.

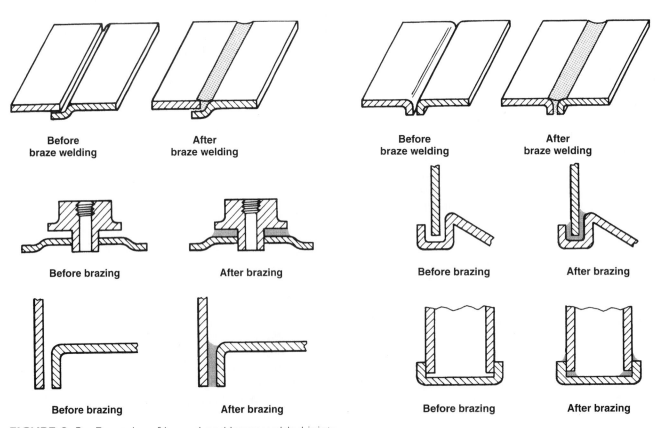

FIGURE 9–3   Examples of brazed and braze-welded joints.

- Because the base metal does not melt, it is possible to join otherwise incompatible metals, such as copper to steel or stainless steel to mild steel (Figure 9-4).
- The soldering and brazing technique takes relatively little time and skill to learn.
- Parts can be easily repositioned or even disassembled later by reheating the joint.
- Parts of varying thicknesses can be joined without burning or overheating.

9.2 BRAZING PRINCIPLES

More often than not, brazing uses brass filler metal. However, silver solder, aluminum, copper, and many other materials fit just as well into the category of brazing. Under the proper conditions, the filler metal is distributed between the properly fitted surfaces of the joint to be brazed. This process is very similar to that of soldering copper pipes with wire solder.

Although many brazed joints are as strong as welded steel, there is one notable exception—a butt joint on light-gauge metal. Because of its high incidence of failure, this type of joint should be avoided if at all possible. In repairing a brazed butt joint on a classic car, for example, the brass deposited on the base metal must be completely removed and rewelded with steel filler rod. Many inexperienced auto body technicians fail to do this when trying to repair old fenders with cracks caused by stress or vibrations. Use the lap joint technique whenever possible (Figure 9-5).

Automobile assembly plants use arc brazing to join the roof and quarter panels (Figure 9-6). Arc brazing uses essentially the same principles as MIG welding. The major difference is that argon is used as the shielding gas (Figure 9-7). Since the amount of heat applied to the base metal is low, the chances of overheating are greatly minimized, and there is little distortion or warpage of the base metal. Arc brazing shortens both the time for making the weld and the time for finishing.

In today's collision repair facility, the brazing equipment is usually the same equipment that is used for oxyacetylene welding—an oxyacetylene torch, filler rods, flux welding goggles, gloves, and a torch lighter.

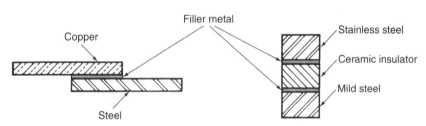

FIGURE 9–4 *Joining dissimilar metals by brazing.*

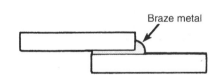

FIGURE 9–5 *Brazed lap joint.*

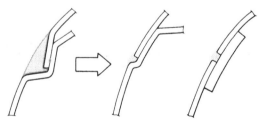

FIGURE 9–6 *Typical body construction using arc brazing.*

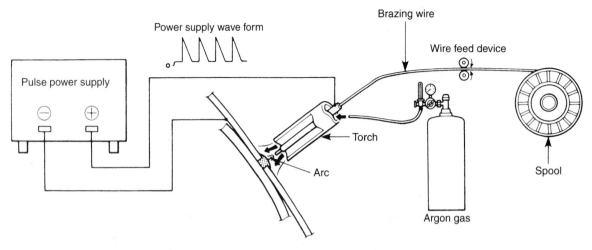

FIGURE 9–7 The arc brazing process.

SOLDERING AND BRAZING FILLER METALS

GENERAL

The type of filler metal used for any specific joint should be selected by considering as many as possible of the criteria listed in Figure 9-8. It would be impossible to consider each of these items to be equally important. Auto body technicians must decide which factors they consider most important and then base their selection on that decision.

Soldering and brazing metals are alloys, that is, a mixture of two or more metals. Each alloy is available in a variety of different percentage mixtures. Some mixtures are stronger, and some melt at lower temperatures than others. Each one has specific properties.

Some manufacturers of brass or bronze rods make brazing rods to specifications. Brass and bronze are available either plain or flux covered; common sizes range from 1/16 to 1/8 inch (1.5 mm to 3 mm) in diameter and larger and lengths of 18 or 36 inches (4.4 cm to 8.8 cm). Some technicians prefer to use a bare brazing rod. A flux-coated rod can produce too much molten flux in the brazing area that can interfere with the proper flow of the brass. Using a powdered flux is an even better way to control the amount of flux.

- Material being joined
- Strength required
- Joint design
- Availability and cost
- Appearance
- Service (corrosion)
- Heating process to be used
- Cost

FIGURE 9–8 Criteria for selecting filler metal.

BRAZING ALLOYS

The American Welding Society's classification system for brazing alloys uses the letter B to indicate that the alloy is to be used for brazing. The next series of letters in the classification indicates the atomic symbol of the metals used to make up the alloy such as CuZn (copper and zinc). There may be a dash followed by a letter or number to indicate a specific alloyed percentage. The letter R may be added to indicate that the braze metal is in rod form. An example of a filler metal designation is BRCuZn-A, which indicates a copper-zinc brazing rod with 59.25 percent copper, 40 percent zinc, and 0.75 percent tin. Table 9-1 is a list of the base metals and the most common alloys used to join the base metals. Not all of the available brazing

TABLE 9–1: BASE METALS AND COMMON BRAZING FILLER METALS USED TO JOIN THE BASE METALS

Base Metal	Brazing Filler Metal
Aluminum	BAlSi, aluminum silicon
Carbon Steel	BCuZn, brass (copper-zinc)
	BCu, copper alloy
	BAg, silver alloy
Alloy Steel	BAg, silver alloy
	BNi, nickel alloy
Stainless Steel	BAg, Silver alloy
	BAu, gold base alloy
	BNi, nickel alloy
Cast Iron	BCuZn, Brass (copper-zinc)
Galvanized Iron	BCuZn, Brass (copper-zinc)
Nickel	BAu, gold base alloy
	BAg, silver alloy
	BNi, nickel alloy
Nickel-Copper Alloy	BNi, nickel alloy
	BAg, silver alloy
	BCuZn, brass (copper-zinc)
Copper	BCuZn, Brass (copper-zinc)
	BAg, silver alloy
	BCuP, copper-phosphorous
Silicon Bronze	BCuZn, Brass (copper-zinc)
	BAg, silver alloy
	BCuP, copper-phosphorus
Tungsten	BCuP, copper-phosphorus

alloys have an AWS classification. Some special alloys are known by registered trade names.

Copper-Zinc

Copper-zinc alloys are the most popular brazing alloys. They are available as regular and low-fuming alloys. The zinc in this braze metal has a tendency to burn out if it is overheated. Overheating is indicated by a red glow on the molten pool that gives off a white smoke. The white smoke is zinc oxide. If zinc oxide is inhaled, it can cause zinc poisoning. Using a low-fuming alloy will help to eliminate this problem. Examples of low-fuming alloys are RCuZn-B and RCuZn-C.

CAUTION: Breathing zinc oxide can cause zinc poisoning. If you think you have zinc poisoning, get medical treatment immediately.

The following material describes the five major classifications of copper-zinc filler rods.

Class BRCuZn is used for the same application as BCu fillers. The addition of 40 percent zinc (Zn) and 60 percent copper (Cu) improves the corrosion resistance and aids in this rod's use with silicon-bronze, copper-nickel, and stainless steel.

CAUTION: Care must be exercised in order to prevent overheating this alloy as the zinc will vaporize, causing porosity and poisonous zinc fumes.

Class BRCuZn-A is commonly referred to as naval brass and can be used to fuse-weld naval brass. The addition of 17 percent tin (Sn) to the alloy adds strength and corrosion resistance. The same types of metal can be joined with this rod as can be joined with BRCuZn.

Class BRCuZn-B is a manganese-bronze filler metal. It has a relatively low melting point and is free flowing. This rod can be used to braze-weld steel, cast iron, brass, and bronze. The deposited metal is higher than BRCuZn or BRCuZn-A in strength, hardness, and corrosion resistance.

Class BRCuZn-C is a low-fuming, high silicon (Si) bronze rod. It is especially good for general-purpose work due to the low-fuming characteristic of the silicon on the zinc.

Class BRCuZn-D is a nickel-bronze rod with enough silicon to be low fuming. The nickel gives the deposit a silver-white appearance and is referred to as white brass. This rod is used to braze and braze-weld steel, malleable iron, cast iron, and to build up wear surfaces on bearings.

Copper-Phosphorus

This alloy is sometimes referred to as phos-copper. It is a good alloy to consider for joints where silver braze alloys may have been used in the past.

Copper-Phosphorus-Silver

This alloy is sometimes referred to as sil-phos. Its characteristics are similar to those of copper-phosphorus, except the silver gives this alloy a little better wetting and flow characteristics.

Silver-Copper

Silver-copper alloys can be used to join almost any metal, ferrous or nonferrous, except aluminum, magnesium, zinc, and a few other low-melting metals. This alloy is often referred to as silver braze and is the most versatile. It is among the most expensive alloys, except for the gold alloys.

ALUMINUM FILLER METALS

Cadmium-Zinc

Cadmium-zinc alloys have good wetting action and corrosion resistance on aluminum and aluminum alloys. The melting temperature is high, and some alloys have a wide paste range (Table 9-2).

Aluminum-Silicon (BAlSi)

These brazing filler metals can be used to join most aluminum sheet and cast alloys.

SOLDERING ALLOYS

Soldering alloys are usually identified by their major alloying elements. In many cases, a base material can be joined by more than one solder alloy. In addition to the considerations for selecting filler metal listed in Figure 9-8, specific factors are listed in the following sections for the major soldering alloys.

The soldering practices that follow will use tin-lead or tin-antimony solders. Both solders have low melting temperatures. If an oxyacetylene torch is used, it is very easy to overheat the solder. Caution is necessary, because most of the fluxes used with this type of solder are easily overheated. The best type of flame to use for this type of soldering is air acetylene, air MAPP®, air propane, or any air fuel-gas mixture. The most popular types are air acetylene or air propane. If galvanized metal is used, additional ventilation should be used to prevent zinc oxide poisoning.

Tin-Lead

This is the most popular solder and is the least expensive one. An alloy of 61.9 percent tin and 38.1 percent lead melts at 362° F (183° C) and has no paste range. (Paste range is the temperature range in which a metal is partly solid and partly liquid.) This is the eutectic composition (lowest possible melting point of an alloy) of the tin-lead solder. An alloy of 60 percent tin and 40 percent lead is commercially available and is close enough to the eutectic alloy to have the same low melting point with only a 12° F (7.8° C) paste range. The widest paste range is 173° F (78° C) for a mixture of 19.5 percent tin and 80.5 percent lead. This mixture begins to solidify at 535° F (289° C) and is totally solid at 362° F (193° C). The closest mixture that is commercially available is a 20 percent tin and 80 percent lead alloy. Table 9-3 lists the percentages, temperatures, and paste ranges for tin-lead solders. Tin-lead solders are most commonly used on electrical connections but must never be used for water piping. Most health and construction codes will not allow tin-lead solders for use on water or food-handling equipment.

CAUTION: Tin-lead solders must not be used where lead could become a health hazard in things such as food or water.

Tin-Antimony

This family of solder alloys has a higher tensile strength and lower creep than the tin-lead solders. The most common alloy is 95/5, 95 percent tin and 5 percent antimony. This is often referred to as "hard solder." This is the most common solder used in plumbing because it is lead-free. The use of C flux, which is a mixture of flux and small flakes of solder, makes it easy to fabricate quality joints. This mixture of flux and solder will draw additional solder into the joint as it is added.

TABLE 9–2:   CADMIUM-ZINC ALLOYS

Cadmium	Zinc	Completely Liquid	Completely Solid	Paste Range
82.5%	17.5%	509° F (265° C)	509° F (265°C)	No paste range
40.0%	60.0%	635° F (335° C)	509° F (265°C)	126° F (52° C)
10.0%	90.0%	750° F (399° C)	509° F (265°C)	241° F (116° C)

TABLE 9–3: MELTING, SOLIDIFICATION, AND PASTE RANGE TEMPERATURES FOR TIN-LEAD SOLDERS

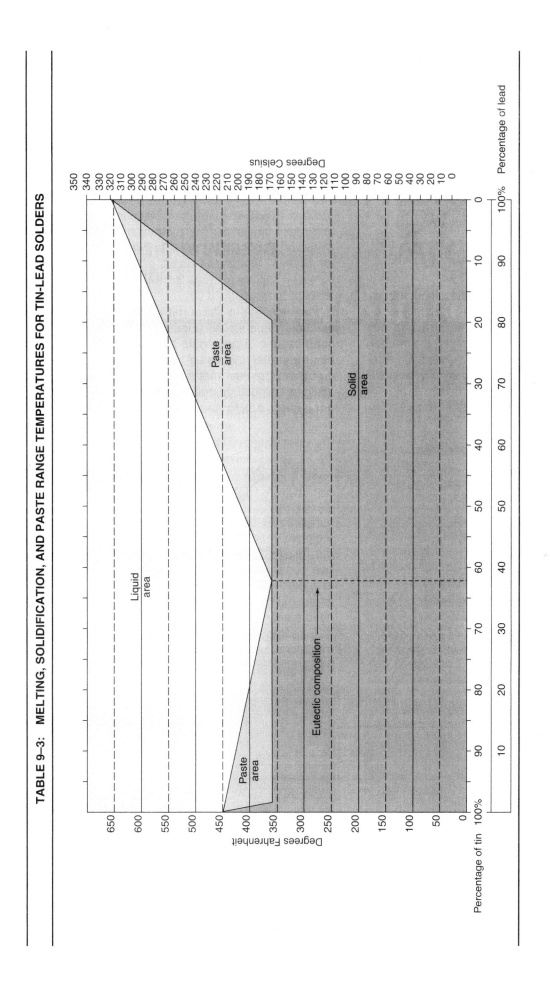

9.4 — FLUXES

GENERAL

Fluxes used in soldering and brazing have three major functions:

- They must remove any oxides that form as a result of heating the parts.
- They must promote wetting.
- They should aid in capillary action.

The flux, when heated to its reacting temperature, must be thin and flow through the gap provided at the joint. As it flows through the joint, the flux absorbs and dissolves oxides, allowing the molten filler metal to be pulled in behind it (Figure 9-9). After the joint is complete, the flux residue should be easily removable.

Fluxes are available in many forms, such as solids, powders, pastes, liquids, sheets, rings, and washers (Figure 9-10). They are also available mixed with the filler metal, inside the filler metal, or on the outside of the filler metal (Figure 9-11). The most common types of fluxes used in collision repair are pastes and powders. They may be brushed on the joint before or after the material is heated. Pastes and powders may also be applied to the end of the rod by heating the rod and dipping it in the flux. Most powders can be made into a paste, and some paste can be thinned by adding distilled water or alcohol. See manufacturers' specifications for details. If water is used, it should be distilled, because tap water may contain minerals that will weaken the flux.

Flux and filler metal combinations are most convenient and easy to use (Figure 9-12). These combinations are more expensive than buying the filler and flux separately. In cases where the flux covers the outside of the filler metal, it may be damaged by humidity or chipped off during storage.

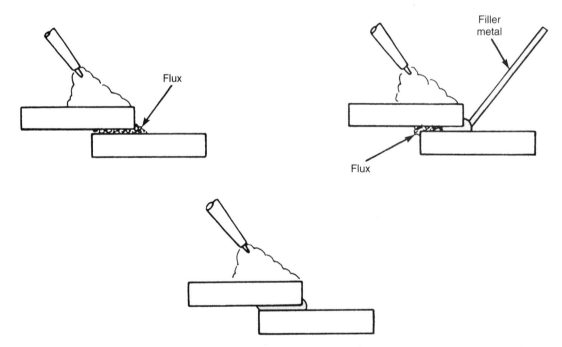

FIGURE 9–9 Flux flowing into a joint reduces oxides to clean the surfaces and gives rise to a capillary action that causes the filler metal to flow behind it.

FIGURE 9–10 Flux chips that can be preplaced in a braze/solder joint.

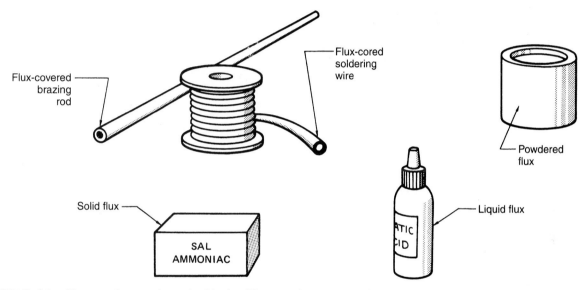

Flux-covered brazing rod

Flux-cored soldering wire

Powdered flux

Solid flux

SAL AMMONIAC

Liquid flux

FIGURE 9–11 Flux can be purchased with the filler metal or separately.

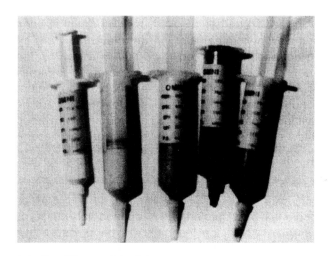

FIGURE 9–12 Tubes that contain flux/filler metal mixtures.

FLUXING ACTION

Soldering fluxes are chemical compounds such as muriatic acid (hydrochloric acid), sal ammoniac (ammonium chloride), or rosin. Brazing fluxes are chemical compounds such as fluorides, chlorides, boric acids, and alkalies. These compounds react to dissolve, absorb, or mechanically break up thin surface oxides that are formed as the parts are being heated. They must be stable and remain active through the entire temperature range of the solder or braze filler metal. The chemicals in the flux react with the oxides as either acids or bases. Some dip fluxes are salts.

The reactivity of a flux is greatly affected by temperature. As the parts are heated to the soldering or brazing temperature, the flux becomes more active. Some fluxes are completely inactive at room temperature. Most fluxes have a temperature range within which they are most effective. Care should be taken to avoid overheating fluxes. If they become overheated or burned, they will stop working as fluxes and become a contamination in the joint. If overheating has occurred, the auto body technician must stop and clean off the damaged flux before continuing. If the part is to be finished in any way, the flux must be completely removed.

CAUTION: Some fluxes give off a very toxic fume when heated. Use them in a well-ventilated place. When removing flux residue, wear safety glasses to prevent eye infection or other damage.

9.5 JOINT STRENGTH

The tensile strength of a joint is its ability to withstand being pulled apart (Figure 9-13). A soldered and brazed joint can be made that has a tensile strength four to five times higher than the filler metal itself. If a few drops of water are placed between two smooth and flat panes of glass and the panes are pressed together, a tensile load is required to pull the panes of glass apart. The water, which has no tensile strength itself, has added tensile strength to the glass joint.

The glass is being held together by the surface tension of the water. As the space between the pieces

FIGURE 9–13 Joint in tension.

TABLE 9–4: TENSILE STRENGTH OF BRAZED JOINT INCREASES AS JOINT SPACE DECREASES

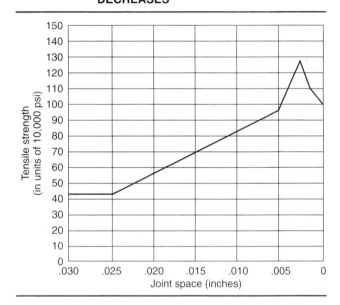

of glass decreases, the tensile strength increases. The same action takes place with a soldered or brazed joint. As the joint spacing decreases, the surface tension increases the tensile strength of the joint (Table 9-4).

Since the strength of the brazing material is less than that of the base metal, the shape and the clearance of the joint are extremely important. Figure 9-14 shows two basic brazing joints. Joint strength depends on the surface area of the pieces to be joined, so make the joint overlap as wide as possible.

Even when the items being joined are of the same material, the brazed surface area must be larger than that of a welded joint (Figure 9-15). As a general rule, the overlapping portion must be at least three times wider than the panel thickness.

The spacing between the parts being joined greatly affects the tensile strength of the finished part. Table 9-5 lists the spacing requirements at the joining temperature for the most common alloys.

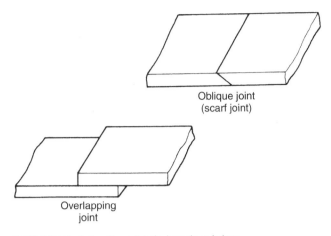

Oblique joint
(scarf joint)

Overlapping
joint

FIGURE 9–14 Two basic brazing joints.

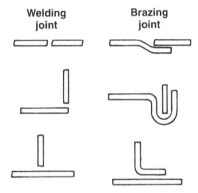

Welding joint   Brazing joint

FIGURE 9–15 Comparing welded and brazed joints.

TABLE 9–5:   BRAZING JOINT CLEARANCES

Filler Metal	Joint Spacing	
	Inches	Millimeters
BAlSi	0.006–0.025	0.15–0.61
BAg	0.002–0.005	0.05–0.12
BAu	0.002–0.005	0.05–0.12
BCuP	0.001–0.005	0.03–0.12
BCuZn	0.002–0.005	0.05–0.12
BNi	0.002–0.005	0.05–0.12

SOLDERING AND BRAZING GENERAL APPLICATIONS

9.6

Soldering and brazing methods are grouped according to the method with which heat is applied. In this text we cover only torch soldering and brazing.

Oxyfuel or air-fuel torches can be used either manually or automatically (Figure 9-16). Acetylene is often used as the fuel gas, but it is preferable to use one of the other fuel gases that has a higher heat level in the secondary flame (Figure 9-17). The oxyacetylene flame has a very high temperature near the inner cone, but it has little heat in the outer flame. This often results in the parts being overheated in a localized area. Such fuel gases as MAPP®, propane, butane, and natural gas have a flame that will heat parts more uniformly. Often torches are used that mix air with the fuel gas in a swirling or turbulent manner to increase the flame's temperature (Figure 9-18). The flame may even completely surround a small-diameter pipe, heating it from all sides at once (Figure 9-19).

Some advantages of using a torch include the following:

- *Versatility.* Using a torch is the most versatile method. Both small and large parts in a wide variety of materials can be joined with the same torch.
- *Portability.* A torch is very portable. Anyplace a set of cylinders can be taken or anywhere the hoses can be pulled into can be soldered or brazed with a torch.
- *Speed.* The flame of the torch is one of the quickest ways of heating the material to be joined, especially on thicker sections.

Some of the disadvantages of using a torch include the following:

- *Overheating.* When using a torch, it is easy to overheat or burn the parts, flux, or filler metal.
- *Skill.* A high level of skill with a torch is required to produce consistently good joints.
- *Fires.* It is easy to start a fire if a torch is used around combustible (flammable) materials.

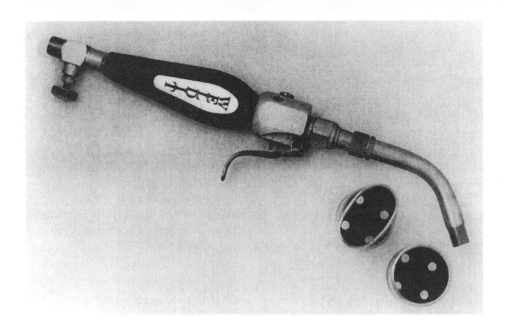

FIGURE 9–16 An air propane torch can be used in soldering joints. (Courtesy of National Torch Tip Co.)

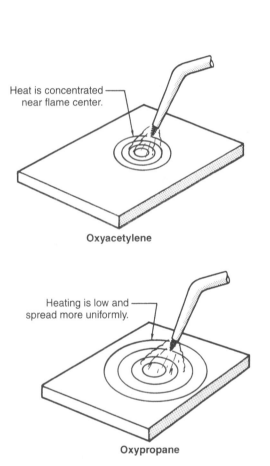

Heat is concentrated near flame center.

Oxyacetylene

Heating is low and spread more uniformly.

Oxypropane

FIGURE 9–17 The high temperature of an oxyacetylene flame may cause localized overheating.

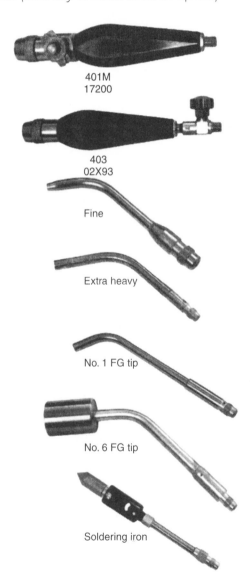

401M
17200

403
02X93

Fine

Extra heavy

No. 1 FG tip

No. 6 FG tip

Soldering iron

FIGURE 9–18 Examples of torch tips and handles that use air-fuel mixtures for brazing. (Courtesy of ESAB Welding and Cutting Products)

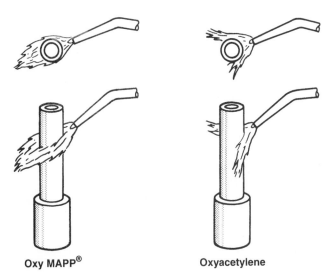

Oxy MAPP®   **Oxyacetylene**

FIGURE 9–19 Heating characteristics of oxy MAPP® compared with oxyacetylene on round materials.

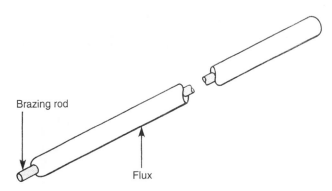

Brazing rod

Flux

FIGURE 9–20 Brazing rod with a flux coating.

As was mentioned earlier, joint preparation is very important to a successful soldered or brazed part. The surface must be free of all oil, dirt, paint, oxides, or any other contaminants. The surface can be either mechanically cleaned (wire brushed, sanded, ground, scraped, or filed) or chemically cleaned (with an acid, alkaline, or salt bath). Soldering or brazing should start as soon as possible after the parts are cleaned to prevent any additional contamination of the joint.

This discussion of the general brazing technique will be geared toward auto body work on 20- to 24 gauge sheet metal. It is wise to remember that the great majority of brazing jobs will not take place under ideal conditions. While it might be very tempting not to bother stripping the surface, the time and effort should be taken to ensure a high-quality brazing job.

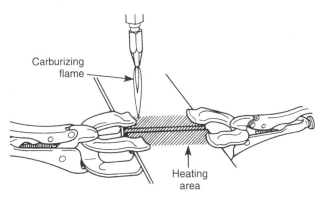

Carburizing flame

Heating area

FIGURE 9–21 Base metal heating.

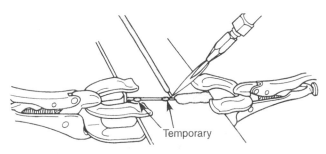

Temporary

FIGURE 9–22 Base metal brazing operation.

The general brazing procedure is as follows:

1. Clean the surface thoroughly with a wire brush, grinder, or scraper.
2. Apply flux uniformly to the brazing surface. If a brazing rod with flux is being used (Figure 9-20), this step can be omitted.
3. Adjust the flame of the welding torch to a slight carburizing flame. Use the melting of the flux to estimate the proper temperature of the brazing material.
4. Heat the base metal to a uniform temperature capable of accepting the brazing material (Figure 9-21).
5. When the base metal has reached the proper temperature, melt the brazing material onto the base metal (Figure 9-22), letting it flow naturally.
6. Stop heating when the brazing material has flowed into the gaps of the base metal. If the surface temperature of the base metal is allowed to get too high, the flux will not clean it properly, and the result will be a poor brazing bond and inferior joint strength.

The following additional precautions should be taken when brazing:

- The brazing temperature must be higher than the melting point of the filler metal but not too hot.
- Preheat the panel until the flux becomes active. This will promote more effective depositing of the brazing filler.
- Secure the panel to prevent the base metal from moving and the joint breaking loose.
- Heat the portion to be welded evenly without melting the base metal.
- Make sure the torch angle is tilted in the direction of the joint being made.
- Brazing time must be as short as possible to prevent the joint strength from being reduced.
- Avoid brazing the same spot twice.

Once the brazed portion has cooled sufficiently, rinse off the remaining flux residue with water and scrub the surface with a stiff wire brush. Baked-on flux can be removed with a sander. If all of the flux is not removed, the paint will not adhere properly, and corrosion and cracks might form in the joint.

PRACTICE 9-1

Brazed Stringer Bead

Using a properly lit and adjusted torch, 6 inches (152 mm) of clean 16 gauge mild steel, brazing flux, and BRCuZn brazing rod, you will make a straight bead the length of the sheet.

Place the sheet flat on a firebrick and hold the flame at one end until the metal reaches the proper temperature. Then touch the flux-covered rod to the sheet and allow a small amount of brazing rod to melt onto the hot sheet (Figure 9-23). Once the molten brazing metal wets the sheet, start moving the torch in

a circular pattern while dipping the rod into the molten braze pool as you move along the sheet. If the size of the molten pool increases, you can control it by reducing the torch angle, raising the torch, traveling at a faster rate, or flashing the flame off the molten braze pool (Figure 9-24). Flashing the torch off a braze joint

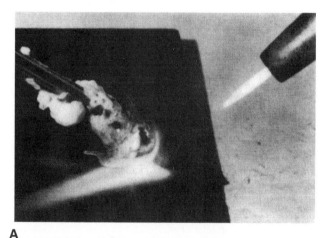

A

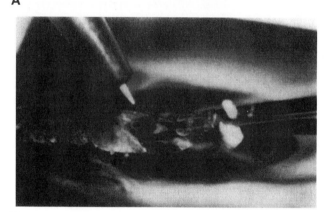

B

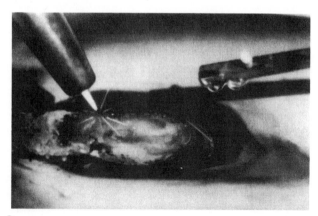

C

FIGURE 9-24 (A) Once the plate is up to temperature, start adding more filler. (B) Dip the brazing rod into the leading edge of the molten weld pool. (C) Remove the rod from the flame area when it is not being added to the molten weld pool.

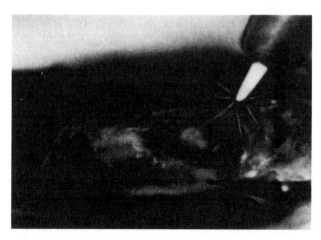

FIGURE 9-23 Checking the surface temperature with a spot of braze metal.

will not cause oxidation problems as it does when welding, because the molten metal is protected by a layer of flux.

As the braze bead progresses across the sheet, dip the end of the rod back in the flux, if a powdered flux is used, as often as needed to keep a small molten pool of flux ahead of the bead (Figure 9-25).

The object of this practice is to learn how to control the size and direction of the braze bead. Controlling the width, buildup, and shape shows that you have a good understanding and control of the process. Keeping the braze bead in a straight line indicates that you have mastered the bead well enough to watch the

bead and the direction at the same time. Turn off the cylinders, bleed the hoses, back out the regulator adjusting screw, and clean up your work area when you are finished.

Brazed Butt Joint

Using the same equipment and setup as listed in Practice 9-1, make a braze butt joint on two pieces of 16 gauge mild steel sheet, 6 inches (152 mm) long.

Place the metal flat on a firebrick, hold the plates tightly together, and make a tack braze at both ends of the joint. If the plates become distorted, they can be bent back into shape with a hammer before making another tack weld in the center. Align the sheets so that you can comfortably make a braze bead along the joint. Starting as you did in Practice 9-1, make a uniform braze along the joint. Repeat this practice until a uniform braze can be made without defects. Turn off the cylinders, bleed the hoses, back out the regulator adjusting screw, and clean up your work area when you are finished.

Brazed Butt Joint with 100 Percent Penetration

Using the same equipment, material, and setup as described in Practice 9-2, make a brazed butt joint with 100 percent penetration (Figure 9-26). To ensure

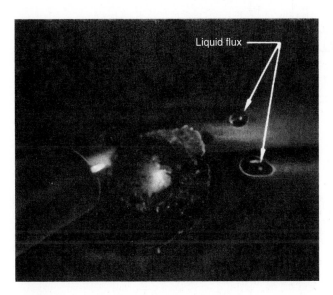

FIGURE 9–25 Observe the molten flux flowing ahead of the molten weld pool.

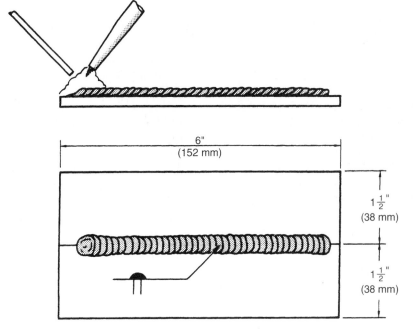

FIGURE 9–26 Brazed butt joint with 100 percent penetration.

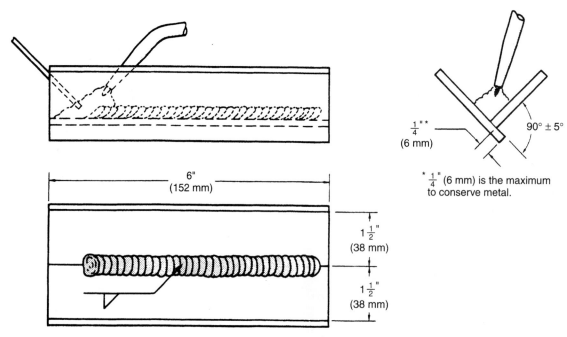

$\frac{1}{4}$" * (6 mm)

90° ± 5°

* $\frac{1}{4}$" (6 mm) is the maximum to conserve metal.

6" (152 mm)

$1\frac{1}{2}$" (38 mm)

$1\frac{1}{2}$" (38 mm)

FIGURE 9–27 *Brazed tee joint.*

that 100% penetration is obtained, a little additional heat is required to flow the braze metal through the joint. Apply the additional heat just ahead of the bead on the base sheets. After the braze is completed, turn the plate over and look for a small amount of braze showing along the entire joint. Repeat this practice until it can be made without defects. Turn off the cylinders, bleed the hoses, back out the regulator adjusting screw, and clean up your work area when you are finished.

PRACTICE 9–4

Brazed Tee Joint

Using the same equipment, material, and setup as listed in Practice 9-2, you will make a brazed tee joint with 100 percent root penetration (Figure 9-27).

Tack the pieces of metal into a tee joint. To obtain 100 percent penetration, direct the flame on the sheets just ahead of the braze bead, being careful not to overheat the braze metal. If the bead has a notch, the root of the joint is still not hot enough to allow the braze metal to flow properly. If the braze metal appears to have flowed properly after you have

completed the joint, look at the back of the joint for a line of braze metal that flowed through. Repeat this practice until the joint can be made without defects. Turn off the cylinders, bleed the hoses, back out the regulator adjusting screw, and clean up your work area when you are finished.

PRACTICE 9–5

Brazed Lap Joint

Using the same equipment, material, and setup as listed in Practice 9-2, you will make a brazed lap joint in the flat position.

Place the pieces of sheet metal on a firebrick so that they overlap each other by approximately 1/2 inch (13 mm). It is important that the pieces be held flat relative to each other (Figure 9-28). Make a small tack braze on both ends and then one or two tack brazes along the joint. Hold the torch so the flame moves along the joint and heats up both pieces at the same time. When the sheets are hot, touch the rod to the sheets and make a bead similar to the butt joint. After completing the brazed joint, it should be uniform in width and appearance. Repeat this practice

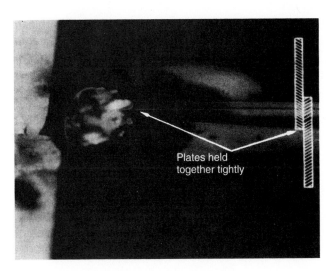

FIGURE 9–28 Tack brazing a lap joint.

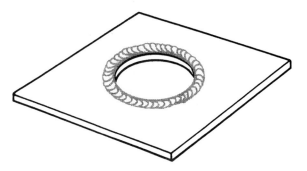

FIGURE 9–29 Filling a hole with braze. First run a bead around the outside of the hole.

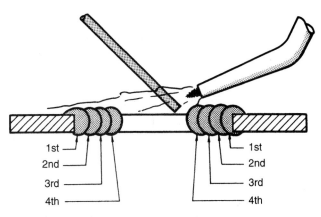

FIGURE 9–30 Keep running beads around the hole until it is closed.

until the joint can be made without defects. Turn off the cylinders, bleed the hoses, back out the regulator adjusting screw, and clean up your work area when you are finished.

PRACTICE 9–6

Braze Welding To Fill a Hole

Using a properly lit and adjusted torch, one piece of 16 gauge mild steel, flux, and BRCuZn filler rod, you will fill a 1-inch (25-mm) hole. Place the piece of metal on two firebricks so that the hole is between them. Start by running a stringer bead around the hole (Figure 9-29). Once the bead is complete, turn the torch at a very sharp angle and point it at the edge of the hole nearest the torch. Hold the end of the filler rod in the flame so that both the bead around the hole and the rod meet at the same time (Figure 9-30). Put the rod in the molten bead and flash the torch off to allow the molten braze pool to cool. When it has cooled, repeat this process. Surface tension will hold a small piece of molten metal in place. If the piece of molten metal becomes too large, it will drop through. Progress around the hole as many times as needed to fill the hole. When the braze weld is complete, it should be fairly flat with the surrounding metal. Repeat this practice until it can be made without defects. Turn off the cylinders, bleed the hoses, back out the regulator adjusting screw, and clean up your work area when you are finished.

9.7 SOLDERING PROCEDURES

The metals to be soldered must be thoroughly cleaned. A chemical cleaner is good to use, provided it is rinsed and dried off completely. The metals can also be cleaned by filing or wire brushing; steel wool also works well. After the surface has been cleaned, proceed as follows:

1. Heat the portion to be soldered. Wipe it with a cloth after heating.
2. Stir the solder paste well. Apply it with a brush to an area 1 to 1-1/2 inches (25 mm to 37 mm) larger than the build-up area.
3. Heat it from a distance.
4. Wipe the solder paste from the center to the outside.

5. Make sure the soldered portion is silver gray. If it is bluish, it has overheated.

6. If any spot is not adequately soldered, reapply the paste for additional soldering.

When soldering, keep the following additional points in mind:

- It is wise to use a special torch for soldering. If a gas welding torch is used, the oxygen and acetylene gas pressures must be 4.3 to 5.0 psi.
- To maintain the appropriate temperature, move the torch so that the flame evenly heats the entire portion to be soldered. When the solder begins to melt, remove the flame and finish with a spatula.
- When additional solder is required, the previously built-up solder must be reheated.

If either piece of metal moves while it is cooling from its flow temperature to its solidification temperature, the solder will form cracks and will probably fail. It is therefore necessary to firmly support the base metals with a fixture, clamps, or the like to make sure they do not move while the soldering operation is in progress.

A common soldering error is to use too much solder; this is wasteful because it does not add any strength to the joint. If too much solder is used, a neat-looking joint can still be achieved by wiping off the excess molten solder using a clean, thick cloth. The solder should be wiped away while it is at its flowing temperature rather than at its melting temperature. It is very important to avoid overheating the metal.

A popular use for soldering is to build up irregular surfaces to secure a smooth finish. This method is used extensively in collision repair work. The irregular surface is mechanically cleaned, then chemically cleaned with a weak acid. A wooden paddle is sometimes used to apply the solder to the torch-heated surfaces. Figure 9-31 shows a joint being filled in by soldering. The solder is then dressed by filing and sanding to match the sheet metal surface.

Soldering copper produces a very concentrated heat while at the same time acting as a means of spreading the solder as it adheres to the base metals. In some parts of the United States, it is known as soldering iron. It has the disadvantage of requiring reheating quite frequently. However, electrically heated soldering coppers and internal flame-heated coppers

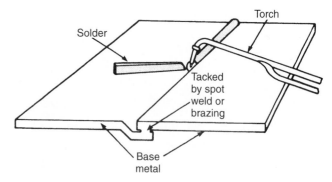

FIGURE 9-31 *Filling a joint using the soldering method.*

do not have this problem. The tip of the soldering copper must be kept clean at all times.

One of the best applications of soldering copper is to use it for sweating a soldered joint. This means that the two metals to be soldered together are lapped at their joint with a previously applied film of solder on the two contacting surfaces. (This application of film is known as *tinning.*) These edges are then lapped, and the copper is slowly moved along the seam, permitting the heat from the soldering copper to penetrate through the metal and fuse the solder films together, as shown in Figure 9-32. The resulting joint is strong and neat. This method is recommended when a high-quality, leakproof joint is desired and for difficult-to-solder joints.

Almost all of the alloys used for soldering or brazing have a paste range. A paste range is the temperature range in which a metal is partly solid and partly liquid as it is heated or cooled. As the joined part cools through the paste range, it is important that the part not be moved. If the part is moved, the solder or braze metal may crumble like dry clay, destroying the bond.

PRACTICE 9-7

Paste Range

This experiment shows the effect on bonding of moving a part as the filler metal cools through its paste range. The experiment also shows how metal

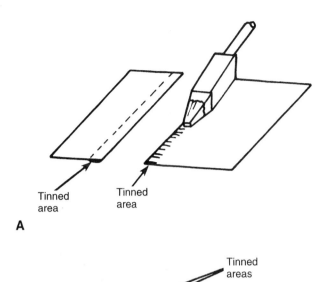

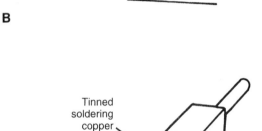

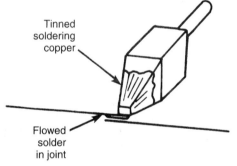

FIGURE 9-32 *Steps required to make a sweated soldering joint: (A) Solder is applied in a thin film: (B) metal surfaces are lapped to form the joint: (C) soldering copper is moved along the joint to flow the solder on the previously tinned surfaces.*

can be "worked" using its paste range. You will need tin-lead solder composed of 20 to 50 percent tin with the remaining percentage being lead. You also will need a properly lit and adjusted torch, a short piece of brazing rod, and a piece of sheet metal. Using a hammer, make a dent in the sheet metal about the size of a quarter (Figure 9-33).

Using the torch, melt a small amount of the solder into the dent and allow it to harden. Remelt the solder slowly, frequently flashing the torch off and touching the solder with the brazing rod until it is evident the solder has all melted. Once it has melted, stick the

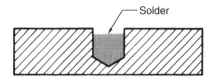

FIGURE 9-33 *Partially fill the drilled hole with solder.*

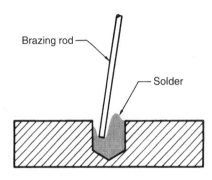

FIGURE 9-34 *Solder being shaped as it cools to its paste range.*

brazing rod in the solder and remove the torch. As the solder cools, move the brazing rod in the metal and observe what happens (Figure 9-34).

As the solder cools to the uppermost temperature of its paste range, it will have a rough surface appearance as the rod is moved. When the solder cools more, it will start to break up around the rod. Finally, as it becomes a solid, it will be completely broken away from the rod.

Now slowly reheat the solder and work the surface with the rod until it can be shaped like clay. If the surface is slightly rough, a quick touch of the flame will smooth it. This is the same way in which "lead" is applied to some body panel joints on a new car so that the joints are not seen on the car when it is finished. The lead used is actually a tin-lead alloy or solder. A large area can be made as smooth as glass without sanding by simply flashing the area with the flame.

PRACTICE 9-8

Soldered Tee Joint

Using a properly lit and adjusted torch, two pieces of 18 to 24 gauge mild steel sheet, 6 inches long (152 mm) flux, and tin-lead or tin-antimony solder wire, you will solder a flat tee joint.

Hold one piece of metal vertical on the other piece and spot-solder both ends. If flux-cored wire is not being used, paint the flux on the joint at this time. Hold the torch flame so it moves down the joint in the same direction you will be soldering. Continue flashing the torch off and touching the solder wire to the joint until the solder begins to melt. Keeping the molten pool small enough to work with is a major problem with soldering. The flame must be flashed off frequently to prevent overheating. When the joint is completed, the solder should be uniform. Repeat this practice until it can be made without defects. Turn off the cylinders, bleed the hoses, back out the regulator adjusting screw, and clean up your work area when you are finished.

PRACTICE 9–9

Soldered Lap Joint

Using the same equipment, materials, and setup as listed in Practice 9-8, you will solder a lap joint in the horizontal position.

Tack the pieces of metal together as shown in Figure 9-35. Apply the flux and heat the metal slowly, checking the temperature by touching the solder wire to the metal often. When the work gets hot enough, flash the flame off frequently to prevent overheating and proceed along the joint. When the joint is completed, the solder should be uniform. Repeat this practice until it can be made without defects. Turn off the cylinders, bleed the hoses, back out the regulator adjusting screw, and clean up your work area when you are finished.

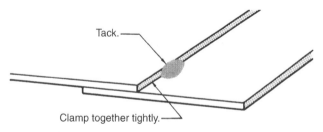

Tack.

Clamp together tightly.

FIGURE 9–35 *Tacking metal together for soldering.*

9.8 — ALUMINUM

Aluminum can be soldered or brazed with an oxyacetylene torch in a way that is very similar to iron. It shares with iron a great affinity for oxygen. Once aluminum is cleaned, it immediately reacts with oxygen to produce a thin, glass-like film of aluminum oxide on the surface. Immediately before brazing aluminum, the oxides must be removed by scraping, wire brushing, sandpapering, or other means. The oxides can also be removed chemically by the use of a flux.

The AWS type number 1 flux must be used when brazing aluminum. It is very easy to overheat the joint. If the flux is burned by overheating, it will obstruct wetting. Use standard torch brazing practices but guard against overheating.

An important factor that makes aluminum different from steel is its high heat conductivity. Aluminum conducts heat almost five times faster than steel or iron, depending on the type of alloy. The result is that aluminum requires more heat to achieve its proper brazing temperature.

For a good look at how aluminum reacts to the heating process, take a piece of steel sheet metal and a piece of aluminum of the same size and thickness and heat one corner of both pieces for a few seconds. On the piece of sheet metal, the heat will be concentrated at the corner, while the heat will travel across the aluminum very rapidly. Another key difference is the color change. The steel will gradually turn redder, then a bright red until it melts. The aluminum will not change color until the melting point is reached, and then it will simply fall away. A good method to determine the proper temperature at which to braze aluminum is to use a special temperature-indicating crayon that will melt or become liquid at a certain temperature.

When brazing aluminum, use a flux to obtain the proper adhesion. Apply the flux to the surface or, if necessary, to the brazing rod. Applying the flux directly to the rod is especially useful if the rod has been in the shop for a long time. There are many types of aluminum brazing rods available, the most common ones being the 1100 and the 4043. Some manufacturers produce a rod with a low melting point that can be used for general aluminum repairs.

When applied to the surface of the repair area, the flux will dry out as the temperature rises, then turn to a glossy liquid. Rub or scrape the aluminum brazing rod in the heated area at 3- to 5-second intervals, removing the rod from the flame each time. When the right temperature is reached, the rod will flow. Keep the torch moving on and off the surface to avoid overheating. Do not melt the rod with the torch before the surface is hot enough. The torch should be held at a flatter angle than when working with steel.

The hardest part of brazing aluminum is its tendency to melt away the edges. By moving the torch continually, there is less chance of this happening. Aluminum is more difficult to braze than brass, but in time it can be mastered.

Cleaning the area after brazing is very important, because the flux is corrosive. If left on the aluminum, the flux will cause oxidation or corrosion. Also, when the component is painted, any remaining flux residue will cause the paint to peel. There are several different chemicals used by manufacturers to remove flux, but boiling hot water and a scrubbing brush will remove most of it. Finally, rinse the aluminum with cold water.

CAUTION: Aluminum flux might give off a toxic fume when heated. Always work in a well-ventilated area. If you are using powdered flux, be very careful to avoid getting it in your eyes.

PRACTICE 9–10

Soldering Aluminum

Using a properly lit and adjusted torch, a piece of aluminum plate, steel wool, flux, and tin-lead or tin-antimony solder wire, you will tin both the pieces of aluminum with solder and then join both together.

The surface of the aluminum must be clean and free of paint, oils, dirt, and coatings such as anodizes. Hold the flame on the aluminum until it warms up slightly. Hold the solder in the flame and allow a small amount to melt and drop on the aluminum plate, but don't add flux (Figure 9-36). Move the flame off the plate and rub the liquid solder with the steel wool. Be careful not to burn your fingers or allow the flame to touch the steel wool. The solder should be stuck in the steel wool when it is lifted off the plate. Alternately heat the plate and rub it with the steel wool solder. When the plate becomes hot enough, it will melt the solder, and the solder will tin the aluminum surface.

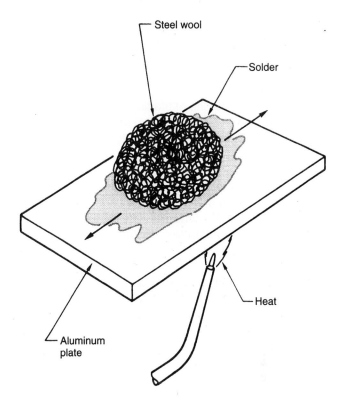

FIGURE 9–36 *Tinning aluminum with solder.*

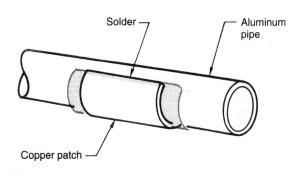

FIGURE 9–37 *Soldering a copper patch on aluminum tubing.*

Once both aluminum plates have been tinned, place them so that they are touching each other. Heat the two until the solder melts and flows out from between the two plates. When the parts cool, to check the bond, try to break the joint apart.

This process will work on other types of metals that have a strong oxide layer that prevents the solder from bonding. By breaking the oxide layer free with the mechanical action of the steel wool, the metals can join. This process can also be used to make a copper patch over an aluminum tube, such as those used in air-conditioning (Figure 9-37). Turn off the cylinders, bleed the hoses, back out the regulator adjusting screw, and clean up your work area when you are finished.

REVIEW QUESTIONS

1. What is the difference between brazing and soldering?
2. List the advantages of brazing and soldering.
3. List the most common filler metals for brazing.
4. What is one notable exception to the principle that many brazed joints are as strong as welded steel?
5. What is the major difference between arc brazing and MIG welding?
6. Explain the AWS classifications for brazing alloys.
7. What is the warning sign that a copper-zinc alloy is being overheated?
8. What is the white smoke given off when a copper-zinc alloy is being overheated?
9. Describe the five major classifications of copper-zinc filler rods.
10. What is the most versatile brazing alloy?
11. What are the most popular types of flames to use when soldering tin-lead or tin-antimony?
12. What shouldn't tin-lead solders be used for?
13. What solder is referred to as "hard solder"?
14. What are the three major reasons that fluxes are used in soldering and brazing?
15. What soldering fluxes are used to react to dissolve, absorb, or mechanically break up thin surface oxides that are formed as the parts are being heated?
16. What is tensile strength?
17. Joint strength is dependent on what?
18. What is the general rule about the overlapping portion of brazed joints?
19. List the advantages and disadvantages of using a torch.
20. What important thermal factors make aluminum different than steel?

Chapter

10

Plasma Arc and Flame Cutting

Objectives

After completing this chapter, you should be able to:

- Describe a plasma torch or describe plasma.
- Explain how a plasma cutting torch works.
- List the advantages and disadvantages of using a plasma cutting torch.
- Demonstrate an ability to set up and use a plasma cutting torch.
- Explain how the flame-cutting process works.
- Demonstrate how to properly set up and use an oxyfuel gas cutting torch.
- Safely use an oxyfuel gas cutting torch to make a variety of cuts.

The plasma process was originally developed in the mid-1950s as an attempt to create an arc, using argon, that would be as hot as the arc created when using helium gas. The early gas tungsten arc welding process used helium gas and was called heliarc. This early GTA welding process worked well with helium, but helium was expensive. The gas manufacturing companies had argon as a by-product from the production of oxygen. There was no good commercial market for this waste argon gas, but gas manufacturers believed there would be a good market if they could find a way to make argon weld similar to helium.

Early experiments found that by restricting the arc in a fast-flowing column of argon, a plasma could be formed. The plasma was hot enough to rapidly melt any metal. The problem was that the fast-moving gas blew the molten metal away. Experimenters could not find a way to control this scattering of the molten metal, so they decided to introduce this as a cutting process, not a welding process (see Figure 10-1).

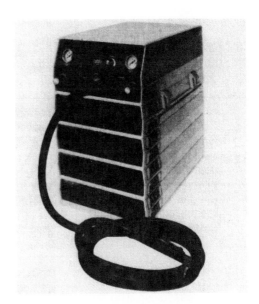

FIGURE 10–1 *Plasma arc cutting machine. This unit can have additional power modules added to the base of its control module to give it more power. (Courtesy of Thermal Dynamics)*

Several years later, with the invention of the gas lens, plasma was successfully used for welding. Today the plasma arc can be used for plasma arc welding (PAW), plasma spraying (PSP), plasma arc cutting (PAC), and plasma arc gouging. Plasma arc cutting is the most often used plasma process.

PLASMA ARC CUTTING

As was mentioned earlier, plasma arc cutting is replacing oxyacetylene cutting as the best way to cut modern car metals. It cuts mangled metal effectively and quickly, but it will not destroy the properties of the base metal. This is important to today's metal shop, which sees more unibody cars with high-strength steel or high-strength alloy steel components. The old method of flame cutting just does not work that well anymore. Plasma's high heat, fast travel speed, and low heat-input qualities, coupled with the fact that plasma cutting will cut rusted, painted, or coated metal with little difficulty, make it an ideal process for the auto body repair field.

Basically, plasma cutting is an extension of the TIG process. It utilizes an electrically ionized gas that is capable of rapidly transferring its heat to the part being cut. To accomplish this, the electric current must pass through compressed air or an inert gas, usually nitrogen because it is a reasonably priced gas. The nitrogen gas (the atmosphere contains 78 percent nitrogen) is ionized, shaped, and then forced through an orifice. The nitrogen or compressed air is superheated by an electrode inside the torch and expands tremendously. The gas itself comes out of the torch at a temperature of up to 2,000° F (1,090° C). The gas or air, because it is so hot and moves so fast, heats up the metal and blows it away.

There are two circuits in plasma cutting (Figure 10-2). The pilot arc starts the ionization process: high frequency transfers from the electrode, which is negative, to the torch, which is positive. When the torch is brought down to the work, the ionized gas or air flows to the grounded workpiece, starting the cutting action.

The secondary gas (or shielded gas) is also drawn from the compressed air. The shielding gas performs exactly the same function as in MIG welding. It protects the cut from outside contaminants while the cutting is in progress.

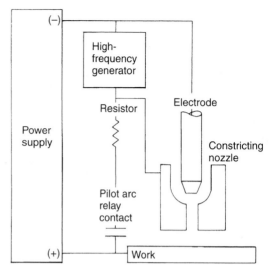

FIGURE 10–2   *Basic plasma arc cutting circuitry.*

CAUTION: Because of the presence of 78 percent nitrogen in the atmosphere, it is very important to maintain the volume and psi ratings recommended by the machine's manufacturer. Also keep in mind that moisture in the air will affect the quality of the cut somewhat.

PLASMA

The word *plasma* has two meanings: it is the fluid portion of blood, and it is a state of matter that is found in the region of an electrical discharge (arc). The plasma created by an arc is an ionized gas that has both electrons and positive ions whose charges are nearly equal to each other. For welding we use the electrical definition of plasma.

A plasma is present in any electrical discharge. A plasma consists of charged particles that conduct the electrons across the gap. Both the glow of a neon tube and the bright fluorescent lightbulb are examples of low-temperature plasmas.

A plasma results when a gas is heated to a high enough temperature to convert into positive and negative ions, neutral atoms, and negative electrons. The temperature of an unrestricted arc is about 11,000° F (6,000° C), but the temperature created when the arc is concentrated to form a plasma is about 43,000° F (24,000° C) (Figure 10-3). This is hot enough to rapidly melt any metal it comes into contact with.

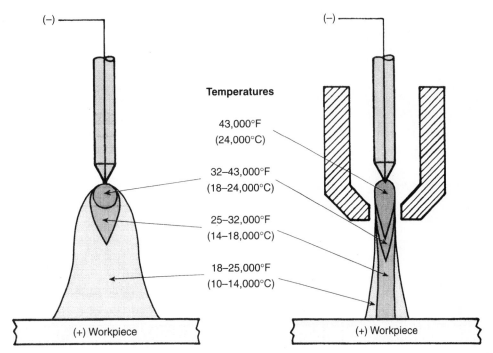

Temperatures

43,000°F
(24,000°C)

32–43,000°F
(18–24,000°C)

25–32,000°F
(14–18,000°C)

18–25,000°F
(10–14,000°C)

(+) Workpiece

(+) Workpiece

FIGURE 10–3 _Approximate temperature differences between a standard arc and a plasma arc. (Courtesy of the American Welding Society)_

ARC PLASMA

The term _arc plasma_ is defined as gas that has been heated to at least a partially ionized condition, enabling it to conduct an electric current (ANSI/AWS A3.0-89 An American National Standard, Standard Welding Terms and Definitions). The term _plasma arc_ is the term most often used in the welding industry when referring to the arc plasma used in welding and cutting processes. The plasma arc produces both the high temperature and the intense light associated with all forms of arc welding and cutting processes.

PLASMA TORCH

The plasma torch is a device, depending on its design, which allows the creation and control of the plasma for welding or cutting processes. The plasma is created in both the cutting and welding torches in the same basic manner, and both torches have the same basic parts. A plasma torch supplies electrical energy to a gas to change it into the high-energy state of a plasma.

Torch Body

The torch body, on a manual-type torch, is made of a special plastic that is resistant to high temperatures, ultraviolet light, and impact. The torch body provides a good grip area and protects the cable and hose connections to the head. The torch body is available in a variety of lengths and sizes. Generally the longer, larger torches are used for the higher-capacity machines; however, sometimes you might want a longer or larger torch simply to give yourself better control or a longer reach. On machine torches the body is often called a barrel and may come equipped with a rack attached to its side. The rack is a flat gear that allows the torch to be raised and lowered manually to the correct height above the work (Figure 10-4).

Torch Head

The torch head is attached to the torch body where the cables and hoses attach to the electrode tip, nozzle tip, and nozzle. The torch head may be connected at any angle—such as 90 degrees, 75 degrees,

FIGURE 10–4 *Machine plasma torch. (Courtesy of Cerametals Inc.)*

or 180 degrees (straight)—or it may be flexible. The 75- and 90-degree angles are popular for manual operations, and the 180 degree straight torch heads are most often used for machine operations. Because of the heat in the head produced by the arc, some provisions for cooling the head and its internal parts must be made. This cooling for low-power torches may be either by air or by water. Higher-power torches must be liquid cooled. The torch head on most torches can be replaced if it becomes worn or damaged.

Power Switch

Most handheld torches have a manual power switch, which is used to start and stop the power source, gas, and cooling water (if used). The switch most often used is a thumb switch located on the torch body, but it may be a foot control or a switch on the panel for machine-type equipment. The thumb switch may be molded into the torch body or it may be attached to the torch body with a strap clamp. The foot control must be rugged enough to withstand the welding shop environment. Some equipment has an automatic system that starts the plasma when the torch is brought close to the work.

Common Torch Parts

The electrode tip, nozzle insulator, nozzle tip, nozzle guide, and nozzle are the parts of the torch that must be replaced periodically as they wear out or become damaged from use (Figure 10-5).

CAUTION: Improper use of the torch or assembly of torch parts may result in damage to the torch body as well as the frequent replacement of these parts.

The metal parts are usually made out of copper, and they may be plated. The plating of copper parts will help them stay spatter-free longer.

Electrode Tip

The electrode is often made of copper with a tungsten tip attached. The use of a copper/tungsten tip in the newer torches has improved the quality of work they can produce. With the use of copper, the heat generated at the tip can be conducted away faster. Keeping the tip as cool as possible lengthens the life of the tip and allows for better-quality cuts for a longer time. The newer torches are a major improvement over earlier torches. Some older torches require you to accurately grind the tungsten electrode into shape. If you are using a torch that requires the grinding of the electrode tip, you must have a guide to insure that the tungsten is properly prepared.

Nozzle Insulator

The nozzle insulator is between the electrode tip and the nozzle tip. The nozzle insulator provides the critical gap spacing and the electrical separation of the parts. The spacing between the electrode tip and the nozzle tip, called electrode setback, is critical to the proper operation of the system.

Nozzle Tip

The nozzle tip has a small, cone-shaped, constricting orifice in the center. The electrode setback space, between the electrode tip and the nozzle tip, is where the electric current forms the plasma. The preset, close-fitting parts provide the restriction of the gas in the presence of the electric current so the plasma can be generated. Nozzle tips come in a variety of types (Figure 10-6). The diameter of the constricting orifice and the electrode setback are major factors in the

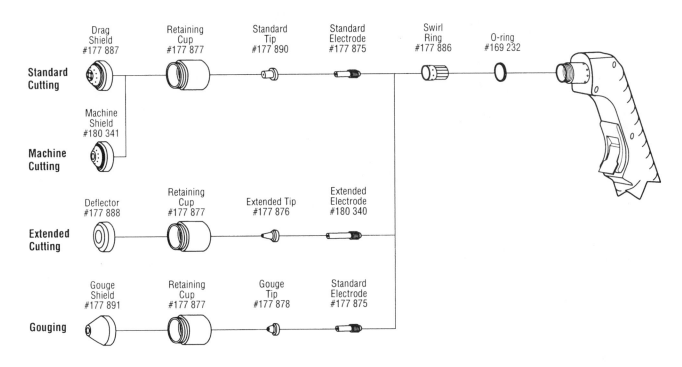

Standard Cutting

Machine Cutting

Extended Cutting

Gouging

FIGURE 10–5   Replaceable torch parts. (Courtesy of Miller Electric)

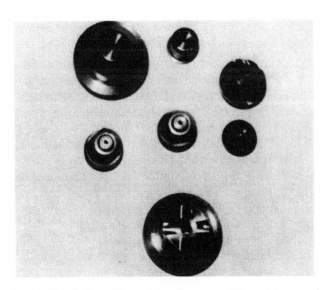

FIGURE 10–6   Different torches use different types of nozzle tips.

operation of the torch. As the diameter of the orifice changes, the plasma jet action will be affected. When the setback distance is changed, the arc voltage and current flow will change.

Nozzle

The nozzle, sometimes called the cup, is made of ceramic or any other high temperature–resistant substance. This helps to prevent the internal electrical parts from accidental shorting and provides control over the shielding gas or water injection if they are used (Figure 10-7).

Torch Guard

A torch guard can be attached to the nozzle both to keep the electrode tip from accidently shorting to the work and as a way of maintaining the proper distance between the work and the torch. The torch guard is allowed to rest on the metal surface and slide along as a cut is made.

FIGURE 10–7 *Nozzles are available in a variety of shapes for different types of cutting jobs.*

POWER AND GAS CABLES

A number of power and control cables and gas and cooling water hoses connect the power supply with the torch (Figure 10-8). These are combined in a multi-part cable that is usually covered to provide some protection to the cables and hoses inside and to make handling the cable easier. This covering is heat resistant but will not prevent damage to the cables and hoses inside if it comes in contact with hot metal or is exposed directly to the cutting sparks.

Power Cable

The power cable must have a high-voltage rated insulation, and it is made of finely stranded copper wire to give the torch maximum flexibility (Figure 10-9). For all non-transfer-type torches and those that use a high-frequency pilot arc, there are two power conductors, one positive and one negative. The size and

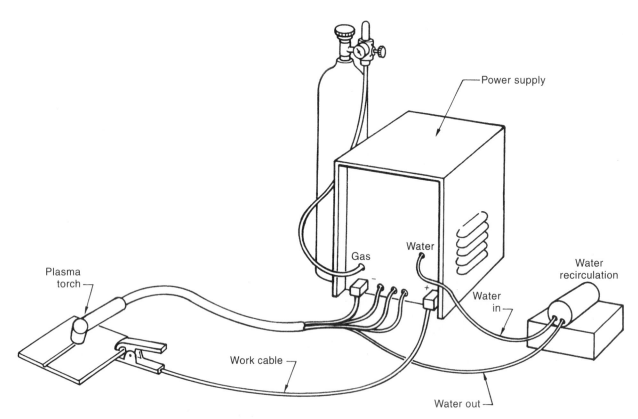

FIGURE 10–8 *Typical manual plasma arc cutting setup.*

FIGURE 10-9 *Portable plasma arc cutting machine.* (Courtesy of the Lincoln Electric Company)

current-carrying capacity of this cable is a controlling factor in the power range of the torch. As the capacity of the equipment increases, the cable must be made large enough to carry the increased current. The larger cables are less flexible and more difficult to manipulate. In order to make the cable smaller on water-cooled torches, the cable is run inside the cooling water return line. By putting the power cable inside the return water line, it allows a smaller cable to carry more current. The water prevents the cable from overheating.

Gas Hoses

There may be two gas hoses running the torch. One hose carries the gas used to produce the plasma, and the other provides a shielding gas coverage. On some small-amperage cutting torches there is only one gas line. The gas line is made of a special heat-resistant, ultraviolet light–resistant plastic. If it is necessary to replace the tubing because of damage, be sure to use the tubing provided by the manufacturer or a welding supplier. The tubing must be sized to carry the required gas flow rate within the pressure range of the torch, and it must be free from solvents and oils that might contaminate the gas. If the pressure of the gas supplied is excessive, the tubing may leak at the fittings or rupture.

Control Wire

The control wire is a two-conductor, low-voltage, stranded copper wire. This wire connects the power switch to the power supply. This allows the auto body technician to start and stop the plasma power and gas as needed during the cut or weld.

Water Tubing

Medium- and high-amperage torches may be water cooled. The water for cooling early-model torches must be deionized. Failure to use deionized water on these torches will result in the torch arcing out internally. This arcing may destroy or damage the torch's electrode tip and the nozzle tip. To see if your torch requires this special water, refer to the manufacturer's manual. If cooling water is required, it must be switched on and off at the same time as the plasma power. Allowing the water to circulate continuously might result in condensation in the torch. When the power is reapplied, the water will cause internal arcing damage.

POWER REQUIREMENTS

Voltage

The production of the plasma requires a direct-current (DC), high-voltage, constant-current (drooping arc voltage) power supply. A constant-current–type machine allows for a rapid start of the plasma arc at the high open circuit voltage and a more controlled plasma arc as the voltage rapidly drops to the lower closed voltage level. The voltage required for most welding operations, such as shielded metal arc, gas metal arc, gas tungsten arc, and flux-cored arc, ranges from 18 volts to 45 volts. The voltage for a plasma arc process ranges from 50 to 200 volts closed circuit and 150 to 400 volts open circuit. This higher electrical potential is required because the resistance of the gas increases as it is forced through a small orifice. The potential voltage of the power supplied must be high enough to overcome the resistance in the circuit in order for electrons to flow (Figure 10-10).

Amperage

Although the voltage is higher, the current (amperage) flow is much lower than it is with most other welding processes. Some low-powered PAC torches will

FIGURE 10–10 *Inverter-type plasma arc cutting power supply. (Courtesy of Miller Electric)*

operate with as low as ten amps of current flow. High-powered plasma cutting machines can have amperages as high as 200 amps, and some very large automated cutting machines may have 1,000-ampere capacities. The higher the amperage capacity, the faster and thicker the machine can cut.

Watts

The plasma process uses approximately the same amount of power, in watts, as a similar nonplasma process. Watts are the units of measure for electrical power. By determining the total watts used for both the nonplasma process and plasma operation, you can make a comparison. Watts used in a circuit are determined by multiplying the voltage times the amperage (Ohm's law; see Figure 10-11). For example, a 1/8-inch diameter E6011 electrode will operate at 18 volts and 90 amperes. The total watts used would be:

W = V x A
W = 18 x 90
W = 1,620

A low-power PAC torch operating with only 20 amperes and 85 volts would be using a total of:

W = V x A
W = 85 x 20
W = 1,700

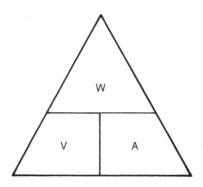

FIGURE 10–11 *Ohm's law.*

BASIC OPERATING PRINCIPLES

The plasma arc cutting process operates on direct current, straight polarity, electrode negative with a constricted transferred arc. Its severing action is produced by melting a specific area with the heat of a constricted arc, then removing the molten material with a high-velocity jet of hot, ionized gas expelled from the nozzle orifice of the cutting torch. In any plasma cutting process, heated gas is transformed into a plasma gas, resulting in extremely high temperatures.

To establish the arc, a low-current pilot arc is initiated by a high-voltage, high-frequency discharge between the electrode and the nozzle of the torch. When this pilot arc comes into contact with the electrically grounded part, it is automatically transferred to the workpiece, and a higher-current arc is initiated. Once the plasma arc cutting current is established, the pilot arc circuit is opened and the unit begins to cut. After the cut is completed, the plasma cutting arc is removed from the workpiece, the electrical cutting circuit is opened, current stops flowing, and a postflow gas cools the torch components. This dissipates damaging heat from the torch, thus prolonging electrode and nozzle life.

Control consoles for plasma arc cutting can contain solenoid valves to turn the gases and cooling water on and off. They usually have flowmeters for the cutting gases and a water flow switch to stop the operation if the cooling water flow falls below a safe limit. Controls for high-powered automatic plasma arc cutting can also contain programming features for upslope and downslope of current and orifice gas flow.

Plasma cutting gas selection depends on the material being cut and the surface quality requirements. Most nonferrous metals are cut using nitrogen,

TABLE 10–1: TYPICAL PAC TORCH PERFORMANCE

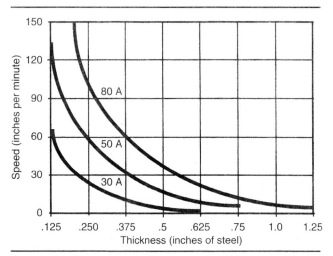

Courtesy of ESAB Welding and Cutting Products.

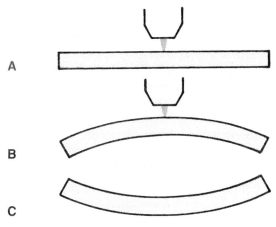

FIGURE 10–12 (A) When metal is heated, (B) it bends up toward the heat. (C) As the metal cools, it bends away from the heated area.

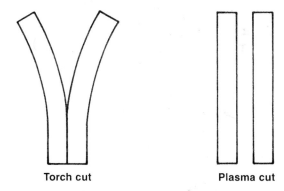

Torch cut   Plasma cut

FIGURE 10–13 Distortion in oxyfuel-cut versus plasma-cut sheets.

nitrogen-hydrogen mixtures, or argon-hydrogen mixtures. Titanium and zirconium are cut with pure argon because of their susceptibility to embrittlement by reactive gases.

Carbon steels are cut using compressed air (80 percent nitrogen, 20 percent oxygen) or nitrogen. Nitrogen is used with the water injection method of plasma arc cutting. Some systems use nitrogen for the plasma-forming gas, with oxygen injected into the plasma downstream of the electrode. This arrangement prolongs the life of the electrode by not exposing it to oxygen. For some nonferrous cutting with this dual-flow system, nitrogen is used as the plasma gas and carbon dioxide (CO_2) is used for shielding. For better quality cuts, argon-hydrogen plasma gas and nitrogen shielding are used.

HEAT INPUT

Although the total power used by both plasma and nonplasma processes is similar, the actual energy input into the work per linear foot is less with plasma. The very high temperatures of the plasma process allow much higher traveling rates, so the same amount of heat input is spread over a much larger area. This has the effect of lowering the joules per inch of heat the weld or cut will receive. Table 10-1 shows the cutting performance of a typical plasma torch. Note the relationship between amperage, cutting speed, and metal thickness. The lower the amperage, the slower the cutting speed or the thinner the metal that can be cut.

A high travel speed with plasma cutting will result in a heat input that is much lower than that of the oxyfuel

cutting process. A steel plate cut using the plasma process may have only a slight increase in temperature following the cut. It is often possible to pick up a part only moments after it is cut using plasma. The same part cut with oxyfuel would be much hotter and require a longer time to cool off.

DISTORTION

Any time metal is heated in a localized zone or spot, it expands in that area; and, after the metal cools, it is no longer straight or flat (Figure 10-12). If a piece of metal is cut, there will be localized heating along the edge of the cut. Unless special care is taken, this effect will make the part unusable. This distortion is a much greater problem with thin metals. By using a plasma cutter, an auto body worker can cut the thin, low-alloy sheet metal of a damaged car with little distortion (Figure 10-13).

On thicker sections, the hardness zone along the edge of a plasma cut is so small that it is not a problem. On the other hand, oxyfuel cutting of thick plate, especially higher alloyed metals, results in a hardness zone large enough to cause cracking and failure if the metal is shaped after cutting (Figure 10-14). Often the plates must be preheated before they are flame cut in order to reduce the heat-affected zone. This preheating adds greatly to the cost of fabrication both in time and fuel costs. By being able to make most cuts without preheating, the plasma process will greatly reduce fabrication costs.

APPLICATIONS

Early plasma arc cutting systems required that either helium or argon gas be used for the plasma and shielding gases. As the process improved, it was possible to start the PAC torch using argon or helium and then switch to less expensive nitrogen. The use of nitrogen as the plasma cutting gas greatly reduced the cost of operating a plasma system. Because of its operating expense, plasma cutting was limited to metals not easily cut using oxyfuel. Aluminum, stainless steel, and copper were the metals most often cut using plasma.

As the process improved, less-expensive gases and even dry compressed air could be used, and the torches and power supplies improved. By the early 1980s, the PAC process had advanced to a point where it was used for cutting all but the thicker sections of mild steel.

ADVANTAGES AND LIMITATIONS

Early plasma arc cutting processes required the use of special gas mixtures and equipment. Today's solid-state machines and air-cooled, lightweight torches use compressed air and electricity to form the required arc. This has served to reduce costs dramatically and has greatly expanded the application of plasma arc cutting. The cost effectiveness is achieved because of plasma arc's very high cutting speed (up to ten times that of oxyfuel cutting on 1-inch-(25 mm) thick steel) and its ease of use, which reduces operator training time. No preheating is required, and air plasma arc cutting is a continuous process—the time-consuming start-and-stop procedures typical of oxyfuel cutting are eliminated. These advantages offset the initially higher investment in plasma arc cutting equipment, permitting the equipment to pay for itself in a short time period.

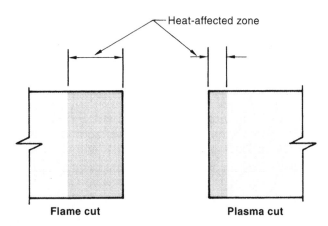

FIGURE 10–14 A smaller heat-affected zone will result in less hardness or weakening along the cut edge.

The economic advantages of plasma arc versus oxyfuel cutting are clear in situations where long, continuous cuts must be made on several pieces. Plasma arc can be used for stack cutting, shape cutting, plate beveling, and piercing. In the case of thicknesses totaling 2-1/2 inches (60 mm) or more, the decision about whether to use plasma or oxyfuel should depend upon such factors as equipment cost, load factors, the material being cut, and specific application considerations.

Cutting Speed

PAC cutting speeds can reach 300 inches per minute. That is 25 feet (7.6 m) per minute, or about 1/4 mile (.5 K/h) an hour. The fastest oxyfuel cutting equipment can cut at only about one-fourth that speed. A problem with early high-speed machine cutting was that the cutting machines could not reliably make cuts as fast as the PAC torch. That problem has been resolved, and the new machines and robots can operate at the upper limits of the plasma torch's capacity. These machines and robots are capable of automatically maintaining the optimum torch standoff distance to the work. Some cutting systems will even follow the irregular surfaces of preformed part blanks (Figure 10-15).

Metals

Any material that is conductive can be cut using the PAC process. In a few applications, nonconductive materials can be coated with conductive

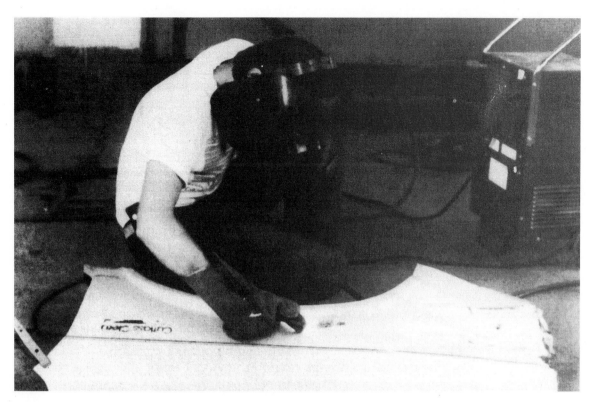

FIGURE 10–15   *Making a quality plasma arc cut. (Courtesy of Century Mfg. Co.)*

material so that they can be cut also. Although it is possible to make cuts in metal as thick as 7 inches (170 mm), it is not cost effective. The most popular materials cut are carbon steel up to 1 inch (25 mm), stainless steel up to 4 inches (200 mm), and aluminum up to 6 inches (150 mm). These are not the upper limits of the PAC process, but beyond these limits other cutting processes may be less expensive. Often a shop may PAC thicker material even if it is not cost effective because they don't have ready access to the alternative process.

Other metals commonly cut using PAC are copper, nickel alloys, high-strength, low-alloy steels, and clad materials. It is also used to cut expanded metals, screens, and other items that would require frequent starts and stops.

Standoff Distance

The standoff distance is the distance from the nozzle tip to the work (Figure 10-16). This distance is very critical to producing quality plasma arc cuts. As the distance increases, the arc force is diminished and

tends to spread out. This causes the kerf to be wider, the top edge of the plate to become rounded, and the formation of more dross on the bottom edge of the plate. However, if this distance becomes too close, the working life of the nozzle tip will be reduced. In some cases, an arc can form between the nozzle tip and the metal that instantly destroys the tip.

On some new torches, it is possible to drag the nozzle tip along the surface of the work without shorting it out. This is a large help when working on metal out of position or on thin sheet metal. Before you use your torch in this manner, you must check the owner's manual to see if it will operate in contact with the work (Figure 10-17). This technique will allow the nozzle tip orifice to become contaminated more quickly.

Starting Methods

Because the electrode tip is located inside the nozzle tip and a high initial resistance to current flow exists in the gas flow before the plasma is generated, it is necessary to have a specific starting method. Two methods are used to establish a current path through the gas.

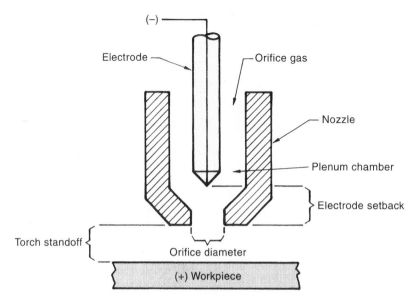

FIGURE 10-16 *Conventional plasma arc terminology. (Courtesy of the American Welding Society)*

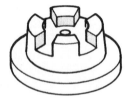

FIGURE 10-17 *A castle nozzle tip can be used to allow the torch to be dragged across the surface.*

The most common method uses a high-frequency alternating current carried through the conductor, the electrode, and back from the nozzle tip. This high-frequency current will ionize the gas and allow it to carry the initial current to establish a pilot arc (Figure 10-18). After the pilot arc has been started, the high-frequency starting circuit can be stopped. A pilot arc is an arc between the electrode tip and the nozzle tip within the torch head. This is a nontransfer arc, so the workpiece is not part of the current path. The low current of the pilot arc, although it is inside the torch, does not create enough heat to damage the torch parts. When the torch is brought close enough to the work, the primary arc will follow the pilot arc across the gap, and the main plasma is started. Once the main plasma is started, the pilot arc power can be shut off.

The second method of starting requires the electrode tip and nozzle tip to be momentarily shorted together. This is accomplished by automatically moving them together and immediately separating them again. The momentary shorting allows the arc to be created without damaging the torch parts.

Kerf

The kerf is the space left in the metal as the metal is removed during a cut. A PAC kerf is often wider than an oxyfuel kerf. Several factors will affect the width of the kerf. A few of the factors are:

- **Standoff distance.** The closer the torch nozzle tip is to the work, the narrower the kerf will be (Figure 10-19).
- **Orifice diameter.** Keeping the diameter of the nozzle orifice as small as possible will keep the kerf smaller.
- **Power setting.** Too high or too low a power setting will cause an increase in the kerf width.
- **Travel speed.** As the travel speed is increased, the kerf width will decrease; however, the bevel on the sides and the dross formation will increase if the speeds are excessive.
- **Gas.** The type of gas or gas mixture will affect the kerf width as the gas change affects travel speed, power, concentration of the plasma stream, and other factors.

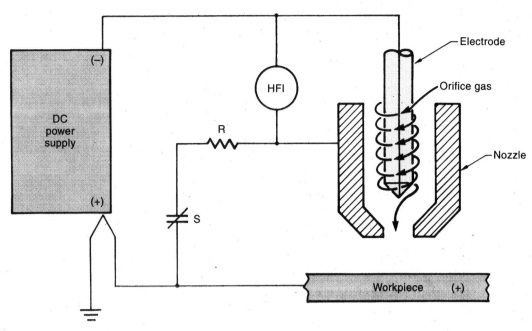

FIGURE 10-18 *Plasma arc torch circuitry. (Courtesy of the American Welding Society)*

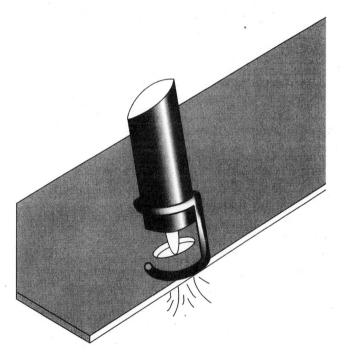

FIGURE 10-19 A wire adapter can be snapped around some shielding cups, which allows the torch to be slid across the surface of the metal as it is cut.

- **Electrode and nozzle tip.** As these parts begin to wear out from use or are damaged, the PAC quality and kerf width will be adversely affected.
- **Swirling of the plasma gas.** On some torches, the gas is directed in a circular motion around the electrode before it enters the nozzle tip orifice. This swirling causes the plasma stream to be more dense with straighter sides. The result is an improved cut quality, including a narrow kerf (Figure 10-20).
- **Water injection.** The injection of water into the plasma stream as it leaves the nozzle tip is not the same as the use of a water shroud. The injection of water into the plasma stream will increase the swirl and further concentrate the plasma. This improves the cutting quality, lengthens the life of the nozzle tip, and makes a squarer, narrower kerf (Figure 10-21).

Table 10-2 lists some standard kerf widths for several metal thicknesses. These are to be used as a guide for the nesting of parts on a plate to maximize the material used and minimize scrap. The kerf size may vary from this depending on a number of variables with your PAC system. You should make test

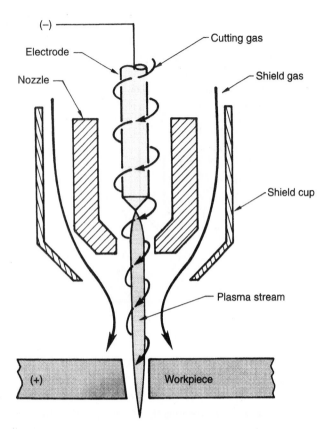

FIGURE 10–20 The cutting gas can swirl around the electrode to produce a tighter plasma column. (Courtesy of the American Welding Society)

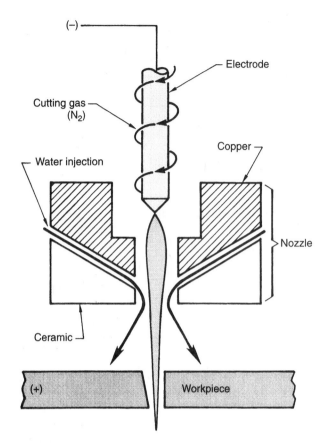

FIGURE 10–21 Water injection plasma arc cutting. Notice that the kerf is narrow, and one side is square. (Courtesy of the American Welding Society)

TABLE 10–2: STANDARD KERF WIDTHS FOR SEVERAL METAL THICKNESSES

| Plate Thickness | | Kerf Allowance | |
inches	mm	inches	mm
1/8 to 1	3.2 to 25.4	+3/32	+2.4
1 to 2	25.4 to 51.0	+3/16	+4.8
2 to 5	51.0 to 127.0	+5/16	+8.0

cuts to verify the size of the kerf before starting any large production cuts.

Because the sides of the plasma stream are not parallel as they leave the nozzle tip, there is a bevel left on the sides of all plasma cuts. This bevel angle is from 1/2 degree to 3 degrees depending on metal thickness, torch speed, type of gas, standoff distance, nozzle tip condition, and other factors affecting a quality cut. On thin metals, the bevel is undetectable and offers no problem in part fabrication or finishing.

With a plasma swirling-type torch, you can plan the direction the cut so that the scrap side has all of the bevel (Figure 10-22). This technique is only effective if one side of the cut is to be scrap.

GASES

Almost any gas or gas mixture can be used today for the PAC process. Changing the gas or gas mixture is one method of controlling the plasma cut. Although the type of gas or gases used will have a major effect on the cutting performance, it is only one of a number of changes that a technician can make to help produce a quality cut. The following are some of the effects on the cut that changing the PAC gas(es) will have:

- **Force.** The amount of mechanical impact on the material being cut. The density of the gas and its ability to disperse the molten metal.
- **Central concentration.** Some gases will have a more compact plasma stream. This factor will greatly affect the kerf width and cutting speed.

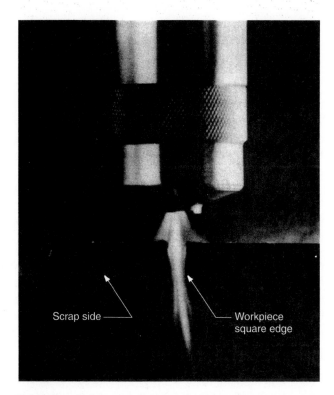

FIGURE 10–22 *Bevel on scrap side of PAC cut.*

Scrap side ——

Workpiece square edge

TABLE 10–3: GASES FOR PLASMA ARC CUTTING AND GOUGING

Metal	Gas
Carbon and low alloy steel	Nitrogen Argon with 0 to 35% hydrogen air
Stainless steel	Nitrogen Argon with 0 to 35% hydrogen
Aluminum and aluminum alloys	Nitrogen Argon with 0 to 35% hydrogen
All plasma arc gouging	Argon with 35% to 40% hydrogen

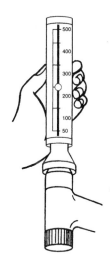

FIGURE 10–23 *Plasma flow measuring kit. (Courtesy of ESAB Welding and Cutting Products)*

- **Heat content.** As the electrical resistance of a gas or gas mixture changes, it will affect the heat content of the plasma it produces. The higher the resistance, the higher the heat produced by the plasma.
- **Kerf width.** The ability of the plasma to remain in a tightly compact stream will produce a deeper cut with less bevel on the sides.
- **Dross formation.** The dross that may be attached along the bottom edge of the cut can be controlled or eliminated.
- **Top edge rounding.** The rounding of the top edge of the plate can often be eliminated by correctly selecting the gas(es) that are to be used.
- **Metal type.** Because of the formation of undesirable compounds on the cut surface as the metal reacts to elements in the plasma, some metals may not be cut with specific gas(es).

Table 10-3 lists some of the popular gases and gas mixtures used for various PAC metals. The selection of a gas or gas mixture for a specific operation to maximize the system performance must be tested with the equipment and setup being used. With constant developments and improvements in the PAC system, new gases and gas mixtures are continuously being added to the list.

In addition to the type of gas, it is important to have the correct gas flow rate for the size tip, metal type, and thickness. Too low a gas flow will result in a cut having excessive dross and sharply beveled sizes. Too high a gas flow will produce a poor cut because of turbulence in the plasma stream and waste gas. A flow measuring kit can be used to test the flow at the plasma torch for more accurate adjustments (Figure 10-23).

STACK CUTTING

Because the PAC process does not rely on the thermal conductivity between stacked parts, like the oxyfuel process, thin sheets can be stacked and cut efficiently. With the oxyfuel cutting of stacked sheets, it is important that there not be any air gaps between layers. Also, it is often necessary to make a weld along the side of the stack in order for the cut to start consistently.

The PAC process does not have these limitations. It is recommended that the sheets be held together for cutting, but this can be accomplished by using standard C-clamps. The clamping needs to be tight because, if the space between layers is excessive, the sheets may stick together. The only problem that will be encountered is that, because of the kerf bevel, the parts near the bottom might be slightly larger if the stack is very thick. This problem can be controlled by using the same techniques as described for making the kerf square.

DROSS

Dross is the metal compound that resolidifies and attaches itself to the bottom of a cut. This metal compound is made up mostly of unoxidized metal, metal oxides, and nitrides. It is possible to make cuts dross-free if the PAC equipment is in good operating condition and the metal is not too thick for the size torch being used. Because dross contains more unoxidized metal than most OFC slag, it is often much harder to remove if it sticks to the cut. The thickness that a dross-free cut can be made depends on a number of factors, including the gas(es) used for the cut, travel speed, standoff distance, nozzle tip orifice diameter, wear condition of the electrode tip and nozzle tip, gas velocity, and plasma stream swirl.

Stainless steel and aluminum are easily cut dross-free. Carbon steel, copper, and nickel-copper alloys are much more difficult to cut dross-free.

MANUAL CUTTING

Manual plasma arc cutting is the most versatile of the PAC processes. It can be used in all positions, on almost any surface, and on most metals. This process is limited to low-power plasma machines; however, even these machines can cut up to 1 1/2-inch-thick metals. The limitation to low power, 100 amperes or less, is primarily for safety reasons.

The higher-powered machines have extremely dangerous open circuit voltages that can kill a person if accidentally touched.

SETUP

The setup of most plasma equipment is similar, but don't ever attempt to set up a system without the manufacturer's owner's manual for the specific equipment.

Be sure all of the connections are tight and that there are no gaps in the insulation on any of the cables. Check the water and gas lines for leaks. Visually inspect the complete system for possible problems.

Before you touch the nozzle tip, be sure that the main power supply is off. The open circuit voltage on even low-powered plasma machines is high enough to kill a person. Replace all parts to the torch before the power is restored to the machine.

PLASMA ARC GOUGING

Plasma arc gouging is a recent introduction to the PAC processes. The process is similar to that of air carbon arc gouging in that a U groove can be cut into the metal's surface. The removal of metal along a joint before the metal is welded or the removal of a defect for repairing can easily be done using this variation of PAC (Figure 10-24).

The torch is set up with a less-concentrated plasma stream. This will allow the washing away of the molten metal instead of thrusting it out to form a cut.

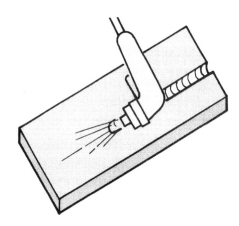

FIGURE 10–24 *Plasma arc gouging a U groove in a plate.*

The torch is held at approximately a 30-degree angle to the metal surface. Once the groove is started, it can be controlled by the rate of travel, torch angle, and torch movement.

Plasma arc gouging is effective on most metals. Stainless steel and aluminum are especially good metals to gouge because there is almost no cleanup required afterwards. The groove is clean, bright, and ready to be welded. Plasma arc gouging is especially beneficial with these metals because there is no reasonable alternative available. The only other process that can leave the metal ready to weld is to have the groove machined, and machining is slow and expensive compared to plasma arc gouging.

It is important not to try to remove too much metal in one pass. The process will work better if small amounts are removed at a time. If a deeper groove is required, multiple gouging passes can be used.

SAFETY

PAC has many of the same safety concerns as do most other electric welding or cutting processes. Some special concerns specific to this process are:

- **Electrical shock.** Because the open-circuit voltage is much higher for this process than for any other, extra caution must be taken. The chance that a fatal shock could be received from this equipment is much higher than from any other welding equipment.
- **Moisture.** Often water is used with PAC torches to cool the torch, to improve the cutting characteristics, or to be part of a water table. Any time water is used it is very important that there be no leaks or splashes. The chance of electrical shock is greatly increased if there is moisture on the floor, cables, or equipment.
- **Noise.** Because the plasma stream is passing through the nozzle orifice at a high speed, a loud sound is produced. The sound level increases as the power level increases. Even with low-power equipment, the decibel (dB) level is above the safety range. Some type of ear protection is required to prevent damage to the operator and to other people in the area of the PAC equipment when it is in operation. High levels of sound can have a cumulative effect on one's hearing. Over time, one's ability to hear will decrease unless proper precautions are taken. See the owner's manual for the recommendations for the equipment in use.

- **Light.** The PAC process produces light radiation in all three spectra. This large quantity of visible light, if the eyes are unprotected, will cause night blindness. The most dangerous of the lights is ultraviolet. As in other arc processes, this light can cause burns to the skin and eyes. The third light, infrared, can be felt as heat, and it is not as much of a hazard. Some type of eye protection must be worn when any PAC is in progress. Table 10-4 lists the recommended lens shade numbers for machines at various power levels.
- **Fumes.** This process produces a large quantity of fumes that are potentially hazardous. A specific means for removing them from the workspace should be in place. A downdraft table is ideal for manual work, but some special pickups may be required for larger applications. The use of a water table and/or a water shroud nozzle will greatly help to control fumes. Often the fumes cannot be exhausted into the open air without first being filtered or treated to remove dangerous levels of contaminants. Before installing an exhaust system, you must first check with local, state, and federal officials to see if specific safeguards are required.
- **Gases.** Some of the plasma gas mixtures include hydrogen. Because this is a flammable gas, extra care must be taken to insure that the system is leak-proof.
- **Sparks.** As with any process that produces sparks, the danger of an accidental fire is always present. This is a larger concern with PAC, because the sparks are often thrown some distance from the work area, and the operator's vision is restricted by a welding helmet. If there is any possibility that sparks will be thrown out of the immediate work area, a fire watch must be present. A fire watch is a person whose sole job

TABLE 10–4: RECOMMENDED SHADE DENSITIES FOR FILTER LENSES

Current Range A	Minimum Shade	Comfortable Shade
Less than 300	8	9
300 to 400	9	12
400 plus	10	14

is to watch for the possible starting of a fire. This person must know how to sound the alarm and have appropriate fire-fighting equipment handy. Never cut in the presence of combustible materials.

- **Operator checkout.** Never operate any PAC equipment until you have read the manufacturer's owner's and operator's manual for the specific equipment to be used. It is a good idea to have someone who is familiar with the equipment go through the operation after you have read the manual.

OXYFUEL GAS CUTTING

Oxyfuel gas cutting (OFC) is a group of processes that uses a high-temperature oxyfuel gas flame to preheat the metal to a kindling temperature at which it will react rapidly with a stream of pure oxygen. The kindling temperature for steel, in oxygen, is 1,625° F (884° C). At this temperature, a molten weld pool need not occur to start a cut. The processes in this group are identified by the types of fuel gases used with oxygen to produce the preheat flame. Acetylene is the most commonly used fuel gas. Table 10-5 lists a number of fuel gases, in addition to acetylene, that are used for cutting.

More people use the oxyfuel cutting torch than any other welding process. The cutting torch is used by workers in virtually all areas, including manufacturing, maintenance, automotive repair, railroad, farming, and more. It is unfortunately one of the most commonly misused processes. Most workers know how to light the torch and make a cut, but their cuts are very poor quality and often unsafe. A good oxyfuel cut should not only be straight and square, but it should require little or no post-cut cleanup. Excessive post-cutting cleanup results in extra cost, which is an expense that cannot be justified.

EYE PROTECTION FOR FLAME CUTTING

Proper filter plates for eye protection during OFC have been identified by the National Bureau of Standards. The recommended filter plates are identified by shade number and are related to the type of cutting operation being performed.

Goggles or other suitable eye protection must be used for flame cutting. Goggles should have vents near the lenses to prevent fogging. Cover lenses or plates should be provided to protect the filter lens. All glass used for lenses should be ground properly so that the front and rear surfaces are smooth. Filter lenses must be marked so that the shade number can be readily identified (Table 10-6).

TABLE 10–5:	FUEL GASES USED FOR FLAME CUTTING	
Fuel Gas	Flame Fahrenheit	Temperature* Celsius
Acetylene	5,589°	3,087°
MAPP®	5,301°	2,927°
Natural Gas	4,600°	2,538°
Propane	4,579°	2,526°
Propylene	5,193°	2,867°
Hydrogen	4,820°	2,660°

*Approximate neutral oxyfuel flame temperature

TABLE 10–6: A GENERAL GUIDE FOR THE SELECTION OF EYE AND FACE PROTECTION EQUIPMENT		
Type of Cutting Operation	Hazard	Suggested Shade Number
Light cutting up to 1 in.	Sparks, harmful	3 or 4
Medium cutting, 1–6 in.	rays, molten metal,	4 or 5
Heavy cutting, over 6 in.	flying particles	5 or 6

CUTTING TORCHES

The oxyacetylene hand torch is the most common type of oxyfuel gas cutting torch used in industry. The hand torch, as it is often called, may be either a part of a combination welding and cutting torch set or a cutting torch only (Figure 10-25). The combination welding-cutting torch offers more flexibility because a cutting head, welding tip, or heating tip can be attached quickly to the same torch body (Figure 10-26). Combination torch sets are often used in schools, automotive repair shops, auto body shops, and small welding shops, or with any job where flexibility in equipment is needed. A cut made with either type of torch has the same quality; however, the dedicated cutting torches are usually longer and have larger gas flow passages than the combination torches. The added length of the dedicated cutting torch helps keep the operator farther away from the heat and sparks and allows thicker material to be cut.

Oxygen is mixed with the fuel gas to form a high-temperature preheating flame. The two gases must be completely mixed before they leave the tip and create the flame. Two methods are used to mix the gases. One method uses a mixing chamber, and the other method uses an injector chamber.

The mixing chamber may be located in the torch body or in the tip (Figure 10-27). Torches that use a mixing chamber are known as equal-pressure torches, because the gases must enter the mixing chamber under the same pressure. The mixing chamber is larger than both the gas inlet and the gas outlet. This larger size causes turbulence in the gases, resulting in the gases mixing thoroughly.

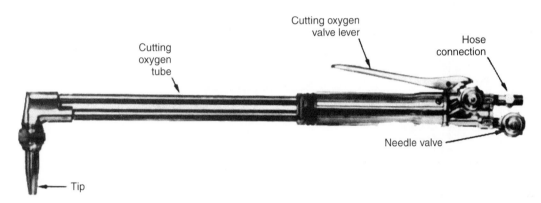

FIGURE 10–25 *Oxyfuel cutting torch. (Courtesy of Victor Equipment Company)*

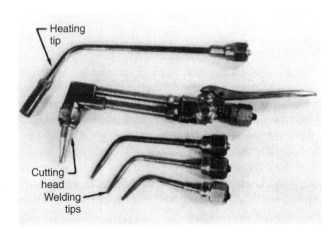

FIGURE 10–26 *The attachments that are used for heating, cutting, welding, or brazing make the combination torch set flexible.*

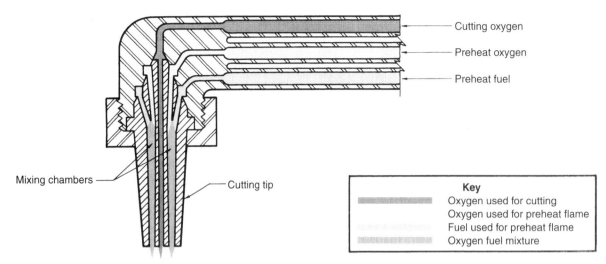

FIGURE 10-27 A mixing chamber located in the tip.

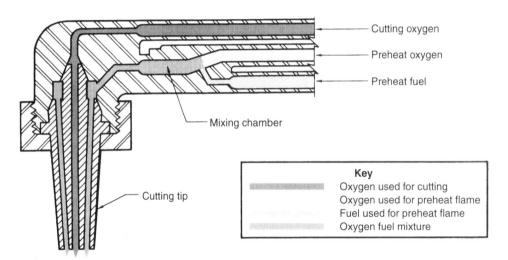

FIGURE 10-28 Injector mixing torch.

Injector torches will work both with equal gas pressures or with low fuel-gas pressures (Figure 10-28). The injector allows the oxygen to draw the fuel gas into the chamber even if the fuel gas pressure is as low as 6 oz./in² (26 g/cm²). The injector works by passing the oxygen through a venturi, which creates a low-pressure area that pulls the fuel gases in and mixes them together. An injector-type torch must be used if a low-pressure acetylene generator or low-pressure residential natural gas is used as the fuel gas supply.

The cutting head may hold the cutting tip at a right angle to the torch body or it may be held at a slight angle. Torches with the tip slightly angled are easier for the auto body technician to use when cutting flat plate. Torches with a right-angle tip are easier for the auto body technician to use when cutting pipe, angle iron, I-beams, or other uneven material shapes. Both types of torches can be used for any type of material being cut, but practice is needed to keep the cut square and accurate.

The location of the cutting lever may vary from one torch to another (Figure 10-29). Most cutting levers pivot from the front or back end of the torch body. Personal preference will determine which one the auto body technician uses.

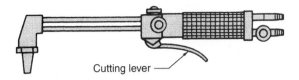
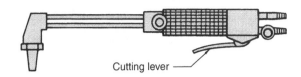

Cutting lever

Cutting lever

FIGURE 10-29 The cutting lever may be located on the front or back of the torch body.

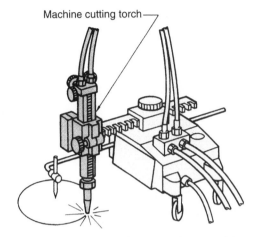

Machine cutting torch

FIGURE 10-30 Portable flame-cutting machine.

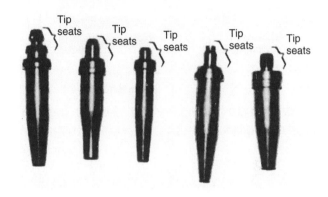

Tip seats   Tip seats   Tip seats   Tip seats   Tip seats

FIGURE 10-32 Five different cutting torch tip designs for different manufacturers' torches.

FIGURE 10-31 Motor-Driver, hand-held cutting torch. (Courtesy of Thermodyne Industries, Inc.)

A machine cutting torch, sometimes referred to as a blowpipe, operates in a similar manner to a hand cutting torch. The machine cutting torch may require two oxygen regulators, one for the preheat oxygen and the other for the cutting oxygen stream. The addition of a separate cutting oxygen supply allows the flame to be more accurately adjusted. It also allows the pressures to be adjusted during a cut without disturbing the other parts of the flame. Two types of machine cutting torches are shown in Figures 10-30 and 10-31.

CUTTING TIPS

Most cutting tips are made of copper alloy, and some tips are chrome plated. Chrome plating prevents spatter from sticking to the tip and thus prolongs its usefulness. Tip designs change for the different types of uses and gases and from one torch manufacturer to another (Figure 10-32).

Tips for straight cutting are either standard or high speed (Figure 10-33). The high-speed cutting tip is designed to allow a higher cutting oxygen pressure, which allows the torch to travel faster. High-speed tips are also available for different types of fuel gases.

The amount of preheat flame required to make a perfect cut is determined by the type of fuel gas used and by the material thickness, shape, and surface condition. Materials that are thick, round, or have surfaces covered with rust, paint, or oil require more preheat flame.

Different cutting tips are available for each of the major types of fuel gases. The differences in the type or number of preheat holes determine the type of fuel gas to be used in the tip.

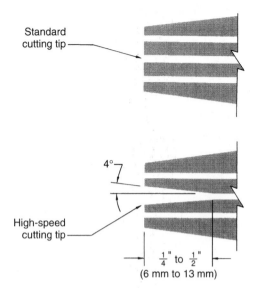

FIGURE 10–33 *Comparison of standard and high-speed cutting tips.*

Gas	Number of Preheat Holes
Acetylene	1 to 6
Mapp®	8, two-piece tip
Propane Natural gas	Two-piece

FIGURE 10–34 *Fuel gas and range of preheat holes.*

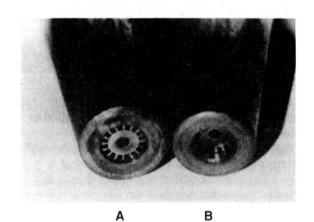

A   **B**

FIGURE 10–35 *Two-piece cutting tips: (A) MAPP, and (B) propane or natural gas.*

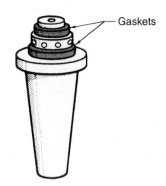

Gaskets

FIGURE 10–36 *Some cutting tips use gaskets to make a tight seal.*

Figure 10-34 lists the fuel gases and the range of preheat holes or tip designs used with each gas. Acetylene is used in tips having from 1 to 6 preheat holes. Some large acetylene cutting tips may have 8 or more preheat holes.

CAUTION: If acetylene is used in a tip that was designed to be used with one of the other fuel gases, the tip may overheat, causing a backfire or causing the tip to explode.

MPS gases are used in tips having eight preheat holes or in a two-piece tip that is not recessed. These gases have a slower flame combustion rate than acetylene. If they are used with tips having fewer than eight preheat holes, there may not be enough heat to start a cut, or the flame may pop out when the cutting lever is pressed.

CAUTION: If MPS gases are used in a deeply recessed, two-piece tip, the tip will overheat, causing a backfire or causing the tip to explode.

Propane and natural gas should be used in a two-piece tip that is deeply recessed (Figure 10-35). The flame burns at such a slow rate that it may not stay lit on any other tip.

Some cutting tips have metal-to-metal seals. When they are installed in the torch head, a wrench must be used to tighten the nut. Other cutting tips have fiber packing seats to seal the tip to the torch. If a wrench is used to tighten the nut for this type of tip, the tip seat may be damaged (Figure 10-36). A torch owner's

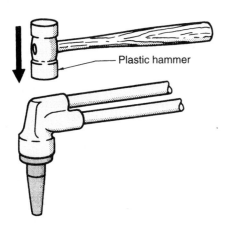

FIGURE 10–37 Tap the back of the torch head to remove a tip that is stuck. The tip itself should never be tapped.

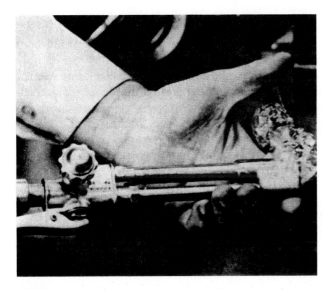

FIGURE 10–38 Checking a cutting tip for leaks. (Courtesy of Albany Calcium Light Co. Inc., Albany, NY)

manual should be checked or a welding supplier should be asked about the best way to tighten various torch tips.

To remove a cutting tip, if the tip is stuck in the torch head, tap the back of the head with a plastic hammer (Figure 10-37). Any tapping on the side of the tip may damage the seat.

To check the assembled torch tip for a good seal, place your thumb over the end of the tip, turn on the oxygen valve, and spray the tip with a leak-detecting solution (Figure 10-38).

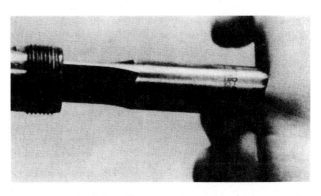

FIGURE 10–39 Damaged torch seats can be repaired by using a reamer.

CAUTION: Carefully handle and store the tips to prevent damage to the tip seats and to keep dirt from becoming stuck in the small holes.

If the cutting tip seat or the torch head seat is damaged, it can be repaired by using a reamer designed for the specific torch tip and head (Figure 10-39), or it can be sent out for repair. New fiber packings are available for tips with packings. The original leak-checking test should be repeated to be sure the new seal is good.

OXYFUEL CUTTING, SETUP, AND OPERATION

CAUTION: NEVER USE A CUTTING TORCH TO CUT OPEN A USED CAN, DRUM, TANK, OR OTHER SEALED CONTAINER. The sparks and oxygen cutting stream may cause even nonflammable residue inside to burn or explode. If a used container must be cut, it must first have one end removed and all residue cleaned out. In addition to the possibility of a fire or explosion, you might be exposing yourself to hazardous fumes. Before making a cut, check the Material Specification Data Sheet (MSDS) for safety concerns.

The setting up of a cutting torch system is exactly like setting up oxyfuel welding equipment except for the adjustment of gas pressures. This chapter covers gas pressure adjustments and cutting equipment operations. In Chapter 8, detailed technical information and instructions are given for oxyfuel systems. The chapter includes:

- Safety
- Pressure regulator setup and operation
- Welding and cutting torch design and service
- Reverse flow and flashback valves
- Hoses and fittings
- Types of flames
- Leak detection

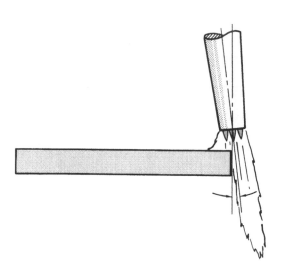

FIGURE 10–40 *Starting a cut on the edge of a plate. Notice how the torch is pointed at a slight angle away from the edge.*

To start a cut on the edge of a plate, hold the torch at a right angle to the surface or pointed slightly away from the edge (Figure 10-40). The torch must also be pointed so that the cut is started at the very edge. The edge of the plate heats up more quickly and allows the cut to be started sooner. Also, fewer sparks will be blown around the shop. Once the cut is started, the torch should be rotated back to a right angle to the surface or to a slight leading angle.

If a cut is to be started in a place other than the edge of the plate, the inner cones should be held as close as possible to the metal. Having the inner cones touch the metal will speed up the preheat time. When the metal is hot enough to allow the cut to start, the torch should be raised as the cutting lever is slowly depressed. When the metal is pierced, the torch should be lowered again (Figure 10-41). By raising the torch tip away from the metal, the amount of sparks blown into the air is reduced, and the tip is kept cleaner. If the metal being cut is thick, it may be necessary to move the torch tip in a small circle as the hole goes through the metal. If the metal is to be cut in both directions from the spot where it was pierced, the torch should be moved backward a short distance and then forward (Figure 10-42). This prevents slag from refilling the kerf at the starting point, thus making it difficult to cut in the other direction. The kerf is the space produced during any cutting process.

The proper alignment of the preheat holes will speed up and improve the cut. The holes should be

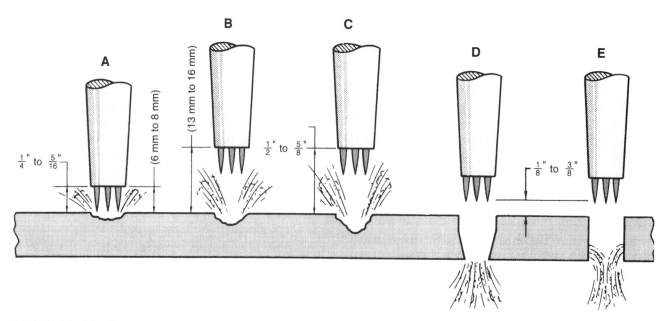

FIGURE 10–41 *Sequence for piercing plate.*

aligned so that one is directly on the line ahead of the cut and another is aimed down into the cut when making a straight line square cut (Figure 10-43). The flame is directed toward the smaller piece and the sharpest edge when cutting a bevel. For this reason, the tip should be changed so that at least two of the flames are on the larger plate and none of the flames are directed on the sharp edge (Figure 10-44). If the preheat flame is directed at the edge, it will be rounded off as it is melted off.

10.3   HAND CUTTING

When you are making a cut with a hand torch, it is important to be steady so that the cut can be as smooth

as possible. You must also be comfortable and free to move the torch along the line to be cut. It is a good idea for an auto body technician to get into position and practice the cutting movement a few times before lighting the torch. Even when you and the torch are braced properly, a tiny movement such as a heartbeat will cause a slight ripple in the cut. Attempting a cut without leaning on the work to brace oneself is tiring and causes inaccuracies.

The torch should be braced with the left hand if the auto body technician is right-handed or with the right hand if the auto body technician is left-handed. The torch may be moved by sliding it toward you over your supporting hand (Figures 10-45 and 10-46). The torch can also be pivoted on the supporting hand. If the pivoting method is used, care must be taken to prevent the cut from becoming a series of arcs.

A slight forward torch angle helps to keep some of the reflected heat off the tip, aids in blowing dirt and

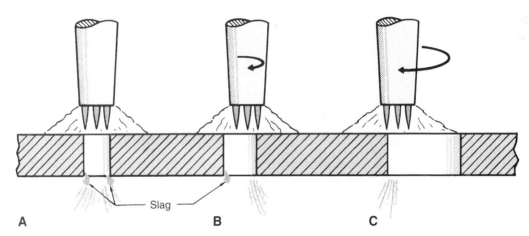

FIGURE 10–42 *A short, backward movement (A) before the cut is carried forward (B) clears the slag from the kerf (C). Slag left in the kerf may cause the cutting stream to gouge into the base metal, resulting in a poor cut.*

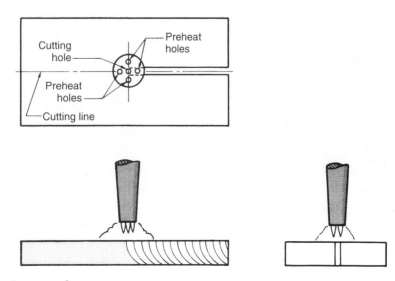

FIGURE 10–43 *Tip alignment for a square cut.*

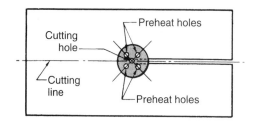

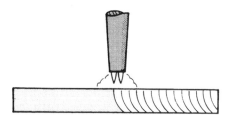

FIGURE 10–44   Tip alignment for bevel cut.

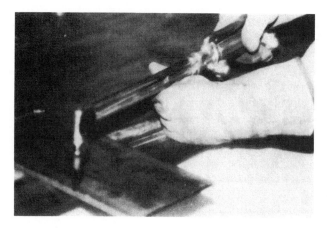

FIGURE 10–45   For short cuts, the torch can be drawn over the gloved hand.

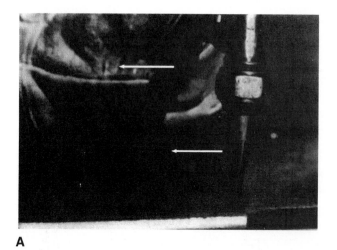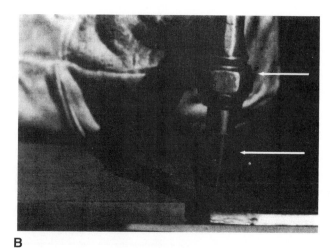

A

B

FIGURE 10–46   For longer cuts, the torch can be moved by sliding your gloved hand along the plate parallel to the cut: (A) start and (B) finish. Always check for free and easy movement before lighting the torch.

oxides away from the cut, and keeps the tip clean for a longer period of time because slag is less likely to be blown back on it (Figure 10-47). The forward angle can be used only for a straight-line square cut. If shapes are cut using a slight angle, the part will have beveled sides.

During the cut, the plasma torch tip or the inner cones of the oxyacetylene flame should be kept 1/8 to 3/8 of an inch (3 to 10 mm) from the surface of the plate (Figure 10-48). This distance is known as the coupling distance.

Starts and stops can be made more easily if one side of the metal being cut is scrap. When it is necessary to stop and reposition oneself before continuing the cut, the cut should be turned out a short distance into the scrap side of the metal (Figure 10-49). The extra space that this procedure provides will allow a smoother and more even restart with less chance that slag will block the cut. If neither side of the cut is to be scrap, the forward movement should be stopped for a moment before releasing the cutting lever. This action will allow the drag, or the distance that the bottom of the cut is behind the top, to

catch up before stopping (Figure 10-50). To restart, use the same procedure that was given for starting a cut at the edge of the plate.

10.4 — LAYOUT

Laying out a line to be cut can be done with a piece of soapstone or a chalk line. To obtain an accurate line, you can use a scribe or a punch. If a piece of soapstone is used, it should be sharpened properly to increase accuracy (Figure 10-51). A chalk line will make

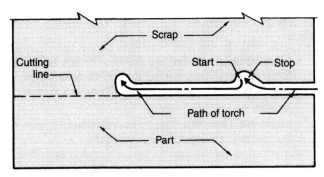

FIGURE 10-49 Turning out into scrap to make stopping and starting points smoother.

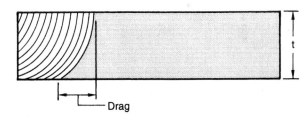

FIGURE 10-50 Drag is the distance by which the bottom of a cut lags behind the top. (Courtesy of Praxair, Inc.)

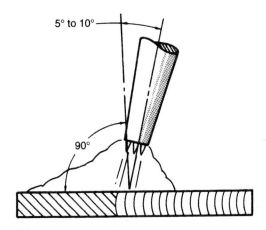

FIGURE 10-47 A slight forward angle helps when cutting thin material.

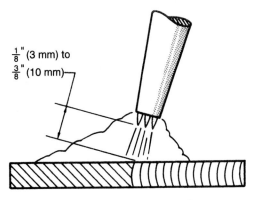

FIGURE 10-48 Inner-cone-to-work distance.

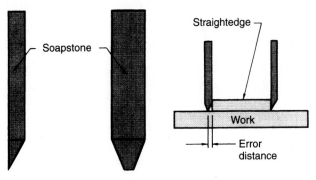

FIGURE 10-51 Proper method of sharpening a soapstone.

a long, straight line on metal and is best used on large jobs. The scribe and punch can both be used to lay out an accurate line, but the punched line is easier to see when cutting. A punch can be held as shown in Figure 10-52 with the tip just above the surface of the metal. When the punch is struck with a lightweight hammer, it will make a mark. If you move your hand along the line and rapidly strike the punch, it will leave a series of punch marks for the cut to follow.

PRACTICE 10-1

Flat, Straight Cuts in Thin Panels

Using a properly set up and adjusted PAC machine, proper safety protection, one or more pieces of mild steel or aluminum, and panels 18 gauge to 24 gauge, you will cut off 1/2-inch- (13 mm) wide strips.

- Starting at one end of the piece of metal, hold the torch as close as possible to a 90-degree angle.
- Lower your hood and establish a plasma cutting stream.
- Move the torch in a straight line down the plate toward the other end (Figure 10-53).
- If the width of the kerf changes, speed up or slow down the travel rate to keep the kerf the same size for the entire length of the plate.

Repeat the cut using both thicknesses of all three types of metals until you can make consistently smooth cuts that are within ±3/32 inch of a straight line and ±5 degrees from square. Turn off the PAC equipment and clean up your work area when you are finished welding.

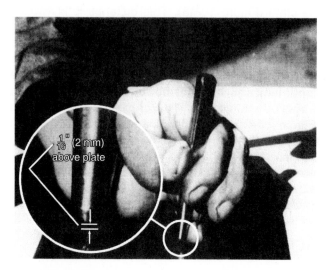

FIGURE 10–52 *Holding the punch slightly above the surface allows the punch to be struck rapidly and moved along a line to mark it for cutting.*

A   **B**   **C**

FIGURE 10–53 *(A) Starting at the edge of a plate, like starting an oxyfuel cut, (B-C) move smoothly in a straight line toward the other end. Note the roughness left along the side of the plate from a previous cut.*

Flat, Curved Cuts in Thin Plate

Using a properly set up and adjusted PAC machine, proper safety protection, one or more pieces of mild steel or aluminum, and panels 18 gauge to 28 gauge thick, you will cut off 1/2-inch (13 mm)-wide strips (Figure 10-54).

Lay out a series of parallel lines on a scrap panel that are spaced 1/2 inch (13 mm) apart.

Follow the same procedure as outlined in Practice 8-1.

Repeat the cut using both thicknesses of all three types of metals until you can make consistently smooth cuts that are within ±3/32 inch (2 mm) of the line and ±5 degrees from square. Turn off the PAC equipment and clean up your work area when you are finished cutting.

Flat Cutting Holes

Using a properly set up and adjusted PAC machine, proper safety protection, one or more pieces of mild steel and aluminum 1/16 inch (1.5 mm) to 24 gauge thick, you will cut 1/2 inch and 1 inch (13 mm and 25 mm) holes.

- Starting with the piece of metal that is 1/16 inch (1.5 mm) thick, hold the torch as close as possible to a 90-degree angle.

- Lower your hood and establish a plasma cutting stream.
- Move the torch in an outward spiral until the hole is the desired size (Figure 10-55).

Repeat the hole-cutting process until both sizes of holes are made using all the thicknesses of all three types of metals and you can make consistently smooth cuts that are within ±3/32 inch (2 mm) of being round and ±5 degrees of being square. Turn off the PAC equipment and clean up your work area when you are finished cutting.

Removing Spot Welds

Using a properly set up and adjusted PAC machine, proper safety protection, and one or more pieces of spot-welded flange from a vehicle, you will cut out the spot weld in order to separate the flange holes.

- Starting at the center of a spot weld, lower your hood and establish a plasma cutting stream.
- The spot weld will immediately be pierced in the center. Now move the torch in a small circle about the size of the spot weld.

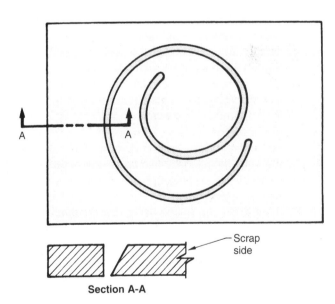

Section A-A

FIGURE 10–55 When cutting a hole, make a test to see which direction to make the cut so that the beveled side is on the scrap piece.

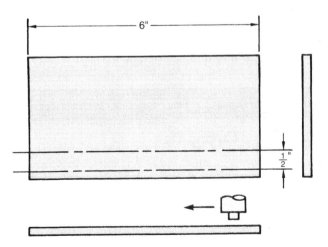

FIGURE 10–54 Flat cutting strips.

- Try to cut out 100 percent of the spot weld. A smaller hole is better, so do not cut an excessively large hole.
- Repeat the process until all of the spot welds are removed.
- Using a hammer and chisel or air chisel, separate the flange.
- Examine the spot welds to see that they separated without excessive distortion of the panel.

Repeat the spot weld cutting process until you can make consistently smooth cuts that are within ±1/16 inch (1.5 mm) of being the same diameter as the spot weld and round. Turn off the PAC equipment and clean up your work area when you are finished cutting.

FIGURE 10–56 *Leak-check all gas fittings. (Courtesy of Albany Calcium Light Co. Inc., Albany, NY)*

PRACTICE 10-5

Setting Up a Cutting Torch

1. The oxygen and acetylene cylinders must be securely chained to a cart or wall before the safety caps are removed.
2. After removing the safety caps, stand to one side and crack (open and quickly close) the cylinder valves, being sure there are no sources of possible ignition that may start a fire. Cracking the cylinder valves is done to blow out any dirt that may be in the valves.
3. Visually inspect all of the parts for any damage, needed repair, or cleaning.
4. Attach the regulators to the cylinder valves and tighten them securely with a wrench.
5. Attach a reverse-flow valve or flashback arrestor, if the torch does not have them built in, to the hose connection on the regulator or to the hose connection on the torch body, depending on the type of reverse-flow valve in the set. Occasionally test each reverse-flow valve by blowing through it to make sure it works properly.
6. If the torch you will be using is a combination-type torch, attach the cutting head at this time.
7. Last, install a cutting tip on the torch.
8. Before the cylinder valves are opened, back out the pressure regulating screws so that when the valves are opened, the gauges will show zero pounds working pressure.
9. Stand to one side of the regulators as the cylinder valves are opened slowly.

10. The oxygen valve is opened all the way until it becomes tight, but don't over-tighten. The acetylene valve is opened no more than one-half turn.
11. Open one torch valve and then turn the regulating screw in slowly until 2 psig to 4 psig (14 kPag to 30 kPag) shows on the working pressure gauge. Allow the gas to escape so that the line is completely purged.
12. If you are using a combination welding and cutting torch, the oxygen valve nearest the hose connection must be opened before the flame adjusting valve or cutting lever will work.
13. Close the torch valve and repeat the purging process with the other gas.
14. Be sure there are no sources of possible ignition that may result in a fire.
15. With both torch valves closed, spray a leak-detecting solution on all connections, including the cylinder valves. Tighten any connection that shows bubbles (Figure 10-56).

PRACTICE 10-6

Cleaning a Cutting Tip

Using a cutting torch set that is assembled and adjusted as described in Practice 10-5 and a set of tip cleaners, you will clean the cutting tip.

1. Turn on a small amount of oxygen (Figure 10-57). This procedure is done to blow out any dirt loosened during the cleaning.

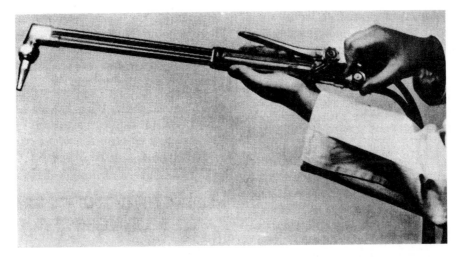

FIGURE 10–57 Turn on the oxygen valve. (Courtesy of Victor Equipment Company)

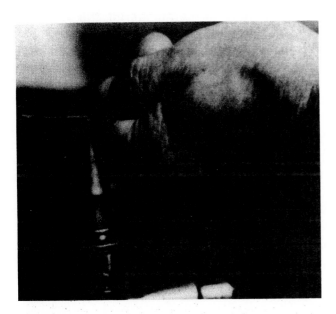

FIGURE 10–58 File the end of the tip flat.

FIGURE 10–59 A tip cleaner should be used to clean the flame and center cutting holes.

2. The end of the tip is first filed flat using the file provided in the tip cleaning set (Figure 10-58).
3. Try several sizes of tip cleaners in a preheat hole until the correct size cleaner is determined. It should easily go all the way into the tip (Figure 10-59).
4. Push the cleaner in and out of each preheat hole several times. Tip cleaners are small, round files. Excessive use of them will greatly increase the orifice (hole) size.
5. Next, depress the cutting lever and, by trial and error, select the correct size tip cleaner for the center cutting orifice.

A tip cleaner should never be forced. If the tip needs additional care, refer back to the section on tip care in Chapter 8.

PRACTICE 10–7

Lighting the Torch

Wearing welding goggles, gloves, and any other required personal protective clothing, and with a cutting torch set that is safely assembled, you will light the torch.

1. Set the regulator working pressure for the tip size. If you don't know the correct pressure for the tip, start with the fuel set at 5 psig (35 kPag) and the oxygen set at 25 psig (170 kPag).
2. Point the torch tip upward and away from any equipment or other students.
3. Turn on just the acetylene valve and use only a sparklighter to ignite the acetylene. The torch may not stay lit. If this happens, close the valve slightly and try to relight the torch.
4. If the flame is small, it will produce heavy black soot and smoke. In this case, turn the flame up to stop the soot and smoke. The auto body technician need not be concerned if the flame jumps slightly away from the torch tip.
5. With the acetylene flame burning smoke free, slowly open the oxygen valve, and by using only the oxygen valve, adjust the flame to a neutral setting.
6. When the cutting oxygen lever is depressed, the flame may become slightly carbonizing. This may occur because of a drop in line pressure due to the high flow of oxygen through the cutting orifice.
7. With the cutting lever depressed, readjust the preheat flame to a neutral setting.

The flame will become slightly oxidizing when the cutting lever is released. Since an oxidizing flame is hotter than a neutral flame, the metal being cut will be preheated faster. When the cut is started by depressing the lever, the flame automatically returns to the neutral setting and does not oxidize the top of the plate.

Extinguish the flame by first turning off the acetylene and then the oxygen.

CAUTION: Sometimes with large cutting tips the tip will pop when the acetylene is turned off first. If that happens, turn the oxygen off first.

SELECTING THE CORRECT TIP AND SETTING THE PRESSURE

Each welding equipment manufacturer uses its own numbering system to designate the tip size. It would be impossible to remember each of the systems. Each manufacturer, however, does relate the tip number to the numbered drill size used to make the holes. On the back of most tip cleaning sets, the manufacturer lists the equivalent drill size of each tip cleaner. By remembering approximately which tip cleaner was used on a particular tip for a metal thickness range, an auto body technician can easily select the correct tip when using a new torch set. Using the tip cleaner that you are familiar with, try it in the various torch tips until you find the correct tip that the tip cleaner fits. Table 10-7 lists the tip drill size, the pressure range, and the metal thickness range for which the tip can be used.

TABLE 10–7:   CUTTING PRESSURE AND TIP SIZE

Metal Thickness		Center Orifice Size		Oxygen Pressure		Acetylene	
inch	(mm)	No. Drill Size	Tip Cleaner No.*	lb/inch²	(kPa)	lb/inch²	(kPa)
1/8	(3)	60	7	10	(70)	3	(20)
1/4	(6)	60	7	15	(100)	3	(20)
3/8	(10)	55	11	20	(140)	3	(20)
1/2	(13)	55	11	25	(170)	4	(30)
3/4	(19)	55	11	30	(200)	4	(30)
1	(25)	53	12	35	(240)	4	(30)
2	(51)	49	13	45	(310)	5	(35)
3	(76)	49	13	50	(340)	5	(35)
4	(102)	49	13	55	(380)	5	(35)
5	(127)	45	**	60	(410)	5	(35)

*The tip cleaner number when counted from the small end toward the large end in a standard tip cleaner set
**Larger than normally included in a standard tip cleaner set

PRACTICE 10–8

Setting the Gas Pressures

Setting the working pressure of the regulators can be done either by following a table or by watching the flame.

1. To set the regulator by watching the flame, first set the acetylene pressure at 2 psig to 4 psig (14 kPag to 30 kPag) and then light the acetylene flame.
2. Open the acetylene torch valve one to two turns and reduce the regulator pressure by backing out the setscrew until the flame starts to smoke.
3. Increase the pressure until the smoke stops, and then increase it just a little more.

This is the maximum fuel gas pressure the tip needs. With a larger tip and a longer hose, the pressure must be set higher. This is the best setting, and it is the safest one to use. With this lowest possible setting, there is less chance of a leak. If the hoses are damaged, the resulting fire will be much smaller than a fire burning from a hose with a higher pressure. There is also less chance of a leak with the lower pressure.

4. With the acetylene adjusted so that the flame just stops smoking, slowly open the torch oxygen valve.
5. Adjust the torch to a neutral flame. When the cutting lever is depressed, the flame will become carbonizing, not having enough oxygen pressure.
6. While holding the cutting lever down, increase the oxygen regulator pressure slightly. Readjust the flame, as needed, to a neutral setting by using the oxygen valve on the torch.
7. Increase the pressure slowly and readjust the flame as you watch the length of the clear-cutting stream in the center of the flame (Figure 10-60). The center stream will stay fairly long until a pressure is reached that causes turbulence, disrupting the cutting stream. This turbulence will cause the flame to shorten in length considerably.
8. With the cutting lever still depressed, reduce the oxygen pressure until the flame lengthens once again. This is the maximum oxygen pressure that this tip can use without disrupting

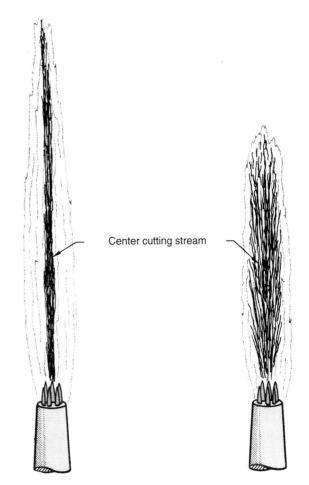

Center cutting stream

FIGURE 10–60 *Watching the length of the clear-cutting stream.*

turbulence in the cutting stream. This turbulence will cause a very poor cut. The lower pressure also will keep the sparks from being blown a longer distance from the work (Figure 10-61).

THE CHEMISTRY OF A CUT

The oxyfuel gas cutting torch works when the metal being cut rapidly oxidizes or burns. This rapid oxidization or burning occurs when a high-pressure stream of pure oxygen is directed on the metal after it has been preheated to a temperature above its kindling point. Kindling point is the lowest temperature at which a material will burn. The kindling temperature of iron is 1,600° F (870° C), which is indicated by a dull red

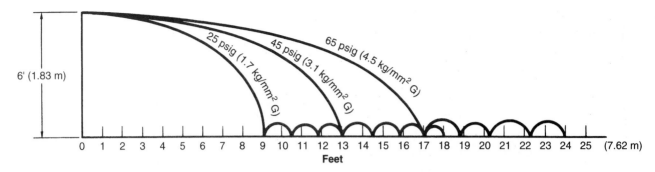

FIGURE 10–61 *The sparks from cutting a 3/8-inch (10 mm) mild steel plate, 6 ft (1.8 m) from the floor will be thrown much farther if the cutting pressure is too high for the plate thickness. These cuts were made with a Victor cutting tip no. 0-1-101 using 25 psig (1.7 kg/mm²) as recommended by the manufacturer and by excessive pressures of 45 psig (3.1 kg/mm²) and 65 psig (4.5 kg/mm²).*

color. Note that iron is the pure element and cast iron is an alloy primarily of iron and carbon. The process will work easily on any metal that will rapidly oxidize, such as iron, low-carbon steel, magnesium, titanium, and zinc.

CAUTION: Some metals release harmful oxides when they are cut. Extreme caution must be taken when cutting used, oily, dirty, or painted metals. They often produce very dangerous fumes when they are cut. You may need extra ventilation and a respirator to be safe.

The process is most often used to cut iron and low-carbon steels because, unlike most of the metals, little or no oxide is left on the metal, and it can easily be welded.

THE PHYSICS OF A CUT

As a cut progresses along a plate, a record of what happened during the cut is preserved along both sides of the kerf. This record indicates to the auto body technician what was correct or incorrect with the preheat flame, cutting speed, and oxygen pressure.

Preheat

The size and number of preheat holes in a tip affect both the top and bottom edges of the metal. An excessive amount of preheat flame results in the top edge of the plate being melted or rounded off. In addition, an excessive amount of hard-to-remove slag is deposited along the bottom edge. If the flame is too small, the travel speed must be slower. A reduction in

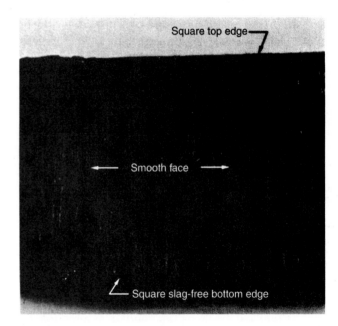

FIGURE 10–62 Correct cut.

speed may result in the cutting stream wandering from side to side. The torch tip can be raised slightly to eliminate some of the damage caused by too much preheat. However, raising the torch tip causes the cutting stream of oxygen to be less forceful and less accurate.

Speed

The cutting speed should be fast enough so that the drag lines have a slight slant backward if the tip is held at a 90-degree angle to the plate (Figure 10-62). If the cutting speed is too fast, the oxygen stream may not have time to go completely through the metal,

FIGURE 10–63   Too fast a travel speed resulting in an incomplete cut; too much preheat and the tip is too close, causing the top edge to be melted and removed.

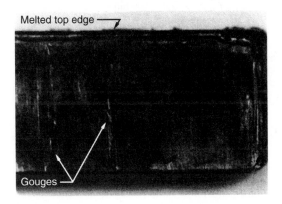

FIGURE 10–64   Too slow a travel speed results in the cutting stream wandering, thus causing gouges in the surface; preheat flame is too close, melting the top edge.

resulting in an incomplete cut (Figure 10-63). Too slow a cutting speed results in the cutting stream wandering, thus causing gouges in the sides of the cut (Figures 10-64 and 10-65).

Pressure

A correct pressure setting results in the sides of the cut being flat and smooth. A pressure setting that is too high causes the cutting stream to expand as it leaves the tip, resulting in the sides of the kerf being

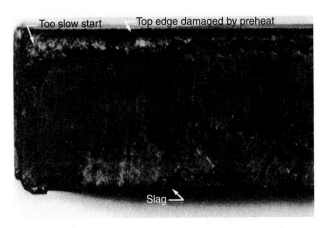

FIGURE 10–65   Too slow a travel speed at the start; too much preheat.

slightly dished (Figure 10-66). When the pressure setting is too low, the cut may not go completely through the metal.

Slag

The two types of slag produced during a cut are soft slag and hard slag. Soft slag is very porous, brittle, and easily removed from a cut. There is little or no unoxidized iron in it. It may be found on some good cuts. Hard slag may be mixed with soft slag. Hard slag is attached solidly to the bottom edge of a cut, and it requires a lot of chipping and grinding to be removed. There is 30 to 40 percent or more unoxidized iron in hard slag. The higher the unoxidized iron content, the more difficult the slag is to remove. Slag is found on bad cuts and is caused by dirty tips, too much preheat, too slow a travel speed, too short a coupling distance, or incorrect oxygen pressure.

The slag from a cut may be kept off one side of the plate being cut by slightly angling the cut toward the scrap side of the cut (Figure 10-67). The angle needed to force the slag away from the good side of the plate may be as small as 2 or 3 degrees. This technique works best on thin sections; on thicker sections the bevel may show.

METHODS OF IMPROVING CUTS

Auto body technicians can use a variety of techniques to improve the quality of the cuts they make. For example, a piece of angle iron can be clamped to the plate being cut. The angle iron can be used to guide the torch for either square or bevel cuts (Figure 10-68).

Correct cut

Top edge square

Face smooth

Bottom square

Oxides, if any, easily removed

Preheat flames too high above the surface

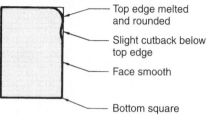

Top edge melted and rounded

Slight cutback below top edge

Face smooth

Bottom square

Travel speed too slow

Top edge rounded

Face gouged

Bottom rough

Oxides hard to remove

Preheat flames too close to the surface

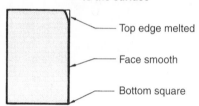

Top edge melted

Face smooth

Bottom square

Travel speed too fast

Top edge sharp

Drag lines pronounced

Bottom rounded

Cutting oxygen pressure too high

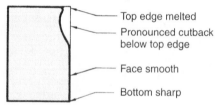

Top edge melted

Pronounced cutback below top edge

Face smooth

Bottom sharp

FIGURE 10–66   Profile of flame-cut plates.

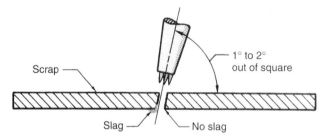

Scrap

1° to 2° out of square

Slag

No slag

FIGURE 10–67   Keeping slag off the good side of a cut.

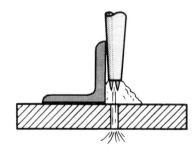
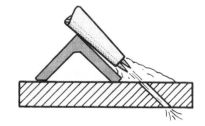

FIGURE 10–68   Using angle irons to aid in making cuts.

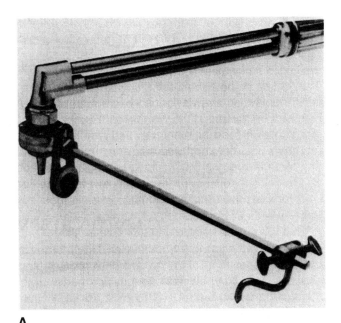

A

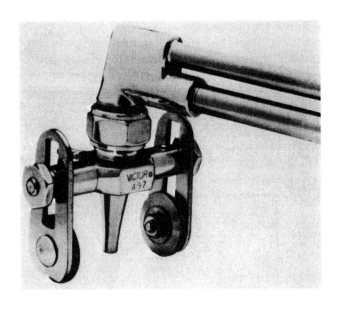

B

FIGURE 10–69 *Devices that are used to improve (A) circular or (B) straight hand cutting. (Courtesy of Victor Equipment Company)*

Devices such as guide rollers for cutting straight lines, circle cutting attachments, and power rollers may be used to reduce operator fatigue and improve the quality of the cuts made (Figure 10-69). These devices work best on new stock when repetitive cuts are being made.

PRACTICE 10–9

Flat, Straight Cut in Thin Plate

Using a properly lit and adjusted cutting torch and one piece of mild steel plate, 6 inches (152 mm) long × 1/4 inch (6 mm) thick, you will cut off 1/2-inch (13-mm) strips. Using a straightedge and soapstone, make several straight lines 1/2 inch (13 mm) apart. Starting at one end, make a cut along the entire length of plate. The strip must fall free, be slag free, and be within ±3/32 inch (2 mm) of a straight line and ±5 degrees from being square. Repeat this procedure until the cut can be made straight and slag free. Turn off the cylinder valves, bleed the hoses, back out the pressure regulators, and clean up your work area when you are finished cutting.

PRACTICE 10–10

Flat, Straight Cut in Thick Plate

Using a properly lit and adjusted cutting torch and one piece of mild steel plate, 6 inches (152 mm) long × 1/2 inch (13 mm) thick or thicker, you will cut off 1/2-inch (13-mm) strips. Note: Remember that starting a cut in thick plate will take longer, and the cutting speed will be slower. Lay out, cut, and evaluate the cut as was done in Practice 10-9. Repeat this procedure until the cut can be made straight and slag free. Turn off the cylinder valves, bleed the hoses, back out the pressure regulators, and clean up your work area when you are finished cutting.

PRACTICE 10–11

Flat, Straight Cut in Sheet Metal

Use a properly lit and adjusted cutting torch and a piece of mild steel sheet that is 10 inches (254 mm) long and 18 gauge to 11 gauge thick. Holding the torch at a very sharp leading angle (Figure 10-70), cut

FIGURE 10–70 *Cut the sheet metal at a very sharp angle.*

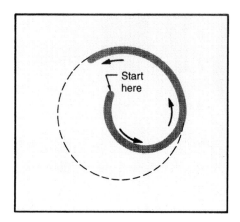

FIGURE 10–71 *Start a cut for a hole near the middle.*

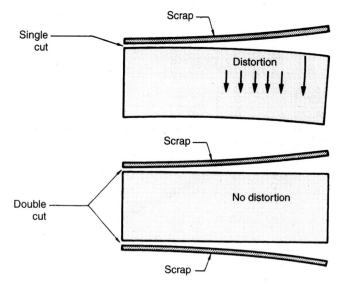

FIGURE 10–72 *Making two parallel cuts at the same time will control distortion.*

the sheet along the line. The cut must be smooth and straight with as little slag as possible. Repeat this procedure until the cut can be made flat, straight and slag free. Turn off the cylinder valves, bleed the hoses, back out the pressure regulators, and clean up your work area when you are finished cutting.

PRACTICE 10–12

Flame-Cutting Holes

Using a properly lit and adjusted cutting torch, welding gloves, appropriate eye protection and clothing, and one piece of mild steel plate 1/4 inch (6 mm) thick, you will cut holes with diameters of 1/2 inch (13 mm) and 1 inch (25 mm). Using the technique described for piercing a hole, start in the center and make an outward spiral until the hole is the

desired size (Figure 10-71). The hole must be within ±3/32 inch (2 mm) of being round and ±5 degrees of being square. The hole may have slag on the bottom. Repeat this procedure until both small and large sizes of holes can be made within tolerance. Turn off the cylinder valves, bleed the hoses, back out the pressure regulators, and clean up your work area when you are finished cutting.

DISTORTION

Distortion occurs when the metal bends or twists out of shape as a result of being heated during the cutting process. This is a major problem when cutting a plate. If the distortion is not controlled, the end product might be worthless. There are two major methods of controlling distortion. One method involves making two parallel cuts on the same plate at the same speed and time. Because the plate is heated evenly, distortion is kept to a minimum (Figure 10-72).

The second method involves starting the cut a short distance from the edge of the plate and skipping other short tabs every 2 feet (0.6 m) to 3 feet (0.9 m) to keep the cut from separating. Once the plate cools, the remaining tabs are cut (Figure 10-73).

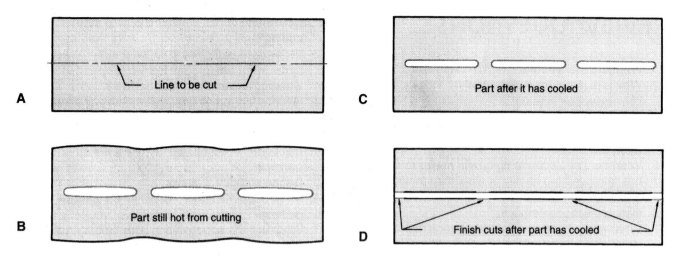

FIGURE 10–73 *Steps used during cutting to minimize distortion.*

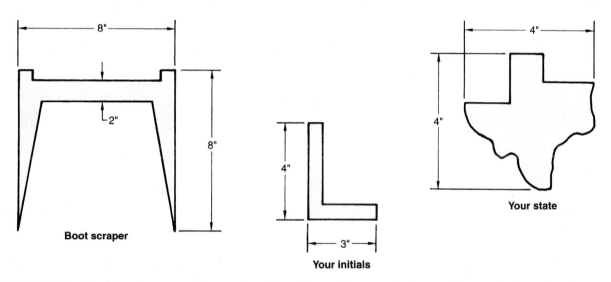

FIGURE 10–74 *Suggested patterns for practice. Note: Sizes of these and other cutting projects can be changed to fit available stock.*

PRACTICE 10-13.

Cutting Out Internal and External Shapes

Using a properly lit and adjusted cutting torch, welding gloves, appropriate eye protection and clothing, and one piece of plate, 1/4 inch (6 mm) to 3/8 inch (10 mm) thick, you may lay out and cut out one of the sample patterns shown in Figure 10-74, or any other design available.

Choose the pattern that best fits the piece of metal you have and mark it using a center punch. The exact size and shape of the layout are not as important as the accuracy of the cut. The cut must be made so that the center-punched line is left on the part and so that there is no more than 1/8 inch (3 mm) between the cut edge and the line. Repeat this practice until the cut can be made within tolerance. Turn off the cylinder valves, bleed the hoses, back out the pressure regulators, and clean up your work area when you are finished cutting.

REVIEW QUESTIONS

1. What are the different types of plasma arcs?
2. Why is plasma arc cutting ideal for today's collision repair industry?
3. Define plasma (electrical form).
4. What is used to cool low- and high-power torches?
5. What type of electrode tip on the newer torches has improved the quality of work they produce, and why?
6. Define electrode setback.
7. What would happen if your torch required deionized water and you didn't use it?
8. What is one of the advantages of using the plasma process over oxyfuel?
9. Describe how to reduce distortion when using the oxyfuel process.
10. Before the use of nitrogen as a shielding gas, what were the only metals cut by the plasma process?
11. How much faster is plasma cutting than oxyfuel cutting?
12. In the case of thicknesses totaling 2-1/2 inches or more, what determines the decision about whether to use plasma or oxyfuel?
13. What is the maximum cost-effective thickness of carbon steel, stainless steel, and aluminum that can be cut with the PAC process?
14. List some of the metals cut using the PAC process.
15. Define standoff distance.
16. What happens if there is too much standoff distance?
17. What are the two methods used to establish a current path through the gas?
18. What are some of the factors that will affect the width of the kerf?
19. What are some of the effects that changing the PAC gas(es) will have on the cut?
20. What problems will too high or too low a gas flow rate cause for a PAC weld?

Chapter

11

Plastic Repair Materials and Equipment

Objectives

After reading this chapter, you should be able to:

- List some of the growing number of plastic automotive applications.
- Explain the difference between the two types of plastics used in automobiles.
- Identify automotive plastics through the use of national identification symbols, the burn test, and by making a trial-and-error weld.
- Describe the basic differences between welding metal and welding plastic.
- Describe the setup and shutdown procedures for a typical hot-air welder.
- Explain the ways airless plastic welding compares with the hot-air method.
- Explain the principles of ultrasonic welding.

In 1975, plastic represented only about 3 percent of the total weight of the average automobile. In the past few years, more and more plastic has been used in various parts of car and light truck bodies, particularly in the front end: in bumper and fender extensions, soft front fascia, fender aprons, grille opening panels, stone shields, instrument panels, trim panels, and elsewhere. The reason is simply that many of the new reinforced plastics are nearly as strong and rigid as steel, and some are even more stable dimensionally. Many such plastics are also extremely corrosion resistant. Figure 11-1 illustrates the difference between today's automobiles and those made just ten years ago. A recent study by Market Search, Inc., of Toledo, Ohio, suggests that this growth is only the beginning. By 1998, the combined use of plastics in vehicle manufacturing is predicted to rise to nearly 3.3 billion pounds, up from 2.2 billion pounds in 1988 (Figure 11-2).

The Saturn Corporation leads the way in the current "plastics boom." The popular Saturns have all of their exterior body panels made from various plastics. Future expected applications for automotive plastics include gas tanks, intake manifolds, composite springs, bumpers, energy absorber systems, and even entire frames.

Because plastic parts are much lighter than sheet metal, they have become an important part of every car manufacturer's fuel-saving, weight reduction program. Because of the high strength-to-weight ratio of plastic, the weight decrease does not mean a decrease in strength. Every indication is that plastic body parts are here to stay, with new applications constantly being found. Therefore, plastic repairs can be expected to be a permanent and increasingly prominent part of life in every collision repair facility. Figure 11-3 shows where plastic parts are commonly used in a typical automobile.

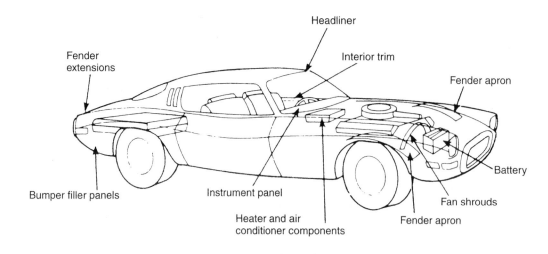

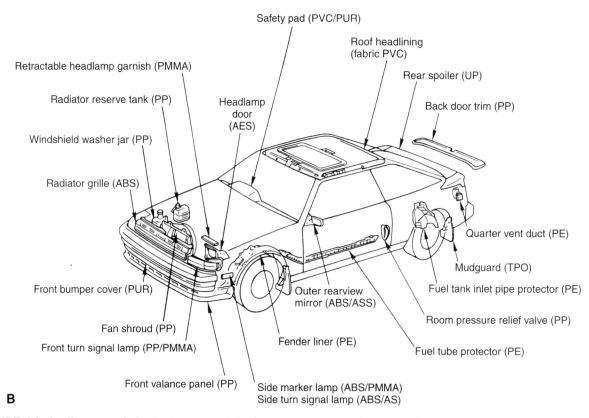

FIGURE 11-1 *The use of plastics in automobiles has grown greatly from (A) 10 years ago to (B) today.*

This increasing use of plastic in the automobile industry has resulted in new approaches to collision damage repair. Many damaged plastic parts can be repaired more economically than they can be replaced, especially if the part does not have to be removed from the vehicle. Cuts and cracks, gouges, tears, and punctures are all repairable, and, when necessary, some plastics can also be re-formed after distortion from their original shape. Repair is quicker as well, since replacement parts are not always available; this means less downtime for the vehicle.

Another factor underlining the importance of learning plastic repair technology is the changing attitude of insurance companies. Simply put, they are much less inclined nowadays to pay for operations such as dash pad removal and replacement because of the considerable time and expense involved. This means that the collision repair technician faced with a badly torn or dented plastic dash has one of two choices: either repair it or make the replacement at his or her own expense.

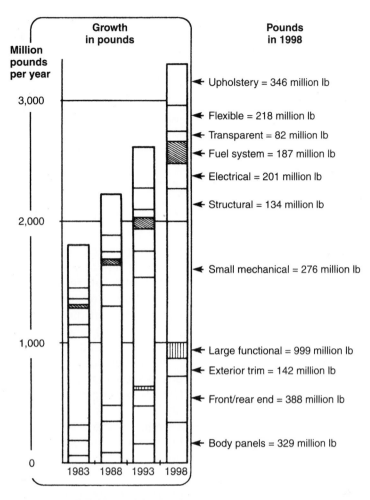

FIGURE 11–2 Projected growth of plastics in automotive applications.

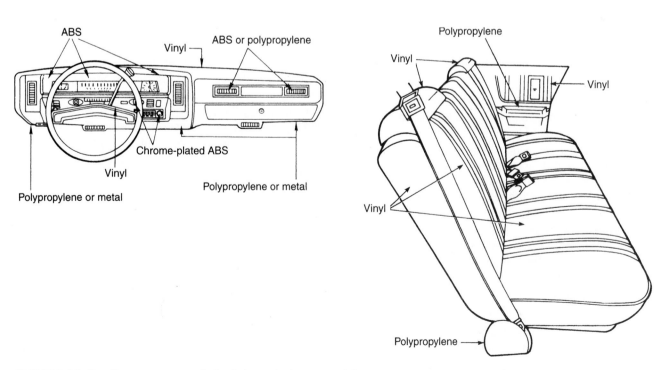

FIGURE 11–3 Common uses of plastic in today's automobile.

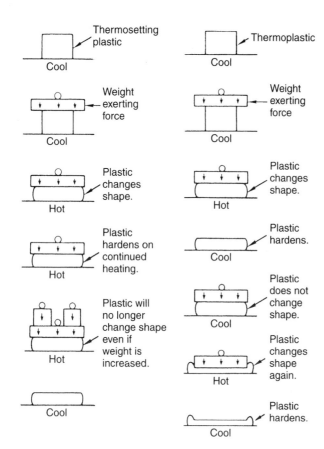

FIGURE 11–4   The effect of heat on plastics.

TYPES OF AUTOMOTIVE PLASTICS

Two types of plastics are used today in automotive production:

1. **Thermoplastics.** These plastics are capable of being repeatedly softened and hardened by heating or cooling, with no change in their appearance or chemical makeup. They soften or melt when heat is applied to them and therefore are weldable with a plastic welder.

2. **Thermosetting Plastics.** These plastics undergo a chemical change by the action of heating, a catalyst, or ultraviolet light, leading to an infusible state. They are hardened into a permanent shape that cannot be altered by reapplying either heat or catalysts. Thermosets are not weldable, although they can be "glued" using an airless welder.

Figure 11-4 explains the relationship of heat to plastics more fully. In general, the recommended

repair techniques are chemical adhesive bonding for thermosetting plastics and welding for thermoplastics. Without a doubt, the ever-growing use of both of these materials has helped to transform plastic welding from a curious technique practiced by a select few into an important skill that has been accepted as an integral part of today's collision repair facility. Following are brief descriptions of the most common automotive plastics.

ACRYLONITRILE/BUTADIENE/ STYRENE (ABS)

Acrylonitrile/butadiene/styrene, or ABS, is a thermoplastic that has excellent forming properties for most applications. ABS is available in normal and high-temperature types, either of which can be welded. Conventional ABS should be welded with air or inert gas at approximately 350 to 400°F (175 to 200°C), and high-temperature ABS at 500 to 550°F (260 to 290°C).

ABS does not have a sharp melting point but softens gradually after reaching its heat distortion point. Normal welding rod pressure is sufficient for holding an approximate 60-degree welding angle. Visual observation of the wavy flow pattern along the edges of the deposited rod is the best way to monitor ABS weld quality.

POLYETHYLENE (PE)

Oxidation is an important consideration when working with this thermoplastic. For this reason, nitrogen is the recommended gas for welding polyethylene. It also helps achieve the maximum weld strength. Because a very thin coating of oxidized film can adversely affect the final welded product, best results are obtained by welding immediately after cutting the polyethylene, then removing the oxidized film from the welding rod with fine sandpaper or by scraping with a knife.

The rod and base material should be of the same composition, since this affects the weld bond. Use large-sized rods for polyethylene wherever possible, and inspect for poor tolerance and stress. The term *stress cracking,* often included in discussions about polyethylene, refers to cracking or splitting of the material under certain conditions, including chemical reaction, heat, or stress. It results from welding materials of slightly different composition, welding at improper temperatures, and subjecting materials to undue stress or chemical attack (Figure 11-5). Correct weld speeds should always be used to avoid stress; maximum polyethylene weld strength is achieved 10 hours after the weld is completed.

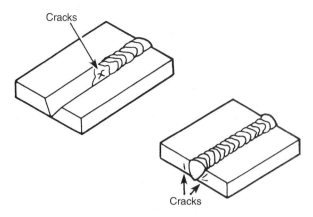

FIGURE 11-5 *Examples of stress cracks.*

FIGURE 11-6 *Most of the panels of a Saturn are made from plastics. (Courtesy of Larry Maupin)*

POLYVINYL CHLORIDE (PVC)

Polyvinyl chloride, or PVC, is another thermoplastic whose use in automobile applications is on the rise. Three important factors influence the welding of this plastic: the type of PVC used (normal or high impact), the amount of plasticizer used, and the quality of the welding rod. Plasticizers are liquids or compounds added during extrusion to make PVC plastic more flexible. They are also used in some welding rods to improve weld quality; under normal conditions, a 10 percent plasticized rod yields improved performance and strength. However, at high temperatures, a plasticized rod decreases in strength, but an unplasticized rod does not.

POLYPROPYLENE (PP)

A higher-temperature thermoplastic that is quite similar to polyethylene, polypropylene is susceptible to stress cracking and oxidation. It requires the use

of nitrogen to obtain maximum weld strength. Some splashing of molten plastic will occur when you are welding polypropylene, but this does not affect the weld and can be eliminated by throttling down the airflow.

POLYURETHANE (PUR)

Polyurethane is unique in that it is available as both a thermoplastic and a thermosetting plastic. Because it is extremely lightweight and formable, it has excellent aerodynamic properties. Polyurethane is also very durable and noncorrosive, thus making it popular with both domestic and foreign carmakers (Figure 11-6).

11.2 PLASTIC IDENTIFICATION

In 1980, General Motors published the names of seventy-nine different plastics used to one extent or another in the auto industry. With such a variety of plastics in existence—and new ones constantly being tested and refined—it is easy for the uninformed technician to be reduced to playing a guessing game when trying to do a repair. This, in turn, jeopardizes the entire job, because a repair that is done based on incorrect identification will likely yield unsatisfactory results. Misidentification increases the chances of using not only the wrong repair products, but also the wrong refinishing techniques and procedures. A repair that initially appears to be sound can quickly delaminate, crack, or discolor if the job was performed based on incorrect identification of the plastic.

There are several ways to identify an unknown plastic. One possibility is the use of national identification symbols, which can often be found on the parts to be repaired. When parts are not identified by these symbols, refer to the manufacturer's technical literature for plastic identification. Domestic manufacturers are using identification symbols more and more; unfortunately, there are still some who do not. Keep in mind that one problem with this system is that it is usually necessary to remove the part to read the letters.

One of the best ways to identify an unknown plastic is to become familiar with the different types of plastics and where they are commonly used. Table 11-1 gives the identification symbols, chemical and common names, and applications of common automotive plastics. Plastic applications information can often be found in shop manuals or in special manufacturer's guides.

TABLE 11–1:   STANDARD SYMBOL, CHEMICAL NAME, TRADE NAME, AND DESIGN APPLICATIONS OF MOST COMMONLY USED PLASTICS

Symbol	Chemical Name	Common Name	Design Applications	Thermosetting or Thermoplastic
ABS	Acrylonitrile-butadiene-styrene	ABS, Cycolac, Abson, Kralastic, Lustran, Absafil, Dylel	Body panels, dash panels, grilles, headlamp doors	Thermoplastic
ABS/MAT	Hard ABS reinforced with fiberglass	—	Body panels	Thermosetting
ABS/PVC	ABS/Polyvinyl chloride	ABS Vinyl	—	Thermoplastic
EP	Epoxy	Epon, EPO, Epotuf, Araldite	Fiberglass body panels	Thermosetting
EPDM	Ethylene-propylene-diene-monomer	EPDM, Nordel	Bumper impact strips, body panels	Thermosetting
PA	Polyamide	Nylon, Capron, Zytel, Rilsan, Minlon, Vydyne	Exterior finish trim panels	Thermosetting
PC	Polycarbonate	Lexan, Merlon	Grilles, instrument panels, lenses	Thermoplastic
PE	Polyethylene	Dylan, Fortiflex, Marlex, Alathon, Hi-fax, Hosalen, Paxon	Inner fender panels, interior trim panels, valances, spoilers	Thermoplastic
PP	Polypropylene	Profax, Olefo, Marlex, Olemer, Aydel, Dypro	Interior mouldings, interior trim panels, inner fenders, radiator shrouds, dash panels, bumper covers	Thermoplastic
PRO	Polyphenylene oxide	Noryl, Olefo	Chromed plastic parts, grilles, headlamp doors, bezels, ornaments	Thermosetting
PS	Polystyrene	Lustrex, Dylene, Styron, Fostacryl, Duraton	—	Thermoplastic
PUR	Polyurethane	Castethane, Bayflex	Bumper covers, front and rear body panels, filler panels	Thermosetting
PVC	Polyvinyl chloride	Geon, Vinylete, Pliovic	Interior trim, soft filler panels	Thermoplastic
RIM	"Reaction injection molded" polyurethane	—	Bumper covers	Thermosetting
R RIM	Reinforced RIM-polyurethane	—	Exterior body panels	Thermosetting
SAN	Styrene-acrylonitrite	Lustran, Tyril, Fostacryl	Interior trim panels	Thermosetting
TPR	Thermoplastic rubber	—	Valance panels	Thermosetting
TPUR	Polyurethane	Pellethane, Estane, Roylar, Texin	Bumper covers, gravel deflectors, filler panels, soft bezels	Thermoplastic
UP	Polyester	SMC, Premi-glas, Selection Vibrin-mat	Fiberglass body panels	Thermosetting

Another technique for identifying various types of plastic is the so-called burn test or melt test. Different plastics have different burn characteristics, and some produce unique odors.

To do a burn test, scrape off a small sliver of the material from the backside or underside of the part, hold it with pliers or on the end of a piece of wire, and try to ignite it with a match or propane torch (Figure 11-7). The burn characteristics of common automotive plastics are given in Table 11-2.

The burn test has fallen out of favor as a technique for identifying plastics for several reasons. For one, having an open flame in a shop environment creates a potential fire hazard. And although the burn test does not generally pose a health threat, it is a good idea to avoid inhaling the fumes. Another reason the burn test is not as popular as it once was is that it is not always reliable. For example,

it is difficult to determine the difference between polypropylene and polyethylene, because both burn with the same characteristics. Furthermore, some parts are now being manufactured from "hybrids," which are blended plastics that use more than one ingredient. A burn test is of no help here.

The best means of identifying an unknown plastic (assuming it is probably a thermoplastic that is potentially weldable) is to make a trial-and-error weld (Figure 11-8). Try several different filler rods until one sticks. Most suppliers of plastic welding equipment offer only half a dozen or so different types of plastic filler rods, so the range of possibilities is not that great. The rods are color coded, so once the rod that works is found, the base material is identified. For more specifics on working with plastics, see Table 11-3.

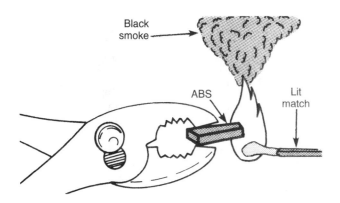

FIGURE 11-7 _Conducting a burn test on ABS plastic._

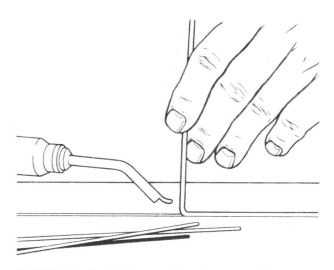

FIGURE 11-8 _Making a trial-and-error weld._

TABLE 11–2:	BURN CHARACTERISTICS OF COMMON PLASTICS
Plastic	**Burn Characteristic**
Polypropylene (PP)	Burns with no visible smoke and continues to burn once the flame source is removed. Produces a smell like burned wax. Bottom of flame is blue and top is yellow.
Polyethylene (PE)	Also smells like burned wax, makes no smoke, and continues to burn once the flame source is removed. Bottom of flame is blue and top is yellow.
ABS	Burns with a thick, black, sooty smoke and continues to burn when the flame source is removed. Produces a sweet odor when burned. Flame is yellowish orange.
PVC	Only chars and does not support a flame when you try to burn it. Gives off gray smoke and an acid-like smell. End of flame is yellowish green.
Thermoplastic Polyurethane (TPUR) and Thermosetting Polyurethane (PUR)	Burns with a yellow-orange sputtering flame and gives off black smoke. The thermoset version of polyurethane, however, will not support a flame.

TABLE 11–3:   HANDLING PRECAUTIONS FOR PLASTICS

Code	Material Name	Heat Resisting Temperature* °F	Resistance To Alcohol or Gasoline	Notes
AAS	Acrylonitrile Acrylic Rubber Styrene Resin	176	Alcohol is harmless if applied only for short time in small amounts (example, quick wiping to remove grease).	Avoid gasoline and organic or aromatic solvents.
ABS	Acrylonitrile Butadiene Styrene Resin	176	Alcohol is harmless if applied only for short time in small amounts (example, quick wiping to remove grease).	Avoid gasoline and organic or aromatic solvents.
AES	Acrylonitrile Ethylene Rubber Styrene Resin	176	Alcohol is harmless if applied only for short time in small amounts (example, quick wiping to remove grease).	Avoid gasoline and organic or aromatic solvents.
EPDM	Ethylene Propylene Rubber	212	Alcohol is harmless. Gasoline is harmless if applied only for short time in small amounts.	Most solvents are harmless, but avoid dipping in gasoline, solvents, etc.
PA	Polyamide (Nylon)	176	Alcohol and gasoline are harmless.	Avoid battery acid.
PC	Polycarbonate	248	Alcohol is harmless.	Avoid gasoline, brake fluid, wax, wax removers, and organic solvents.
PE	Polyethylene	176	Alcohol and gasoline are harmless.	Most solvents are harmless.
POM	Polyoxymethylene (Polyacetal)	212	Alcohol and gasoline are harmless.	Most solvents are harmless.
PP	Polypropylene	176	Alcohol and gasoline are harmless.	Most solvents are harmless.
PPO	Modified Polyphenylene Oxide	212	Alcohol is harmless.	Gasoline is harmless if applied only for quick wiping to remove grease.
PS	Polystyrene	140	Alcohol and gasoline are harmless if applied only for short time in small amounts.	Avoid dipping or immersing in alcohol, gasoline, solvents, etc.

* Above this temperature plastic will begin to soften.

PRINCIPLES OF PLASTIC WELDING

The welding of plastics is not unlike the welding of metals. Both methods use a heat source, welding rod, and similar types of finished welds (butt joints, fillet welds, lap joints, and the like). Joints are prepared in much the same manner, and similarly evaluated for strength. Due to differences in the physical characteristics of each material, however, there are notable differences between welding metal and welding plastics.

When one is welding metal, the rod and base metal are made molten and puddled into a joint. While metals have a sharply defined melting point, plastics have a wide melting range between the temperature at which they soften and the temperature at which they char or burn. Also unlike metals, plastics are poor conductors of heat and thus difficult to heat uniformly. Because of this, the plastic filler rod and the surface of the plastic will char or burn before the material below the surface becomes fully softened. The decomposition time at welding temperature is shorter than the time required to completely soften many plastics for fusion welding. The result is that a plastic welder must work within a much smaller temperature range than a metal welder.

Because a plastic welding rod does not become completely molten and appears much the same before and after welding, a plastic weld might appear incomplete to the shop technician who is used to welding only metal. The explanation is simple: Since only the outer surface of the rod has become molten and the inner core has remained hard, the welder can exert pressure on the rod to force it into the joint and create a permanent bond. When heat is taken away, the rod reverts to its original form. Thus, even though a strong and permanent bond has been obtained between the rod and base material, the appearance of the rod is much the same as before the weld was made, except for molten flow patterns on either side of the bead.

During a plastic weld, the materials are fused together by the proper combination of heat and pressure. With the conventional hand-welding method, this combination is achieved by applying pressure on the welding rod with one hand while at the same time applying heat and a constant fanning motion to the rod and base material with hot gas from the welding torch (Figure 11-9). Successful welds require that both pressure and heat be kept constant and in proper balance. Too much pressure on the rod tends to stretch the bead and produce unsatisfactory results; too much heat will char, melt, or distort the plastic. With practice, plastic welding can be mastered as completely as metal welding.

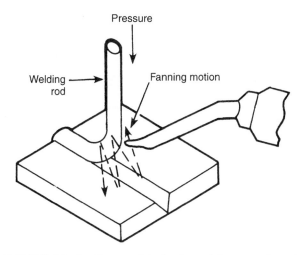

FIGURE 11-9 *Successful plastic welding requires the proper combination of heat and pressure.*

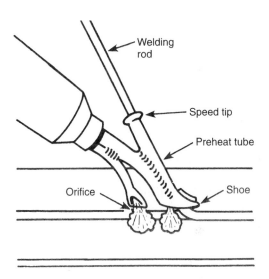

FIGURE 11-10 *Plastic welder utilizing a high-speed tip.*

11.4 HIGH-SPEED WELDING

High-speed welding incorporates the basic methods of hand welding. Its primary difference lies in the use of a specially designed and patented high-speed tip (Figure 11-10), which enables the welder to produce more uniform welds and work at a much higher rate of speed. As with hand welding, constant heat and pressure must be maintained.

The increased efficiency of high-speed welding is made possible by the preheating of both the rod and the base material before the point of fusion. The rod is preheated as it passes through a tube in the speed tip; the base material is preheated by a stream of hot air passing through a vent in the tip ahead of the fusion point. A pointed shoe on the end of the tip applies pressure on the rod, thus eliminating the need for the operator to apply pressure. At the same time, it smooths out the rod, creating a more uniform appearance in the finished weld.

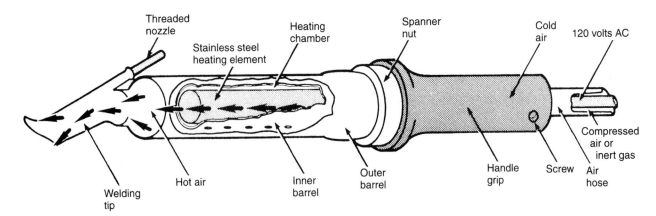

FIGURE 11-11 Typical hot-air welder.

In high-speed welding, the conventional two-hand method becomes a faster and more uniform one-hand operation. Once started, the rod is fed automatically into the preheating tube as the welding torch is pulled along the joint. High-speed tips are designed to provide the constant balance of heat and pressure necessary for a satisfactory weld. The average welding speed is about 40 inches per minute.

High-speed welding does have its disadvantages. Because increased speeds must be maintained in order to achieve the best possible weld, the high-speed welding torch is not suited for small, intricate work. Also, when the operator is new to this technique, the position in which the welder is held might seem clumsy and difficult. However, experience will enable the operator to successfully make all welds that can be made with a hand welder, including butt welds, V-welds, corner welds, and lap joint welds. Speed welds can be made on circular as well as flat work. In addition, inside welds on tanks can be speed-welded, provided the working space is not too small to manipulate the torch.

HOT-AIR PLASTIC WELDING

There are a number of manufacturers who make plastic hot-air welding equipment for the collision repair industry, and all use the same basic technology. A ceramic or stainless steel electric heating element is used to make hot air 450 to 650° F (230 to 340° C), which in turn blows through a nozzle and onto the plastic. The air supply comes from either the shop's air compressor or a self-contained portable compressor that is mounted in a carrying case that comes with the unit. Most hot-air welders use a working pressure at the tip of around 3 psi. A pair of pressure regulators is required to reduce the air pressure first to around 50 psi, and then finally to the working pressure of 2 1/2 to 3 1/2 psi. A typical hot-air welder is illustrated in Figure 11-11.

The barrel of the torch itself gets sufficiently hot so that skin contact could cause a burn if the hot air is directed against the skin long enough. The torch is used in conjunction with the welding rod, which is normally 3/16 inch (4.5 mm) in diameter and made from the same material as the plastic being repaired. This will ensure that the strength, hardness, and flexibility of the repair are the same as the damaged part. Use of the proper welding rod is very important; an adequate weld is impossible if the wrong rod is used.

One of the problems with hot-air welding is that the 3/16-inch (4.5 mm) diameter rod is often thicker than the panel to be welded. This can cause the panel to overheat before the rod has melted. Using a 1/8-inch (3 mm) rod with the hot-air welder can often correct such warpage problems. This works on the same principle as using a 1/64-inch (0.4 mm) wire instead of a 1/32-inch wire (0.7 mm) for MIG welding.

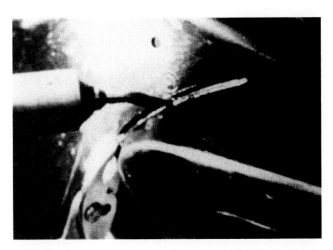

FIGURE 11-12   Tacking welding tip. (Courtesy of Seelye Inc.)

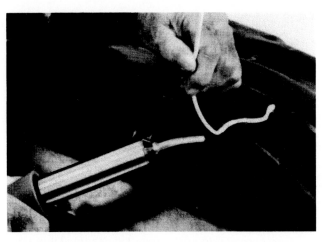

FIGURE 11-13   Round welding tip. (Courtesy of Seelye Inc.)

Three shapes of welding tips are available for use with most plastic welding torches. They are:

- **Tacking Welding Tips.** These are used to tack broken sections together before welding (Figure 11-12). If necessary, tack welds can be easily broken apart for realigning.
- **Round Welding Tips.** These tips are used to fill small holes and make short welds, welds in hard-to-reach places, and on tight or particularly sharp corners (Figure 11-13).
- **Speed-Welding Tips.** These are used for long, fairly straight welds. They hold the filler rod, automatically preheat it, and feed the rod into the weld, thus allowing for faster welding rates (Figure 11-14).

FIGURE 11-14   Speed-welding tip. (Courtesy of Seelye Inc.)

11.6   SETUP, SHUTDOWN, AND SERVICING

Naturally, no two hot-air welders are exactly alike; their design can vary from one manufacturer to another. The setup, shutdown, and service procedures that follow, while typical of all hot-air welders, should nonetheless be regarded as general guidelines only. For specific instruction, always refer to the owner's manual and other material provided by the welder manufacturer. Keep in mind that some manufacturers advise against using their welder on plastic that is any thinner than 1/8 inch (3 mm) because of the likelihood of distortion. In other cases, it is acceptable to weld plastics as thin as 1/16 inch (1.5 mm), provided they are supported from underneath during the operation. Again, it is very important to read and follow the specific directions for the welder being used.

SETUP AND SHUTDOWN

To set up a typical hot-air welder, proceed as follows:

1. Close the air pressure regulator valve by turning the control handle counterclockwise until it is loose. This will prevent possible damage to the gauge from a sudden surge of excess air pressure.

2. Connect the regulator to a supply of either compressed air or inert gas. The standard rating for an air pressure regulator is 200 pounds of line pressure. If inert gas is used, a pressure-reducing valve is needed.

3. Turn on the air supply. The starting air pressure depends on the wattage of the heating element and the air pressure. The operating air pressure requires slightly less air.

4. Connect the welder to a common 120-volt AC outlet. A three-prong grounded plug or temporary adapter must be used with the welder at all times, as shown in Figure 11-15.

5. Allow the welder to warm up at the recommended pressure. It is essential that either air or inert gas flows through the welder at all times, from warm-up to cool down, to prevent the burnout of the heating element and further damage to the gun.

6. Select the proper tip and insert it with pliers to avoid touching the barrel while hot.

7. After the tip has been installed, the temperature will increase slightly due to back pressure. Allow 2 to 3 minutes for the tip to reach the required operating temperature.

8. Check the air temperature by holding a thermometer 1/4 inch (6 mm) from the hot air end of the torch. For most thermoplastics, the temperature should be in the 450 to 650° F range. Information supplied with the welder usually includes a chart of welding temperatures.

9. If the temperature is too high to weld the material, increase the air pressure slightly until the temperature decreases. If the temperature is too low for the particular application, decrease the air pressure slightly until the temperature rises. When increasing and decreasing the air pressure, allow at least 1 to 3 minutes for the temperature to stabilize at the new setting.

10. Damage to the welder or heating element will not occur from too much air pressure; however, the element can become overheated by too little air pressure. When decreasing the air pressure, never allow the round nut that holds the barrel to the handle of the welder to become too hot to touch. This is an indication of overheating.

11. A partial clogging of the dirt screen on the regulator or a fluctuation in the line voltage can also cause over- or underheating. Watch for these symptoms.

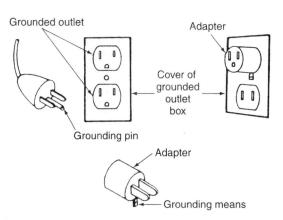

FIGURE 11-15 *Methods of grounding a plastic welder.*

12. If the threads at the end of the barrel become too tight, clean them with a good high-temperature grease to prevent seizing.

13. When the welding is finished, disconnect the electric supply and let the air flow through the welder for a few minutes or until the barrel is cool to the touch. Then disconnect the air supply.

MAINTENANCE

If it becomes necessary to change the heating element, proceed as follows:

1. While pushing the end of the barrel against a solid object, hold the handle tightly and push in. The pressure on the barrel will compress the element spring.

2. Use a spanner wrench to loosen the spanner nut. Keep the pressure on the handle and back off the nut all the way by hand.

3. Hold the barrel and place the entire welder on a bench. Remove the barrel.

4. Gently pull the element out of the handle. At the same time, unwind the cable (which has been spiraled into the handle) until it is completely out of the handle.

5. Grasp the socket at the end of the wire tightly. Rock the element while pulling until the element is disconnected.

6. To install the new element, reverse the procedure. Turn the element clockwise (about 1-1/2 turns) while pushing the wire gently back into the handle. This prevents kinking of the wire.

Other more involved servicing procedures are best left to qualified repair technicians. Many manufacturers make it clear that disassembling the welder automatically invalidates the warranty.

11.7 AIRLESS PLASTIC WELDING

Although only in existence for a relatively short time, airless welding has become very popular with the collision repair industry. Compared to the hot-air method, it is less expensive, easier to learn, simpler to use, and more versatile. Although the hot-air method utilizes 3/16-inch-diameter welding rods, 1/8-inch (3 mm) diameter rods are used with the airless method. This not only provides a quicker rod melt, it also helps to eliminate two troublesome problems: panel warpage and excess rod buildup.

When setting up the typical airless welder, the first and most important step is to put the temperature dial at the appropriate setting, depending on the specific thermosetting plastic being worked on. It is crucial that the temperature setting be correct; otherwise, the entire welding operation will be jeopardized. It will normally take about 3 minutes for the welder to fully warm up. As for the selection of the welding rod, there is another factor to consider besides size—namely, compatibility. Make sure the rod is the same as the damaged plastic, or the weld will more than likely not hold. To this end, many airless welder manufacturers provide rod application charts. When the correct rod has been chosen, it is good practice to run a small piece of it through the welder to clean out the tip before beginning.

11.8 ULTRASONIC PLASTIC WELDING

The technology of ultrasonic plastic welding was originally developed solely as a means of assembly, but it is also suitable for making repairs. Handheld systems (Figure 11-16) are available in 20 and 40 kHz; they are equally adept at welding large parts and tight, hard-to-reach areas. Welding time is controlled by the power supply. Although the use of

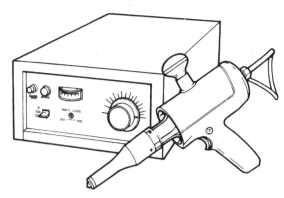

FIGURE 11–16 *Typical handheld ultrasonic welder and power supply.*

ultrasonics in collision repair is still in its infancy, the strong, clean, precise welds the process produces will likely make it increasingly popular in the years to come.

Most of the commonly used injection-molded plastics can be ultrasonically welded without the use of solvents, heat, or adhesives. Ultrasonic weldability depends on the plastic's melting temperature, elasticity, impact resistance, coefficient of friction, and thermal conductivity. Generally, the more rigid the plastic, the easier it is to weld ultrasonically. Thermoplastics such as polyethylene and polypropylene are ideal for ultrasonic welding, provided the welder can be positioned close to the joint area.

11.9 ULTRASONIC STUD WELDING

Ultrasonic stud welding, a variation of the shear joint, is a reliable technique that can be used to join plastic parts at a single point or at numerous locations. In many applications requiring permanent assembly, a continuous weld is not required. Often the size and complexity of the parts seriously limit attachment points or weld locations. With similar materials, this type of assembly can be effectively and economically accomplished using ultrasonic stud welding. The power requirement is low because of the small weld area, and the welding cycle is short—almost always less than half a second.

Figure 11-17 shows the basic stud weld joint before, during, and after welding. The weld is made along the circumference of the stud; its strength is a function of the stud's diameter and the depth of the weld. Maximum tensile strength is achieved when the depth of the weld equals half the diameter of the stud. The radial interference (dimension A) must be uniform and should generally be 0.008 to 0.012 inch for studs having a diameter of 0.5 inch or less. The hole should be a sufficient distance from the edge to prevent breakout; a minimum of 0.125 inch is recommended.

In the joint, the recess can be on the end of the stud or in the mouth of the hole, as shown in the examples. When using the latter, a small chamfer can be used for rapid alignment. To reduce stress concentration, a good-sized fillet radius should be incorporated at the base of the stud. Recessing the fillet below the surface allows flush contact of the parts.

Other ways in which the ultrasonic stud weld can be used are shown in Figure 11-18. A third dissimilar material can be locked in place, as in view A. View B shows separate molded rivets in lieu of metal self-tapping screws. Unlike metal fasteners, they produce a relatively stress-free assembly.

Figure 11-19 shows a variation that can be used where appearance is important or an uninterrupted surface is required. The stud is welded into a boss, whose outside diameter can be no less than twice the stud diameter. When welding into a blind hole, it might be necessary to provide an outlet for air. Two possibilities are shown: a center hole through the stud (view A), or a small, narrow slot in the interior wall of the boss (view B).

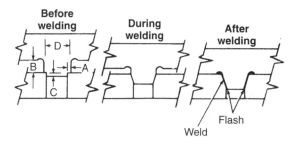

FIGURE 11–17   *Basic ultrasonic stud welding joint.*

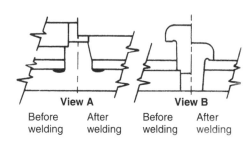

FIGURE 11–18   *Variations of the basic stud joint.*

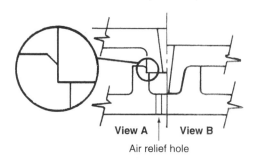

FIGURE 11–19   *Welding a stud in a blind hole.*

REVIEW QUESTIONS

1. Name some of the areas that plastics are used in cars and light truck bodies.
2. Why are the new reinforced plastics such a good choice for today's cars?
3. What are some of the anticipated uses of the new reinforced plastics?
4. What kinds of problems in plastics are more economical to repair than to replace?
5. What are the advantages to repairing a damaged plastic part rather than replacing it?
6. How is the insurance industry affecting plastic repair technology?
7. What two types of plastics are used today in automotive production?
8. What are the recommended repair techniques for plastics?
9. Define stress cracking and its causes.
10. When is the maximum polyethylene weld strength achieved?
11. List the important factors that influence welding on polyvinyl chloride.

12. List the most common automotive plastics, their advantages, and their uses.
13. What are some of the ways to identify an unknown plastic?
14. List the three shapes of welding tips available for use with most plastic-welding torches.
15. List the steps in the setup and shutdown of hot-air welders.
16. List the steps in changing the heating element of a hot-air welder.
17. List the advantages of airless plastic welding compared to hot-air welding.
18. What is the most important factor in setting up for an airless plastic weld?
19. Ultrasonic weldability depends upon what?
20. What kinds of thermoplastics are ideal for ultrasonic welding?

Chapter

12

Plastic Repair Methods

Objectives

After reading this chapter, you should be able to:

- List the key factors that must be considered when deciding whether to repair or replace a damaged part.
- Choose which repair method—adhesive bonding or welding—should be used under any given circumstances.
- Describe the various chemical adhesive bonding techniques.
- Describe the basics of hot-air and airless welding.
- Explain tack-welding, hand-welding, and speed-welding techniques.

Plastic repair, like any other kind of body repair work, begins with an estimate. That is, it must be determined if the part should be repaired or replaced. Often a comparison of costs between the two alternatives dictates the final decision. However, cost is not the only factor that must be taken into consideration. For instance, a small crack, tear, gouge, or hole in a nose fascia or large panel that is difficult to replace or that is not readily available probably indicates that a repair is in order. On the other hand, extensive damage to the same component or damage to a part that is easy to replace (for example, a fender extension or trim piece) would dictate replacement. Another factor that cannot be overlooked when making this decision is whether or not the repairer has the skill required to make this particular job a quality repair.

When all things are taken into consideration, figuring out the comparative costs of repair versus replacement is not as easy as it looks. For instance, suppose a late-model car has a cracked plastic fan shroud. A replacement shroud would cost 25 dollars. To weld it would take about 20 minutes, including the time it takes to prepare the surface. Repair versus replacement might seem like equally acceptable

choices based on the facts related so far, but there are other factors to consider: one is the labor required to remove the shroud (this could be 20 minutes if the fan has to be pulled to get the shroud off); also, a replacement part might not be readily available. Welding the shroud in place would eliminate the labor required to remove and replace it, thus tipping the economic scale in favor of repair. Repairing the existing shroud would also save the time and trouble of back ordering a shroud through a dealer.

Another example is a damaged radiator. The radiator core is in good condition, but the radiator will not hold water because the nylon end tank has a crack in it. In addition, one of the plastic mounting tabs on the end tank has been broken off. The conventional approach for someone who does not make plastic repairs would be to pull the radiator and send it to the local radiator shop. The shop would either replace the damaged end tank or, like an increasing number of shops today, would simply weld up the crack in the existing tank, weld the broken tab back in place, and send a bill for the labor. However, this type of repair could have been done in the original shop if the repairer had learned how to work with plastics.

TABLE 12–1:   PLASTIC PARTS REPAIR SYSTEMS

Key			
AR	Adhesive repair	S	Anaerobic (instant) adhesive
FGR	Fiberglass repair	PC	Patching compound
HAW	Hot-air welding	AW	Airless welding

ISO Code	Name	Repair System
ABS	Acrylonitrile-butadiene-styrene (hard)	HAW, S, FGR, AW
ABS/PVC	ABS/Vinyl (soft)	PC, AW
EPI II or TPO	Ethylene propylene	AR, AW
PA	Nylon	S, FGR, AW
PC	Lexan	S, FGR
PE	Polyethylene	HAW, AW
PP	Polypropylene	HAW, AW
PPO	Noryl	FGR, AW
PS	Polystyrene	S
PUR, RIM, or RRIM	Thermoset polyurethane	AR, AW
PVC	Polyvinyl chloride	PC, AW
SAN	Styrene acrylonitrile	HAW, AW
SMC	Sheet molded compound (polyester)	FGR
TPR	Thermoplastic rubber	AR, AW
TPUR	Thermoplastic polyurethane	AR, AW
UP	Polyester (Fiberglass)	FGR

The factors that must be considered when deciding to replace or repair a part are:

- Cost
- Extent of damage
- Ease of replacement
- Availability of replacement parts
- Skill of the repairer

If repair is the answer, it must be determined if the part has to be removed from the vehicle. The entire damaged area must be accessible for a quality repair job to be performed; if it is not, the part must be removed. Although parts do not always have to be removed, in many cases it is the only way to do the job right. Keep in mind that refinishing will have to be done as well.

12.1 ADHESIVE BONDING VERSUS WELDING

There are two methods of repairing plastics: chemical adhesives and welding. Table 12-1 indicates the best repair systems for the plastics most often used in the automotive industry.

CAUTION: Keep in mind that although the techniques described in this chapter are proven repair methods, no technique is 100 percent reliable when used to repair a fuel tank or in other critical structural applications.

Although most types of plastics can be repaired with adhesive materials, welding is usually preferred for thermoplastics because of its speed and ease. A plastic welder takes only a couple of minutes to heat up to operating temperatures; once hot, it can weld at speeds ranging from 5 to 40 inches (120 mm to 900 mm) per minute, depending on the application. It is also possible to go directly from welding to sanding and painting without waiting for the repair to cure. At most, only a few minutes are needed for the welded plastic to cool.

Many adhesives require mixing and can take anywhere from 30 minutes to several hours to cure, but heat can be used to shorten the curing time. This is not necessarily a major disadvantage, since the repair technician can always find something else to do while the adhesive cures.

As for surface preparation, both welding and adhesives require some grinding, sanding, and/or trimming in order to achieve good adhesion. More surface preparation is generally needed with adhesives, however, because the edges of the damaged area must often be featheredged before the adhesive is applied.

Some adhesives require reinforcing or support patches behind the damaged area as well, making the repair a several-step process. With plastic welding, however, there is no need to feather out the edges; V-grooving along the cracked or damaged area is generally all that is needed. Recommended V-grooves for one- and two-sided welds are shown in Figure 12-1.

Keep in mind, however, that hot-air plastic welding cannot be used on thermosetting plastic parts. Only thermoplastics soften when heated. This includes such plastics as polypropylene (PP), polyethylene (PE), soft thermoplastic polyurethane (TPUR), acrylonitrile butadiene styrene (ABS), polyamide (PA or nylon), polycarbonate (PC or Lexan), polyphenylene oxide (PPO or Noryl), polyvinyl chloride (PVC), and acrylics such as plexiglass (PMMA). As a rule, all thermoplastics—with the exception of ethylene propylene diene (EPDM)—are weldable, although some are more easily welded than others.

For plastics in the thermosetting category, an alternative to hot-air plastic welding is needed. Thermosetting plastics undergo a chemical change when they cure that prevents them from softening when heated. This makes thermosetting plastics impossible to weld, because the parent material will not melt and mix

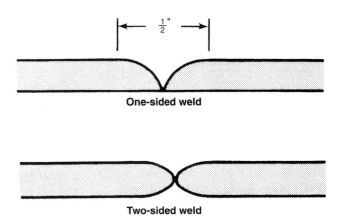

FIGURE 12-1 _Proper V-grooves._

with the filler material. Trying to heat such a plastic will only result in charring the edges. Members of this category include polyurethane (PUR), polyester (UP or fiberglass), and fiberglass-reinforced ABS and PUR.

One alternative to hot-air plastic welding is the more versatile airless welding method. As shown in Table 12-1, airless welding can be used to repair a number of thermosetting plastics, including nylon (PA), thermoset polyurethane (PUR), and thermoplastic rubber (TPR). As was mentioned in the previous chapter, one difference between hot-air plastic welding and airless plastic welding is that the airless method more closely resembles a gluing process. As a result, the airless method is more versatile.

Of course, the other alternative to hot-air plastic welding is adhesive bonding. It is suitable for nearly every type of thermosetting plastic. A surface gouge on a rigid thermoset part where structural strength is not an issue can be more economically repaired with polyester body filler. On the other hand, puncture damage that requires a backup or a structural repair that involves reinforcing the backside is best done, from a practical standpoint, with a combination of structural adhesive and polyester body filler. Since epoxy resins possess superior adhesive properties, all repair work done on the backside of the part should be done with fiberglass cloth and structural adhesive. The cosmetic repair on the front side of the part can then be completed with polyester body filler. Again, as stated earlier, information about which repair method is preferred for a particular repair can be found in shop manuals or special guides.

CHEMICAL ADHESIVE BONDING TECHNIQUES

The main advantage of adhesives is their versatility. Not all plastics can be welded, but most can be repaired with adhesives. In this section, various chemical adhesive bonding techniques are presented.

EPOXY AND URETHANE REPAIR MATERIALS

Both epoxy and urethane adhesive repair systems are two-component systems in which repair materials must be mixed in equal parts just prior to application. Some types of urethane are available in various degrees of flexibility that can be matched to the part being repaired.

Both epoxy and urethane applications are essentially the same. Whenever adhesives are used, keep the following points in mind:

- As with welding, surface preparation and cleanliness are extremely important. The part must be washed and cleaned with plastic cleaner before you start any repair.
- Both the damaged part and the repair material must be at room temperature (at least 60°F (15° C)) to achieve proper curing and adhesion.
- Both parts of the repair material must be equally and completely mixed before application.
- Whenever possible, particularly when repairing cuts or tears that go all the way through the paint, a fiberglass mat reinforcement should be used.

TWO-PART EPOXY REPAIR OF POLYOLEFIN COMPONENT

Until recently, adhesive repairs were limited to urethane plastics only. With the introduction of adhesion promoters, polyolefins can now be repaired using virtually the same adhesive bonding techniques (Table 12-2). The only difference is in the use of the adhesion promoter before each application of repair material; the rest of the procedure is the same. Automobile manufacturers are using polyolefins for large exterior parts such as fascias and bumper covers. Common polyolefins found in automotive applications include TPO, PP, and EP. To determine whether the base material is a polyolefin, grind the damaged area. If it grinds cleanly, an adhesion promoter is not needed. If the material melts or smears, it is a polyolefin and must have an adhesion promoter.

Use the following procedure to make a two-part epoxy repair of a polyolefin bumper cover.

1. Clean the entire cover with soap and water, wipe or blow-dry, then clean with a good plastic cleaner.
2. V-groove and taper the damaged area using a slow-speed grinder and 3-inch 36-grit disc. Grind about a 1-1/2-inch (37 mm) taper around the damage for good adhesion.
3. Use a sander with 180-grit paper to featheredge the paint around the damaged area; then blow dust-free. Depending on the extent of the damage, the backside might need reinforcement. To do this, follow steps 4 through 6:
4. Clean the backside with plastic cleaner, then apply a coat of adhesion promoter according to the manufacturer's recommendations.
5. Dispense equal amounts of both parts of the flexible epoxy adhesive and mix to a uniform color. Apply the material to a piece of fiberglass cloth using a plastic squeegee.
6. Attach the saturated cloth to the backside of the bumper cover and fill in the weave with additional adhesive material.
7. With the backside reinforcement in place, apply a coat of adhesion promoter to the sanded repair area on the front side. Let the adhesion promoter dry completely.
8. Fill in the groove with adhesive material. Allow it to cure according to the manufacturer's recommendations.
9. Rough-grind the repair area with 80-grit paper, then sand with 180-grit, followed by 240-grit. If additional adhesive material is needed to fill in a low spot or pinhole, be sure to apply a coat of adhesion promoter first, followed by a skim coat of adhesive material.

Never try to use flexible putty on polyolefins—it will not work.

TABLE 12–2: ADHESIVE REPAIR SYSTEMS

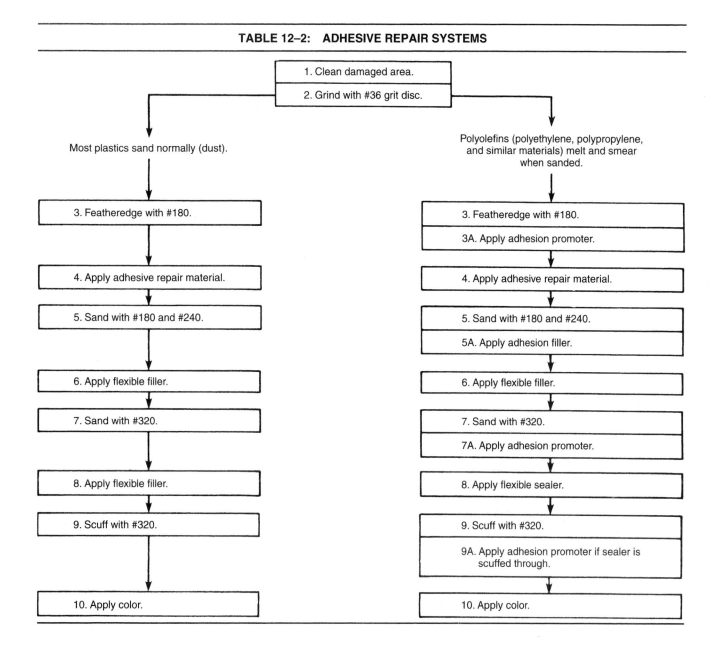

ADHESIVE REPAIR OF OTHER FLEXIBLE PARTS

For repairs to other flexible parts with adhesive materials, the procedure to be followed is similar to the one just presented with these exceptions:

1. Omit the adhesion promoter.
2. Use either a skim coat of adhesive material or a flexible filler to fill in imperfections, not both.

Minor Cut and Crack Repairs

Adhesives are usually used for repairing minor cuts and cracks. First, the repair area must be wiped clean with water and a plastic cleaner. It is very important for the mating surfaces to be clean and free of wax, dust, grease, or any other substance. It is not necessary to use solvents (other than a plastic cleaner) for cleaning plastic parts. Allow the part(s) to warm to 70° F (21° C) before applying adhesives.

After cleaning, the next step is to prepare the damage using an adhesive kit. The kit should have two elements: an accelerator and an adhesive. Spray one side of the crack with the accelerator, then apply the adhesive to the same side of the crack.

Carefully position the two sides of the cut or crack in their original position and press them together with firm pressure. Hold for a full minute to achieve good bond strength. Cure for 3 to 12 hours for maximum strength, following the instructions and precautions on the label.

If the original paint was not damaged and the repair was properly positioned, painting might not be required. Where painting is required, be sure to read the manufacturer's instructions that appear on the paint container.

Gouge, Tear, and Puncture Repair

The procedure for repairing gouges, tears, and punctures is a chemical bonding process; it requires no special tools or skills. As always, the first step is to clean around the damaged area thoroughly with a wax, grease, and silicone-removing solvent. Apply it with a water-dampened cloth, then wipe dry.

To prepare for the structural adhesive, bevel the edges of the hole back about 1/4 to 3/8 inch (6 mm to 10 mm). The beveling will produce a coarse surface for good adhesion; in any repair, the mating surfaces should be scuffed. Use a slow speed when grinding (2,000 RPM or less). If the sanded area has a "greasy" appearance, it would be a good idea to apply a coat of adhesion promoter. Also, apply more adhesion promoter after each sanding step.

The next step is to featheredge the paint around the repair area with a finer-grit disc. Remove the paint, but very little of the urethane plastic, and blend the paint edges into the plastic.

Continue removing paint until there is a paint-free band about 1 to 1 1/2 inches (25 mm to 37 mm) wide around the hole. The repair material must not overlap the painted surface. Carefully wipe off all paint and urethane dust in preparation for the next step. The repair area must be absolutely clean before proceeding.

Next, flame-treat the beveled area of the hole. This is done to improve the adhesion of the structural adhesive. Use a torch with a controlled flame, and develop a 1-inch (25 mm) cone tip. Direct the flame onto the beveled area carefully and keep it moving until the area is slightly brown. Be extremely careful not to warp the urethane or burn the paint.

The next step is to apply auto backing tape to the repair area; an aluminum foil with a strong adhesive and a moisture-proof backing is best. Clean the inner surface of the repair area with silicone and wax remover, then install the tape. Cover the hole completely with about a 1-inch adhesion surface around the edges.

Before you apply the structural repair adhesive, the back of the opening should be thoroughly cleaned. Tape it with aluminum auto body repair tape to provide support for the repair. For best results, slightly dish the tape so that the adhesive overlaps the repair area on the backside. It might be possible to install the tape without loosening or removing any parts from the car. On some makes, partial disassembly is necessary.

Prepare the adhesive repair material as directed by the manufacturer. Most adhesive compounds come in two tubes. On a clean, flat, nonporous surface such as metal or glass, squeeze out equal amounts from each tube. Then, with an even paddling motion to reduce air bubbles, completely mix the two components until a uniform color and consistency are achieved.

Paddle the adhesive into the hole, using a squeegee or plastic spreader. This must be done quickly and carefully; the adhesive material will begin to set in about 2 to 3 minutes. Two applications are usually required. The first application simply fills the bottom of the hole, so it is not necessary to worry about contour at this time. However, an attempt should be made to fill most of the hole's volume. Cure for approximately 1 hour at room temperature or 20 minutes with a heat lamp or gun at approximately 200° F (93° C).

Before the final application of adhesive, use a fine-grit disc to grind down the high spots. Wipe the dust from the repair area. After the first application has been ground and wiped clean, mix the second application of the adhesive as before, and paddle it into an overfill contour of the repair area. A flexible squeegee or spatula is useful in approximating the panel contours.

When the adhesive material has dried, establish a rough contour to the surrounding area with an 80-grit abrasive on a sanding block. Then feather-sand with a disc sander and 180-grit sandpaper, followed by 240-grit sandpaper to achieve levelness with the surface of the part. Check the repair for any low areas, pits, or pinholes. Additional material can be spread over any defects.

Final feathering and finish sanding can be done with a disc sander and 320-grit disc. When final sanding is completed, remove all dust and loose material. The surface is now ready for priming and painting.

12.3 — PLASTIC WELDING

The basic procedures for hot-air and airless welding are very similar. To make a good plastic weld with either procedure, keep the following factors in mind:

- **Welding rod material.** If it is not compatible with the base material, the weld will not hold.
- **Temperature.** Too much heat will char, melt, or distort the plastic.
- **Pressure.** Too much pressure stretches and distorts the weld.
- **Angle between rod and part.** If too shallow, a proper weld will not be achieved.
- **Speed.** If the torch movement is too fast, it will not permit a good weld; too slow a speed can char the plastic.

The basic repair sequence is generally the same for both procedures:

1. Prepare the damaged area.
2. Align the damaged area.
3. Make the weld.
4. Allow it to cool.
5. Sand. If pinholes, voids, and the like exist, bevel the edges of the defective area and add another bead of weld. Resand.
6. Paint or finish.

BASICS OF HOT-AIR WELDING

Following is a typical procedure for hot-air plastic welding:

1. Set the welder to the proper temperature (if a temperature adjustment is provided).
2. Wash and clean the part with plastic cleaner; do not use conventional gap solvents or dewaxers. To remove silicone-type materials, use a conventional cleaner first, making sure to completely remove all residue.
3. V-groove the damaged area.
4. Bevel the part 1/4 inch (6 mm) beyond the damaged area.
5. Tack-weld or tape the break line with aluminum body tape.
6. Select the welding tip best suited to the type of damage, and the proper welding rod.
7. Make the weld. Allow it to cool and cure for about 30 minutes.
8. Grind, sand, or scrape the weld to the proper contour and shape.

BASICS OF AIRLESS PLASTIC WELDING

The typical airless plastic welding procedure is as follows:

1. Wash the damaged part with soap and water and wipe or blow-dry.
2. Clean the damaged part with plastic cleaner.
3. Align and tape the broken or split sections with aluminum body tape.
4. V-groove at least 50 percent of the way through the panel for a two-sided weld and 75 percent of the way through for a one-sided weld.
5. Use a slow-speed grinder and a 60- or 80-grit disc to remove the paint from around the damaged area. Blow dust free.
6. After setting the welder to the proper temperature (Figure 12-2), slowly feed the rod into the melt tube.
7. Apply light pressure to the rod to slowly force it out into the grooved area.
8. As the rod melts, start to move the torch tip very slowly in the direction of the intended weld. Overlap the edges of the groove with melted plastic while progressing forward.
9. After completing the weld, use the flat "shoe" part of the torch tip to smooth it out.

FIGURE 12–2 Setting the temperature dial on an airless welder. (Courtesy of Urethane Supply Co.)

GENERAL WELDING TECHNIQUES

Welding plastic is not difficult when done in a careful and thorough manner. Keep in mind the following general tips:

- The welding rod must be compatible with the base material in order for the strength, hardness, and flexibility of the repair to be the same as the part. To this end, the rods are color-coded for easy identification.
- Always test a welding rod for compatibility with the base material. To do this, melt the rod onto a hidden side of the damaged part, let the rod cool, then try to pull it from the part. If the rod is compatible, it will adhere.
- Pay close attention to the temperature setting of the welder; it must be correct for the type of plastic being welded.
- Never use oxygen or other flammable gases with a plastic welder.
- Never use a plastic welder, heat gun, or similar tool in wet or damp areas. Remember, electric shock can kill.
- Become proficient at horizontal welds before attempting the more difficult vertical and overhead types.
- Make welds as large as they need to be. The greater the surface area of a weld, the stronger the bond.
- Before beginning an airless weld, run a small piece of the welding rod through the welder to clean out the torch tip.
- Consult a supplier for the brands of tools and materials that best fit the shop's needs. Always read and follow the manufacturer's instructions carefully.

For plastic welding, single- or double-V butt welds (Figure 12-3) produce the strongest joints; lap fillet welds are also good. When one is using a round or V-shaped filler rod, the damaged area is prepared by slowly grinding, sanding, or shaving the adjoining surfaces with a sharp knife to produce a single or double V. For flat-ribbon filler rods, V-grooving is not necessary. Wipe any dust or shavings from the joint with a clean, dry rag. The use of cleaning solvents is not generally recommended because they can soften the edges and cause poor welds.

TACK WELDING

The edges to be joined must be aligned. For long tears for which backup is difficult, small tack welds

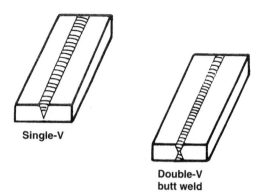

Single-V

Double-V butt weld

FIGURE 12–3 *Single- and double-V butt welds produce strong joints.*

can be made along the length of the tear to hold the two sides in place while you are doing the weld. For larger areas, a patch can be made from a piece of plastic and tacked in place. To tack weld, proceed as follows:

1. Hold the damaged area in its correct position with clamps or fixtures.
2. Using a tacking welding tip, fuse the two sides to form a thin hinge weld along the root of the crack. This is especially useful for long cracks, because it allows easy adjustment and alignment of the edges.
3. Start the tacking by drawing the point of the welding tip along the joint. Press the tip in firmly, making sure to contact both sides of the crack; draw the tip smoothly and evenly along the line of the crack.
4. The point of the tip will fuse both sides in a thin line at the root of the crack. The fused parts will hold the sides in alignment, although they can be separated and retacked if adjustment is necessary. Fuse the entire length of the crack.

HAND WELDING

Several techniques must be mastered for quality hand welding to be performed.

Starting the Weld

Prepare the rod for welding by cutting the end at an angle of approximately 60 degrees. When you start a weld, the tip of the welder should be held about 1/4 to 1/2 inch above and parallel to the base material. The

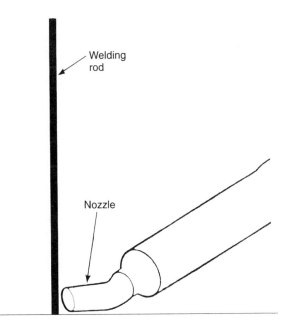

FIGURE 12–4 *Keep the nozzle parallel to the base material and the rod at a right angle to the surface.*

filler rod is held at a right angle to the work as shown in Figure 12-4, with the cut end of the rod positioned at the beginning of the weld.

Direct the hot air from the tip alternately at the rod and the base, but concentrating more on the rod. Always keep the filler rod in line with the V while pressing it into the seam. Light pressure (about 3 psi) is sufficient for achieving a good bond. Once the rod begins to stick to the plastic, start to move the torch and use the heat to control the flow. Be careful not to melt or char the base plastic or to overheat the rod. As the welding continues, a small bead should form ahead of the rod along the entire weld joint. A good start is essential, because this is where most weld failures begin. For this reason, starting points on multiple-bead welds should be staggered whenever possible.

Continuing the Weld

Once the weld has been started, the torch should continue to fan from rod to base material. Because the rod now has less bulk, a greater amount of heat must be directed at the base material. Experience will help you to develop the proper technique.

Feeding the Rod

In the welding process, the rod is gradually being used up, making it necessary for the welder to renew

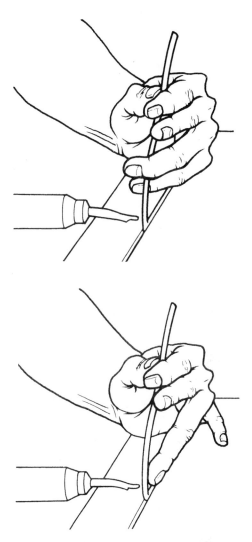

FIGURE 12–5 *Methods of repositioning a grip on the rod.*

his or her grip on the rod. Unless this is done carefully, the momentary release of pressure might cause the rod to lift away from the weld and allow air to become trapped under the weld. As a result, the weld is weakened. To prevent this from happening, the welder must develop the ability to continuously apply pressure on the rod while repositioning the fingers. This can be done by applying pressure with the third and fourth fingers while moving the thumb and first finger up the rod (Figure 12-5). Another way is to hold the rod down into the weld with the third and fourth fingers while repositioning the thumb and first finger. The rod is cool enough to do this because only the bottom of it should be heated. However, care should be observed in touching new welds or aiming the torch near the fingers.

Finishing the Weld

At the end of the weld, maintain pressure on the rod as the heat is removed. Hold it still for a few seconds to make sure the rod has cooled enough so it will not pull loose, then carefully cut the rod with a sharp knife or clippers. Do not try to pull the rod from the joint. About 15 minutes' cooling time is needed for rigid plastic and 30 minutes for thermoplastic polyurethane.

Rough Grinding the Weld

The welded area can be made smooth by grinding with coarse (36-grit) emory or sandpaper. A 9- or 10-inch (220 mm to 250 mm) disc on a low-speed electric grinder will remove large weld beads. Excess plastic can be removed with a sharp knife before grinding. Care must be taken not to overheat the weld area, because it will soften. To speed up the work without damaging the weld, periodically cool it with water.

Checking the Weld

After rough grinding, the weld should be checked visually for defects. Any voids or cracks will make the weld unacceptable. Bending should not produce any cracks, because a good weld is as strong as the part itself. Table 12-3 shows some typical welding defects, their causes, and corrections.

Finishing

The weld area can be finish sanded by using 220-grit sandpaper followed by a 320-grit. Either a belt or orbital sander may be used, plus hand sanding as required. If refinishing is required, follow the procedure designed specifically for plastics.

SPEED WELDING

On panel work, speed welding is very popular. Here are some techniques that are essential for quality speed welding.

Starting the Weld

With the high-speed torch held like a dagger and the hose on the outside of the wrist, bring the tip over the starting point about 3 inches from the material so the hot air will not affect the material (Figure 12-6). Cut the welding rod at a 60-degree angle, insert it into the preheating tube, and immediately place the pointed shoe of the tip on the material at the starting point. Hold the welder perpendicular to the material and push the rod through until it stops against the material at the starting point. If necessary, lift the torch slightly to allow the rod to pass under the shoe. Keeping a slight pressure on the rod with the left hand and only the weight of the torch on the shoe, pull the torch slowly toward you. The weld is now started.

CAUTION: Once the weld is started, do not stop. If the forward movement must stop for any reason, pull the tip off the rod immediately; if this is not done, the rod will melt into the tip's feed tube. Clean the feeder foot with a soft wire brush as soon as the welding is completed.

Continuing the Weld

In the first inch or two of travel, the rod should be helped along by pushing it into the tube with slight pressure. Once the weld has been properly started, the torch is brought to a 45-degree angle; the rod will now feed automatically without further help. As the torch moves along, visual inspection will indicate the quality of the weld being produced.

The angle between the welder and base material determines the welding rate. Since the preheater hole in the speed tip precedes the shoe, the angle of the welder to the material being welded determines how close the hole is to the base material and how much preheating is being done. It is for this reason that the torch is held at a 90-degree angle when starting the weld and at 45-degrees thereafter (Figure 12-7). When a visual inspection of the weld indicates a welding rate that is too fast, the torch should be brought back to the 90-degree angle temporarily to slow down the welding rate, then gradually moved back to the desired angle for proper welding speed. It is important that the welder be held in such a way that the preheater hole and the shoe are always in line with the direction of the weld, so that only the material in front of the shoe is preheated. A heat pattern on the base material will indicate the area being preheated. The rod should always be welded in the center of that pattern.

TABLE 12–3:   PLASTIC WELDING TROUBLESHOOTING GUIDE

Problem	Cause	Remedy
Porous Weld	1. Porous weld rod 2. Balance of heat on rod 3. Welding too fast 4. Rod too large 5. Improper starts or stops 6. Improper crossing of beads 7. Stretching rod	1. Inspect rod. 2. Use proper fanning motion. 3. Check welding temperature. 4. Weld beads in proper sequence. 5. Cut rod at angle, but cool before releasing. 6. Stagger starts and overlap splices 1/2″.
Poor Penetration	1. Faulty preparation 2. Rod too large 3. Welding too fast 4. Not enough root gap	1. Use a 60-degree bevel. 2. Use small rod at root. 3. Check for flow liners while welding. 4. Use tacking tip or leave 1/32″ root or gap and clamp pieces.
Scorching	1. Temperature too high 2. Welding too slowly 3. Uneven heating 4. Material too cold	1. Increase air flow. 2. Hold constant speed. 3. Use correct fanning motion. 4. Preheat material in cold weather.
Distortion	1. Overheating at joint 2. Welding too slowly 3. Rod too small 4. Improper sequence	1. Allow each bead to cool. 2. Weld at constant speed; use speed tip. 3. Use larger-sized or triangular-shaped rod. 4. Offset pieces before welding. 5. Use double V or backup weld. 6. Backup weld with metal.
Warping	1. Shrinkage of material 2. Overheating 3. Faulty preparation 4. Faulty clamping of parts	1. Preheat material to relieve stress. 2. Weld rapidly—use backup weld. 3. Too much root gap. 4. Clamp parts properly; back up to cool. 5. For multilayer welds, allow time for each bead to cool.
Poor Appearance	1. Uneven pressure 2. Excessive stretching 3. Uneven heating	1. Practice starting, stopping, and finger manipulation on rod. 2. Hold rod at proper angle. 3. Use slow, uniform fanning motion, heating both rod and material (for speed welding: use only moderate pressure, constant speed, keep shoe free of residue).
Stress Cracking	1. Improper welding temperature 2. Undue stress on weld 3. Chemical attack 4. Rod and base material not same composition 5. Oxidation or degradation of weld	1. Use recommended welding temperature. 2. Allow for expansion and contraction. 3. Stay within known chemical resistance and working temperatures of material. 4. Use similar materials and inert gas for welding. 5. Refer to recommended application.

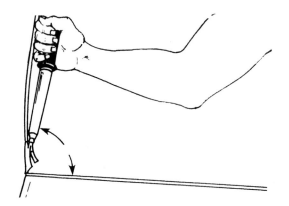

FIGURE 12–6   *Starting a speed weld.*

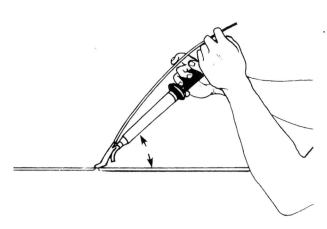

FIGURE 12–7   *Continuing a speed weld.*

Finishing the Weld

It is important to remember that, once started, speed welding must be maintained at a fairly constant rate of speed. The torch cannot be held still. To stop welding before the rod is used up, bring the torch back past the 90-degree angle and cut off the rod with the end of the shoe. This can also be accomplished by pulling the speed tip off the remaining rod. When you are cutting the rod with the shoe, the remaining rod must be removed promptly from the preheater tube. A rod not removed promptly from the preheater tube will char or melt, clogging the tube and making it necessary for the tube to be cleaned out by inserting a new rod in the tube.

A good speed weld in a V-joint will have a slightly higher crown and more uniformity than the normal hand weld. It should appear smooth and shiny, with a slight bead on each side. For best results and faster welding speed, the shoe on the speed tip should be cleaned occasionally with a wire brush to remove any residue that might cling to it and create drag on the rod.

REVIEW QUESTIONS

1. List the factors that must be considered when deciding to replace or repair a part.
2. What are the two methods of repairing plastics?
3. What is the preferred method of repair for thermoplastics and why?
4. Why is more surface preparation generally needed with adhesives?
5. List the thermoplastics that are weldable.
6. What thermoplastic is not weldable?
7. List the thermosetting plastics.
8. Why won't thermosetting plastics melt?
9. Which type of welding will work on thermosetting plastics?
10. Which of the two plastic welding techniques is more versatile?
11. How do you determine whether the base material is a polyolefin?
12. Name two types of adhesive materials used for body repair work.
13. When sanding to prepare for the application of a structural adhesive, what should you do if the sanded area has a "greasy" appearance?
14. When should adhesion promoter be applied?
15. List the basic factors that affect a good plastic weld with either the airless or hot-air welding methods.
16. List the general tips for welding with plastic.
17. Describe the finger position used when hot-air welding to keep pressure on the rod while welding.
18. How long should the rod cool for rigid plastic and thermoplastic polyurethane?
19. When performing speed welding, why shouldn't you stop, and what should you do if you have to?
20. What determines the welding rate for speed welding?

Chapter

13

Special Plastic and Fiberglass Repair Techniques

Objectives

After reading this chapter, you should be able to:

- Perform a one-sided weld on ABS.
- Perform a two-sided weld on polypropylene.
- Restore a distorted bumper cover by reshaping it and a stretched bumper cover by shrinking it.
- Repair a urethane bumper that has sustained damage in the mounting area.
- Repair a cut or tear in a urethane bumper cover.
- Repair vinyl-clad urethane foam that has sustained surface dents, cuts, tears, and cracks.
- List the safety points unique to working with fiberglass.
- Perform fiberglass repairs, including fixing holes in fiberglass panels, fixing shattered fiberglass panels, and replacing fiberglass panels.

As stated earlier, the use of plastic in cars and light trucks is on the rise with each new model year. The actual number of automotive plastics is also expanding, as new hybrids are constantly refined for use in the vehicle manufacturing process. Not all such plastics can be repaired using the conventional methods outlined in the previous chapter; some require special repair materials and procedures to restore them to their original condition. This chapter will examine the following techniques: one- and two-sided welds; reshaping, shrinking, and repairing bumper covers; repairing high-stress areas of a bumper; repairing and refinishing vinyl-clad urethane foam; and repairing and replacing fiberglass. In the case of the one- and two-sided welds, although the airless method is used here because of its previously mentioned advantages, the hot-air method can of course be used instead.

In all cases, follow the procedures exactly as they are presented. Do not skip any of the steps, and do not perform them out of order. And remember: always be sure to read and follow the instructions carefully for the repair products used.

13.1 ONE-SIDED ABS WELD

ABS is one of the most common hard plastics found in today's vehicles. A cosmetic filler is not required when working with hard plastics; the welding rod serves as the filler. Although two-sided welds are preferred when the damaged part can be removed from the vehicle, the one-sided weld is perfectly adequate if done properly. Use the following step-by-step procedure to do a one-sided ABS weld using an airless welder:

1. Set the temperature dial on the welder for ABS. Let the welder warm up while the part is being prepared.

CAUTION: Never set up a welder in wet or damp areas. Electrical shock can kill.

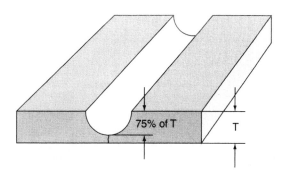

FIGURE 13-1 *V-groove the damaged area 75 percent of the material's thickness for one-sided repairs.*

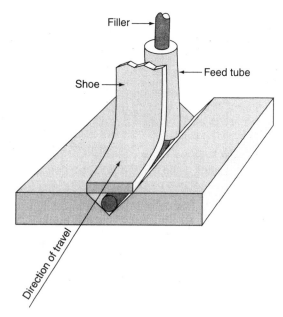

FIGURE 13-2 *Gently feed the rod as welding begins.*

2. As always, proper weld preparation is very important. Wash the part with soap and water, then wipe or blow-dry.
3. Wash the part with plastic cleaner to remove grease, wax, and other contaminants that can cause problems with the repair. Several applications of plastic cleaner might be necessary.
4. Align the break with aluminum body tape; this will hold the plastic securely for welding. Press the tape in place with a plastic squeegee or paint paddle.
5. Because the welding will be done from one side only, it will be necessary to V-groove the damaged area 75 percent of the way through the base material. Use a die grinder, 1/4-inch drill with a cutter bit or grinding pad, a small grinder, or a rotary file (Figure 13-1). The part is now ready to be welded.
6. Begin the weld by inserting the rod into the melt tube. Place the flat shoe part of the tip over the V-groove and very gently feed the rod through the tube as the melting begins (Figure 13-2). Move the tip along the groove slowly for good melt-in and heat penetration.
7. For optimum strength, use the stitch-tamp procedure; this technique is used solely on ABS plastic. To do this, move the pointed end of the tip slowly into the weld area. This will force the rod into the base material and bond the two together for a strong weld. Stitch-tamp the entire length of the weld. The rod and base material should be well mixed and become a uniform color.

8. Resmooth the weld area using the flat shoe part of the tip, again working slowly for good heat penetration.
9. Shape the excess rod buildup to a smooth contour using a slow-speed grinder and 80-grit disc. This can also be accomplished by scraping with a single-edged razor blade.
10. When the weld is sufficiently smooth, scuff sand the area with 320-grit paper; this will provide good adhesion for refinishing. Blow dust-free.

13.2 — **TWO-SIDED POLYPROPYLENE WELD**

Polypropylene is a hard plastic used in interior trim panels, inner fender liners, bumper covers, fan shrouds, and heater housings. The two-sided weld is recommended whenever the damaged part can be removed from the vehicle, because it restores the total strength of the part. It is especially important that high-stress parts such as mounting tabs and edge reinforcements get two-sided welds. Naturally, one-sided welds are fine for solely cosmetic repairs. V-groove 50 percent of the way through the base material on each side of the cut or tear so that the welds touch each other.

Following is the step-by-step procedure for doing a two-sided polypropylene weld using an airless welder:

1. Set the temperature dial on the welder for polypropylene. Let the welder warm up while the part is removed from the vehicle.
2. Wash the part with soap and water. Wipe or blow-dry.
3. Wash the part with a good plastic cleaner to remove grease, wax, and other contaminants.
4. Align the break with aluminum body tape; proper alignment of the joint is critical. Press the tape into place with a plastic squeegee or paint paddle.
5. V-groove 50 percent of the way through the backside of the panel, using a die grinder, 1/4-inch (6 mm) drill with a butter bit or grinding pad (Figure 13-3), a small grinder, or a rotary file.
6. The backside weld should be done using the melt-flow method. With the rod inserted in the melt tube, place the flat shoe part of the tip directly over the V-groove and hold it in place until the melted rod begins to flow out around the shoe (Figure 13-4).
7. Let the rod melt out on its own; do not force it. You should feel the rod begin to collapse as it melts.
8. Move the shoe very slowly and crisscross the groove as it fills with melted plastic. Do not move too fast, or the welder will not have sufficient time to properly heat the base material and melt the rod.
9. Quick-cool the weld with a damp sponge or cloth. Remove the tape in preparation for the front side repair.
10. V-groove 50 percent of the way through the front of the panel and into the backside weld. This will enable the two welds to be tied together for a strong repair.

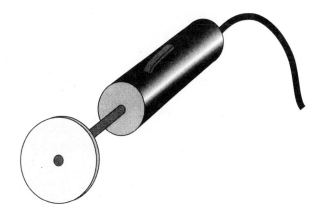

FIGURE 13-3 Die grinder with pad.

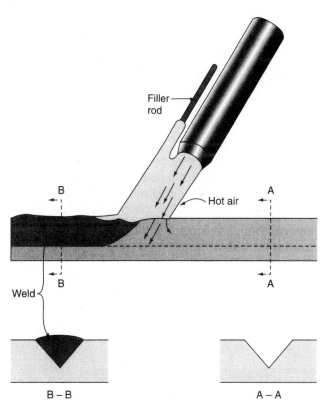

FIGURE 13-4 Using the melt-flow method of polypropylene.

11. Use the same welding and smoothing procedures done on the ABS repair for the front side weld. A slow-speed grinder, a single-edge razor blade, or a drum sander attachment on a die grinder can be used to reshape the contour.

12. Scuff sand the area to be refinished with 320-grit paper and blow dust-free. Prime and color coat to complete the repair (Figure 13-5).

BUMPER COVER REPAIRS

RESHAPING A DISTORTED BUMPER COVER

A concentrated heat source, such as a torch lamp or high-temperature heat gun, is used to reshape a distorted bumper cover. It is important that the heat totally penetrate the plastic. Use the following procedure:

1. Thoroughly wash the cover with soap and water, then clean with plastic cleaner to remove all road tar, oil, grease, and undercoating. Wipe or blow-dry.

2. Apply the heat directly to the distorted area. Check for adequate heat penetration by feeling the surface on the opposite side of the cover. When it becomes too uncomfortable to touch (about 140° F [60° C]), the heat has sufficiently penetrated the plastic.

3. Heat alone is rarely enough to reshape a bumper cover. A paint paddle or squeegee is ideal for helping to shape the cover.

4. When the reshaping is complete, quick-cool the cover with a damp sponge or cloth; an ice cube also works well.

SHRINKING A STRETCHED BUMPER COVER

Flexible plastics do not always tear in an impact. Sometimes they stretch, much like metal. Use the following procedure to shrink a stretched bumper cover:

1. Clean the cover with soap and water, followed by plastic cleaner. Wipe or blow-dry.

FIGURE 13–5 *Grinder reshaping part. (Courtesy of Larry Maupin)*

2. Use a torch lamp or high-temperature heat gun to "hot spot" the stretched area.

3. Quick-cool the repair area immediately with a damp sponge or cloth; an ice cube works even better. Whatever cooling method is used, the point is to cool it as quickly as possible.

REPAIRING HIGH-STRESS AREAS ON A URETHANE BUMPER

When a bumper cover is ripped away from a vehicle, the likely result will be broken mounting tabs. Any damage to the mounting area of a bumper must have a quality repair, or it will break loose when the bumper is remounted. If an airless welder is being used to make the repair, there is no need to determine whether the urethane is a thermoset or a thermoplastic, because both are repairable with this type of welder. The urethane must be a thermoplastic in order for you to use a hot-air system.

Use the following step-by-step procedure to make a two-sided repair:

1. Wash the bumper with soap and water, wipe or blow-dry, then wash it with a good plastic cleaner. Use a cleaner that will remove grease, wax, and other foreign matter.

2. Bevel back the torn edges of the mounting tabs at least 1/4 inch (6 mm) using a drum sander.

3. With the small grinder, rough up the plastic as far as the weld will extend. Blow the area dust-free.

4. Use aluminum body tape to build a form in the shape of the missing tab (Figure 13-6). Turn up the edges of the tape to form the thickness of the weld.
5. Set the temperature dial on the welder to accommodate a clear urethane rod and allow it time to warm up.
6. Push the rod slowly through the melt tube, slightly overfilling the form with melted plastic. Feather the plastic into the beveled area. Make sure the mold cavity is full and the melted plastic is feathered out over the beveled edges of the broken mounting tabs.
7. Use the flat shoe part of the tip to smooth and shape the weld. Quick-cool with a damp sponge or cloth, then remove the tape.
8. Use a die grinder and small cutter bit to make a V-groove, then bevel the backside of the torn edges. This will allow the edges to be welded on both sides for optimum strength.
9. Fill in the groove with melted plastic. Use the flat shoe part of the tip to smooth out the plastic, then quick-cool with a damp sponge or cloth.
10. Finish the weld to the desired contour using a slow-speed grinder and a 60- or 80-grit disc. The bumper can be reinstalled immediately, but it will not regain its total strength until it has cooled completely.

REPAIRING A CUT OR TEAR IN A URETHANE BUMPER COVER

In most cases, a one-sided weld is sufficient to repair a cut or tear in a bumper cover. However, if the damage is in a high-stress area, a two-sided weld is required for greater strength, followed by application of a flexible filler. The following step-by-step procedure outlines a typical two-sided weld and filler application:

1. Wash the cover with soap and water, wipe or blow-dry, then wash with a good plastic cleaner.
2. Align the cut or tear by applying aluminum body tape to the painted side of the cover. The tape must be applied to the painted side because it provides better adhesion.
3. Use a die grinder and a smaller cutter bit to V-groove the cut or tear on the backside. Be

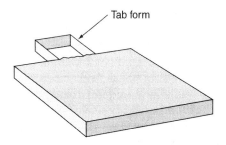

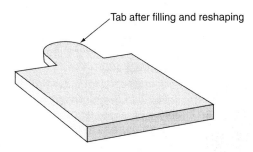

FIGURE 13–6 Build a form for the broken tab with aluminum body tape.

sure to penetrate at least 50 percent of the way through the bumper.
4. Use a drum sander or similar tool to bevel back the edges of the V-groove at least 1/4 inch on each side.
5. Set the temperature dial on the welder to accommodate a clear urethane rod. Give it sufficient time to warm up.
6. Start the weld by slowly feeding the rod into the melt tube, applying only enough pressure to force the melted rod out into the grooved area. As the rod begins to melt, very slowly move the tip in the direction of the intended weld, and overlap the edges of the groove with melted plastic.
7. After completing the full length of the weld, remove the rod from the melt tube. Use the flat shoe part of the tip to smooth out the weld area. Proceed slowly to give the plastic sufficient time to melt under the shoe.
8. Quick-cool the weld with a damp sponge or cloth, then remove the tape.

9. Wash the front side with soap and water and plastic cleaner. Wipe or blow-dry.

10. V-groove the front side, making sure the groove is deep enough to penetrate the backside weld so the two welds can be blended together for added strength.

11. Using a drum sander or similar tool, bevel the sides of the groove. This will permit the weld to feather out over the edges of the base material.

12. Use a slow-speed grinder and a 60- or 80-grit disc to remove the paint from around the weld area. Do not overgrind; 2 or 3 inches (25 mm to 74 mm) is plenty. Then blow the area dust-free with an air hose.

13. Use the same welding technique that was used on the backside, making sure to get a good melt-in. Quick-cool the weld.

14. Smooth the weld using a slow-speed grinder and a 60- or 80-grit disc. Because heat buildup from the grinder can cause the edges of the weld to peel up, grind for only a few seconds at a time, then allow the plastic to cool. Plastic reacts to heat from a grinder very quickly and can experience warpage.

15. Mix and apply the flexible filler as per the manufacturer's instructions. One thin application is sufficient; overfilling will reduce flexibility.

16. When the filler has cured according to the package instructions, use a slow-speed grinder and a 60- or 80-grit disc to grind the filled area to a smooth contour. Follow with 180-grit paper to featheredge the paint and put the final touches on the filler.

17. If any imperfections remain that must be filled, first apply a skim coat of flexible putty. Allow it to dry, then sand it smooth with 220-grit paper. Do not use body putty or primer to fill imperfections.

13.4 REPAIRING VINYL-CLAD URETHANE FOAM

The use of vinyl-clad urethane foam in vehicle interiors is commonplace these days. Most padded instrument panels, or crash pads, are sections of urethane

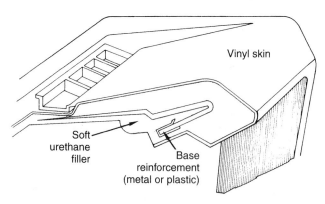

FIGURE 13–7   *Cutaway of a typical crash pad.*

foam covered with a thin vinyl overlay. The foam is molded to a plastic or melt reinforcement base plate and can be repaired easily using an airless welder. Although the basic welding technique is fine for repairing cuts, tears, and cracks, it is also necessary to be able to repair dents and similar deformities.

REMOVING SURFACE DENTS

Besides being unsightly, a surface dent in effect stretches the plastic—similar to a hail dent in metal. When done carefully and precisely, the following technique is very effective in removing dents in crash pads, armrests, and other padded interior panels. Whenever possible, remove the part from the vehicle before beginning. If the part remains in the vehicle, be sure to cover the seats for protection. Figure 13-7 shows a cutaway view of a typical crash pad.

1. Apply a damp sponge or cloth to the dent; soak it for about 30 seconds. The vinyl is porous and will absorb some of the water. Do not wipe it dry; let the area remain moist.

2. Use a high-temperature heat gun or torch lamp to heat the dented area. Hold the heat source 10 to 12 inches (250 mm to 300 mm) from the surface, and keep it moving in a circular motion at all times. Gradually move the heat inward into the center of the dent, but be careful not to overheat the vinyl or it will blister.

3. Continue heating until the dented area is uncomfortable to touch (about 140° F [60° C]), then turn off the heat source and set it outside the vehicle to prevent accidental damage.

4. If necessary, massage the pad with a forcing motion toward the center of the dent. Some dents will not require massaging because the heat is enough to remove them. It might be necessary to heat and massage more than once.

5. After reshaping, quick-cool the area with a damp sponge or cloth.

REPAIRING AND REFINISHING CUTS, TEARS, AND CRACKS

A cut, tear, or crack in a crash pad detracts greatly from a car's appearance. Many such defects are the direct result of abuse; others are caused by weather and age. The basic repair procedure is as follows:

1. Set the temperature dial on the welder to accommodate a clear urethane rod. Let the welder warm up.

2. Use the same cleaning procedure as for the ABS and polypropylene welds. Wash the entire pad with soap and water, blow it dry, then clean it with plastic cleaner.

3. Examine the damaged area:
   • If it is brittle, warm it with a heat gun to soften the vinyl before doing the repair.
   • If the edges are curled or jagged, cut them away with a small cutter bit before making the V-groove.

4. V-groove at least 1/4 inch (6 mm) deep in the foam padding. This will enable the melted rod to flow into the inner pad and provide extra strength.

5. The vinyl is very thin; despite this, bevel the edges as much as possible. Then rough them up at least 1/4 inch (6 mm) on both sides for good weld penetration.

6. Start the weld at the bottom of the groove and rotate the welder so the flat shoe part of the tip is up and the melt tube is down. Slowly feed the rod through the melt tube until the groove is filled with melted plastic flush with the surface.

7. Rotate the welder so the shoe is turned toward the weld; smooth out the excess rod buildup. Feather it out over the beveled edges at least 1/4 inch (6 mm) on each side of the groove.

8. Use a die grinder with a drum sander attachment to remove any remaining excess rod buildup. Rough up the vinyl about 2 inches beyond the weld on each side for good filler adhesion. Be careful not to grind away the entire weld.

9. The filler used must be flexible and designed specifically for vinyl. Follow the manufacturer's mixing and application instructions. With a plastic squeegee, apply the filler to the smooth area only. Spread it to the desired contour.

10. Allow the filler to cure, then begin contour sanding with 40-grit paper to remove the tacky glaze and rough up the contour. Follow up with 80-grit and finish with 180-grit to remove the deeper sanding marks. If any of the filler is accidentally sanded through, apply a skim coat and resand.

11. To retexture, follow the same steps used in the ABS and polypropylene repairs. Remember to apply the texture material only to the repair area at first. Then blend it out to a natural break line or retexture the entire panel.

12. Nib sand the newly textured area with 220-grit paper and blow dust-free. Follow the manufacturer's recommendations when applying the color coat. Remember that weather cracks are caused by prolonged exposure to the sun, which causes the plasticizer to evaporate from the flexible PVC material. However, they can be repaired if the panel is not too badly deteriorated. To test for this, press on the panel to flex the vinyl covering; if more cracks appear, the panel is probably beyond repair.

13.5 — FIBERGLASS REPAIRS

The fiberglass plastics used on car and truck bodies are made from molten glass drawn out into tiny, threadlike fibers. These fibers are reinforced with plastic resin and catalyst, then gel coated to a shiny finish. The combination of fiberglass and resin forms an extremely tough, durable, and corrosion-resistant material that is used in both OEM construction and repair work. However, in the case of the latter, it should be approached with caution; the techniques required are quite different from those in metalwork and even from those plastic repairs discussed previously.

WORKING WITH FIBERGLASS

Working with fiberglass requires thinking safety at all times. The resin and related ingredients can irritate the skin and stomach lining. The curing agent or hardener is generally a methyl ethyl ketone peroxide, which produces harmful vapors. Read and understand these safety points before using any fiberglass products:

- Read all label instructions and warnings carefully.
- Wear rubber gloves when working with fiberglass and resin or hardener. Long-sleeved shirts with buttoned collar and cuffs are helpful in preventing sanding dust from getting on the skin.
- A protective skin cream should be used on any exposed areas of the body.
- If the resin or hardener comes in contact with the skin, wash with borax soap and water or denatured alcohol.
- Always work in a well-ventilated area of the shop.
- Wear a respirator to avoid inhaling sanding dust and resin vapors.
- When making fiberglass repairs, mask the surrounding areas to avoid spilling resin on them.
- Clean all tools and equipment with lacquer thinner immediately after use. Dispose of the leftover mixed material in a safe container.

REPAIRING HOLES IN FIBERGLASS PANELS

Repairing holes in fiberglass is a simpler operation than most people think; it is actually easier than filling holes in metal body panels. If the underside of the damaged panel is easily accessible, use the following repair procedure:

1. Clean the surface surrounding the damage with a good commercial grease and wax remover. Use a 36-grit grinding disc to remove all paint and primer at least 3 inches beyond the repair area.

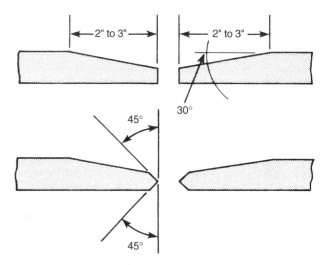

FIGURE 13-8 *Beveling the inside and outside of the repair area permits better adhesion.*

2. Grind, file, or use a hacksaw to remove all cracked or splintered material from the hole on both the inside and outside of the repair area.
3. Remove any dirt, sound deadener, and the like from the inner surface of the repair area. Clean with reducer, lacquer thinner, or a similar solvent.
4. Scuff around the hole with 80-grit paper to provide a good bonding surface.
5. Bevel the inside and outside edges of the repair area about 30 degrees to permit better patch adhesion (Figure 13-8).
6. Clean the repair surface thoroughly with reducer or thinner.
7. Cut several pieces of fiberglass cloth or mat large enough to cover the hole and the scuffed area. The exact number will vary depending on the thickness of the original panel, but five is usually a good number to start with.
8. Prepare a mixture of resin and hardener following the label recommendations. Using a small paintbrush, saturate at least two layers of the fiberglass cloth with the activated resin mix and apply it to the inside or back surface of the repair area. Make sure the cloth fully contacts the scuffed area surrounding the hole.

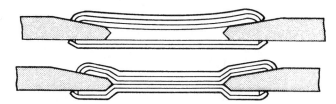

FIGURE 13-9 The layers of saturated glass cloth should contact the outside repair area.

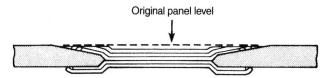

FIGURE 13-10 Sand the patch slightly below the contour of the panel.

FIGURE 13-11 Hardened patching material.

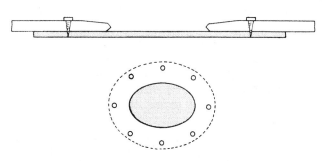

FIGURE 13-12 Sheet metal attached to the backside of a damaged panel.

9. Saturate three more layers of cloth with the mix. Apply it to the outside surface, making certain that these layers fully contact the inner layers and the scuffed outside repair area (Figure 13-9).
10. With all of the layers of cloth in place, form a saucerlike depression in them. This is necessary to increase the depth of the repair material. Use a rubber squeegee to work out any air bubbles.
11. Clean all tools with lacquer thinner immediately after use.
12. Let the saturated cloth become tacky. An infrared heat lamp can be used to speed up the process; if one is used, keep it 12 to 15 inches (300 mm to 365 mm) away from the surface. Do not heat the repair area above 200° F (90° C), because too much heat will distort the material.
13. With 50-grit paper, disc-sand the patch slightly below the contour of the panel (Figure 13-10).
14. Prepare more fiberglass resin mix. Use a plastic spreader to fill the depression in the repair area, leaving a sufficient mound of material to grind down smooth and flush.
15. Allow the patch to harden (Figure 13-11). Again, a heat lamp can be used to speed the curing process.
16. When the patch is fully hardened, sand the excess material down to the basic contour, using 80-grit paper and a sanding block or pad. Finish sand with 120-grit or finer grit paper.

This repair can also be made by attaching sheet metal to the backside of the panel with sheet metal screws (Figure 13-12). Sand the metal and both the inner and outer sides of the hole to provide good adherence for the repair material. Before fastening the sheet metal, apply resin mix to both sides of the rim of the hole. Follow the previous procedure for the remainder of the repair.

When the inner side of the hole is not accessible, apply a fiberglass patch to the outer side only. After the usual cleaning and sanding operations, apply several layers of fiberglass cloth to the outer side of the hole. Before it dries, make a saucerlike depression in the cloth to provide greater depth for the repair material.

REPLACING FIBERGLASS PANELS

The fiberglass bodies used on Corvettes and some large truck noses are made of individual panels. These panels are joined together in such a way that the seams and joints are invisible. Bonding strips are often used at the joints to provide a sound bonding surface, as seen in Figure 13-13. Most kit cars are also made of fiberglass panels with invisible seams. It is important that any replacement be performed to exacting standards so as to preserve the visual integrity of the vehicle body.

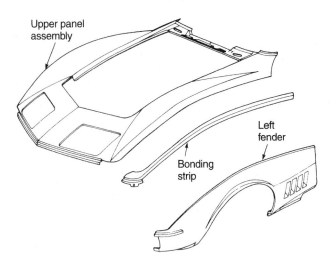

FIGURE 13–13 *Bonding strips are useful at fiberglass panel joints.*

If struck hard enough, a fiberglass panel could crack or break away, depending on the force of the blow. A judgment must be made regarding repair versus replacement; the extent of the damage determines whether or not the panel can be salvaged. The following procedure details specifically the replacement of an outer rear quarter panel on a Corvette, but it is a good example of the general fiberglass panel replacement technique. It can also be applied to the replacement of the fiberglass parts used on truck cab bodies. Keep in mind that the car cannot be jacked up or resting on safety stands for the fitting of the new panels; it must be in its normal resting position until you are satisfied with the alignment and fit.

1. Because there are no visible seams between the panel and the body, the first step is to determine where to make the separation. This can be done using the new panel as a guide or by following along the crack created when the panel was struck.
2. Use a disc sander and a coarse-grit disc to remove the finish along the seam. This will expose the seam and leave no question as to where the separation is to be made.

CAUTION: For personal safety, always wear safety goggles, a respirator, a long-sleeved shirt, and gloves when sanding fiberglass.

---

3. To separate the panel from the body, a hammer and wide-blade chisel can be used, though an air chisel fitted with a wide blade makes the job easier.
4. Use light, intermittent finger pressure on the air chisel's trigger, just enough to crack the seam. Too much force on the chisel can break the bonding strip, which provides the underneath surface for joining the two panels.
5. Continue separating the panels at the seams. If possible, work from the underside of the panel and force the chisel between the seams to separate and remove it.
6. Use a disc sander with a coarse-grit disc along all seam areas; sand 3 or 4 inches (74 mm to 100 mm) beyond the seam. Remove all the finish and reduce the thickness of the panel where it meets the lower bonding strip. This will allow space for the adhesive and filler that will be applied later.
7. Continue sanding all the surfaces to which the new panel will be bonded, in order to remove all the remaining adhesive and fiberglass particles.
8. Hold the new panel in place to check that all the bonding areas have been sanded and properly prepared for the new panel.
9. While still holding the new panel in place, carefully check the fit at the door opening and the areas where it will be joined. Proper fit is necessary for good appearance, as well as to provide sufficient allowance for the adhesive filler.
10. To be absolutely sure that the panel will be in correct alignment, use clamps and sheet metal screws to temporarily hold it in position. Set a clamp to hold the panel and align the fit at the door.
11. Drill a hole near the top edge of the panel and install a sheet metal screw to hold it in place. Check the fit once again and install screws wherever a separation between the panel and the bonding strip appears. If desired, clamps or other types of quick-release clips can be used in place of screws.
12. Double-check all the joints to be sure there is clearance for the adhesive and filler. Also, make sure that all the places where there might be a gap between the panel and bonding strip have been clamped.

13. With the panels held in place with screws or clamps, closely check the alignment and fit of the panels. Step back and view the assembly from a distance; all panels must be straight and level, because once they are cemented in place they cannot be readjusted.

14. Cut strips of fiberglass matting approximately 3 inches wide. The lengths should be short and easy to handle. Prepare enough so that their length equals the length of all the areas to which the new panels will be joined. When mixed with the resin, the fiberglass matting forms the adhesive that joins the panels together.

15. Apply several layers of masking tape to the adjacent areas to protect them from adhesive.

16. Raise the vehicle and place it on safety stands. Remove the wheel, sheet metal screws, clamps, and panel.

17. Pour a small amount of fiberglass resin into a container, add the hardener, and stir thoroughly. The amount of hardener used affects the hardening rate of the resin. Because of the time needed for fitting, adjusting, and aligning the panels, prepare the mix so that it will not harden too fast.

18. On a clean, disposable work surface, apply resin to the mat strips. Thoroughly saturate the strips. Coat the bonding areas with resin, then apply the mat to the bonding strips and brush out all air bubbles.

19. Brush resin onto the bonding areas of the panel. With the help of an assistant, position the panel onto the bonding area and reinstall the first of the sheet metal screws into its original hole. Do not tighten the screw yet.

20. Check the alignment of the panel with the door, then reinstall the screws and clamps. After rechecking the alignment, tighten the screws just enough for the resin to ooze from the seams.

21. Working from the underside of the panel, brush the excess fiberglass mat so that it contacts the panel; this will add strength to the joint. Install the rear panel and brush out the resin and matting in the same way.

22. Fill any remaining holes or broken areas with resin and matting, and allow the resin to harden completely. Depending on the temperature and the amount of hardener used, the length of the curing process varies.

CAUTION: It is best not to rush the hardening process. If the panel is moved before the adhesive has set fully, the bond will be broken and an early failure will result.

23. After the adhesive has cured completely, remove the clamps and sheet metal screws. Disc-sand all the seam areas.

24. The strength of the joint depends on the bond between the bonding strip and each panel. All spaces and screw holes must be filled with a very strong fiberglass-reinforced plastic filler. Mix the hardener and filler thoroughly.

25. Wipe the panel with a clean cloth. Apply the filler to all seams and joints, and let it harden slightly.

26. Use a cheese grater to level the filler so that it conforms to the panel's shape. A long-line air sander can be used to smooth the filler areas. Difficult-to-reach corners and joints can be hand sanded.

27. Wipe the surface clean and apply a level coat of common plastic filler to fill any minor imperfections or irregularities. Use the long-line air sander to smooth the filler. Finally, check the filled areas and sand out any imperfections.

With the seams filled and smoothed, the repair area must be thoroughly cleaned and prepared for refinishing. The restored vehicle now shows no signs of the severe collision damage it sustained.

REPAIRING SHATTERED FIBERGLASS PANELS

Many times, an impact will cause part of a fiberglass panel to shatter or break into several pieces. Rather than replace the whole panel, it might be possible to reassemble the pieces, much like putting a puzzle together. This procedure is not only less expensive than replacing the panel, it also saves the time usually spent waiting for new replacement parts. The following procedure can be used to repair most shattered fiberglass panels:

1. Clean the repair area with a wax and grease remover. Use a 36-grit sanding disc to remove the finish 3 inches around the damage.

2. Grind 3 inches around the backside of the repair area to remove any undercoating, dirt, and other foreign material.

3. Prepare a mixture of fiberglass resin and hardener. Follow the manufacturer's instructions for mixing and applying. Add fiberglass fibers to give the mixture greater bridging strength, and mix only no more filler than can be applied before it begins to set up and harden.

4. When you are assembling a large number of pieces, some cure time might be needed periodically before subsequent pieces can be added.

5. If most of the pieces are available, reassemble the panel one piece at a time. Begin by smearing a thin coat of filler on the mating edges. Use a C-clamp or vise grip to hold the piece in its original position on the panel.

6. Smear a coating of filler over the panel and the piece; the aim is to build a "bridge" over the joint. Repeat this process with each of the pieces until they are all clamped and glued together.

7. If the shattered area is large or if some of the pieces are missing, fiberglass bonding strips or a piece of sheet metal can be fastened to the backside of the panel to provide support for the repair area. Bonding strips can be obtained from other wrecked panels or from a salvage yard.

8. To apply the bonding strip, hold it in position and drill holes through both the panel and the strip. Roughen the face of the strip with a 36-grit sanding disc and coat the panel-to-strip mating surfaces with filler.

9. Use sheet metal screws to screw the outside of the panel into the bonding strip (Figure 13-14). After the repair is finished, these screws will be removed. If a piece of sheet metal is being used as backing, wrap it with plastic before screwing it to the panel; this will allow it to be easily removed after the repair is finished.

10. After the backing is in place, coat the panel edges and the edges of the pieces with the resin/fiber mixture. Assemble the pieces, then coat the repair area with the filler. Be sure to completely fill all voids.

11. If fiberglass bonding strips are not used to support the pieces, the rigidity of the repair area can be increased with fiberglass cloth. Cut two or three pieces of cloth or mat approximately 3 inches larger (74 mm) than the repair area.

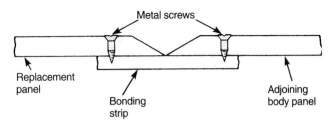

FIGURE 13-14 *The outside of the panel screwed into the bonding strip.*

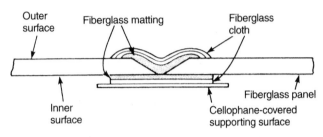

FIGURE 13-15 *The repair area covered with fiberglass cloth.*

12. Saturate the cloths with fiberglass filler and place them over the repair area. Use a plastic spreader to press the cloth into any voids or low spots and to work out air bubbles.

13. After the panel has been reassembled and the filler has hardened, remove any sheet metal screws and grind the repair area with a 36-grit sanding disc to remove all high spots.

14. Cut three pieces of fiberglass cloth or mat 3 inches larger than the repair area. Thoroughly saturate the cloth pieces with filler and place them over the repair (Figure 13-15). Carefully smooth them flat and work out any air bubbles.

15. After the filler has cured, sand to the correct contour and prime.

USING MOLDED CORES

Naturally, holes are much more difficult to repair in a curved portion of a fiberglass panel than on a flat surface. Basically, the only solution in such a case (short of purchasing a new panel section) is to use the mold core method of replacement. This is often the quickest and cheapest way to repair a curved fiberglass surface; the entire process takes almost an hour. While the procedure illustrated in Figure 13-16 and described here relates to a rear fender section, the principle can be applied to any type of curved fiberglass panel damage. The mold core is made as follows:

1. First off, locate an undamaged panel on another vehicle that matches the damaged one; this will be used as a model. A new or used car can be employed, since the model vehicle will not be harmed if care is taken.

2. On the model vehicle, mask off an area slightly larger than the damaged area. Apply additional masking paper and tape to the surrounding area, especially on the low side of the panel. This will prevent any resin from getting on the finish.

3. Coat the area being used as a mold with paste floor wax. Leave a wet coat of wax all over the surface. A piece of waxed paper can be substituted for the coat of wax; be sure that it is taped firmly in place (Figure 13-16A).

4. Cut several pieces of special fiberglass mold veil (thin fiberglass mat material) in sizes ranging from 2 by 4 inches (50 mm by 100 mm) to 4 by 6 inches (100 mm by 150 mm). Standard fiberglass mats can be used if the panel does not have reverse curves.

5. Mix the fiberglass resin and hardener following the label instructions on both products.

6. Starting from one corner of the mold area, place pieces of veil on the waxed area so that each edge overlaps the next one; use just one layer of veil (Figure 13-16B).

7. Apply the resin/hardener to the veil material with a paintbrush (Figure 13-16C). Force the mixture into the curved surfaces and around corners with the tips of the bristles.

8. Use the smaller pieces of veil along the edges and on difficult curves. Additional resin can be applied if needed, brushing in one direction only to force the material into the indentations. In all cases, use only one layer of veil.

CAUTION: Be sure that the resin does not get on any part of the model vehicle that is not coated with wax.

---

9. After the veil pieces have been applied to the entire waxed area, allow the mold core to cure a minimum of 1 hour.

10. Once the mold core has hardened, gently work the piece loose from the model vehicle (Figure 13-16D). The core should be an exact reproduction of this section of the panel.

11. Remove the floor wax protecting the model vehicle's paint finish using a wax and grease remover. Then polish this section of the panel.

12. Since the mold core is generally a little larger than the original panel, place it under the damaged panel and align. If necessary, trim down the edges, the core, and the damaged panel slightly where needed for better alignment. The edges of the damaged panel and core must also be cleaned.

13. Using fiberglass adhesive, cement the mold core in place (Figure 13-16E). Allow the core and panel to cure.

14. Grind back the original damaged edges to a taper or bevel, maintaining the desired contour.

15. Lay the fiberglass mat, soaked in resin/hardener, on the taper or bevel and over the entire inner core. Once the mat has hardened, level it with a coat of fiberglass filler. Then prepare it for painting (Figure 13-16F).

In some instances, it might not be possible to place the core on the inside of the damaged panel. In this case, the damaged portion must be cut out to the exact size of the core. After the panel has been trimmed and its edges beveled, tabs must be installed to support the core from the inside. These tabs can be made from pieces of the salvaged fiberglass panel or from fiberglass cloth strips saturated in resin/hardener.

After cleaning and sanding the inside sections, attach the tabs to the inside edge of the panel and bond with fiberglass adhesive. Vise grips can be used to hold the tabs in place during bonding. Taper the edge of the opening and place the core on the tabs. Fasten the core to the tabs with fiberglass adhesive. Grind down any high spots so that layers of fiberglass mat can be added.

Place the saturated mats over the core, extending about 1-1/2 to 2 inches (37 mm to 50 mm) beyond the hole in all directions. Work each layer with a spatula or squeegee to remove all air pockets. Additional resin can be added with a paintbrush to secure the layers. Allow the resin/hardener time to cure sufficiently, then sand the surface level. For a smooth surface, use fiberglass filler to finish the job.

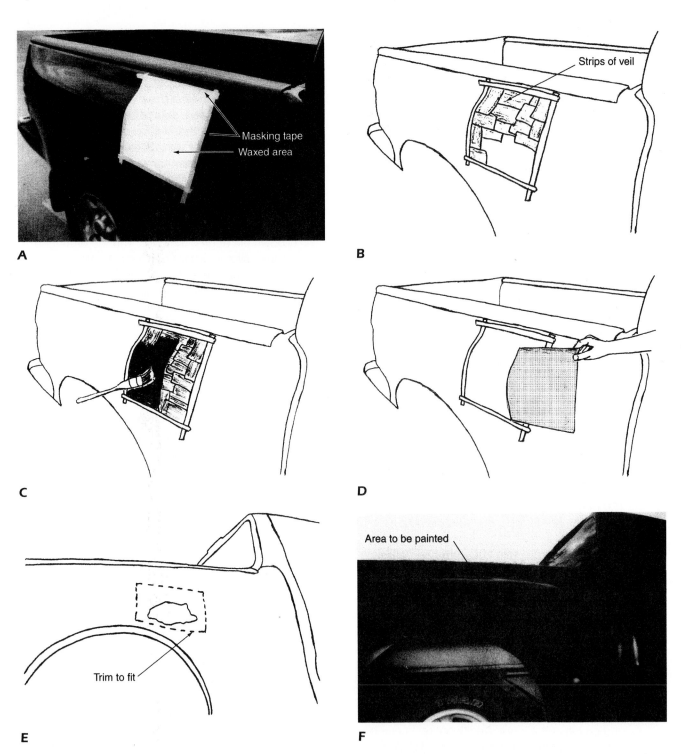

A

B

C

D

E

F

FIGURE 13–16 Steps in making a fiberglass core: (A) coat the area being used as a mold model with paste floor wax on a piece of waxed paper; (B) place pieces of fiberglass veil over the waxed or waxed paper surface; (C) apply resin/hardener to the veil material; (D) remove the mold core from the model; (E) cement the core piece in place; (F) the completed job. (Courtesy of Unican Corp.)

REVIEW QUESTIONS

1. What type of weld should be used for optimum strength on ABS plastic?
2. Where is polypropylene used?
3. What high-stress parts made of polypropylene should have two-sided welding?
4. What is used to reshape a distorted bumper cover?
5. What should be used to reshape a bumper cover besides heat?
6. What should be used to heat-shrink a stretched bumper cover?
7. Define quick-cool.
8. Describe how to use flexible filler.
9. What is the best way to repair cuts, tears, and cracks in the vinyl-clad urethane foam of vehicle interiors?
10. Describe how to repair dents in vinyl-clad urethane.
11. List the causes of damage to the crash pad.
12. How can you tell if a panel is too deteriorated to repair?
13. What makes fiberglass extremely tough, durable, and corrosion-resistant?
14. Describe how fiberglass is made.
15. List the safety procedures used when working with fiberglass.
16. When replacing fiberglass panels the car must be in what position?
17. What is the best way to temporarily affix a panel in place while aligning it?
18. What affects the hardening rate of fiberglass resin?
19. Describe how a mold core is made.
20. What should body panel tabs be made from?

GLOSSARY

Abrasive Material such as sand, crushed steel grit, aluminum oxide, silicon carbide, or crushed slag used for cleaning or surface roughening.

ABS (Acrylonitrile/Butadiene/Styrene) A common thermoplastic.

Acceptable Weld A weld that meets all the requirements prescribed by the welding specifications.

Acetone Flammable, volatile liquid used in acetylene cylinders to dissolve and stabilize acetylene under high pressure.

Acetylene Highly combustible gas composed of carbon and hydrogen. Used as a fuel gas in the oxyacetylene welding process.

Adhesion Promoter A spray material that, when applied to a polyolefin, enables it to be repaired using the adhesive bonding method.

Adhesive Bonding Material-joining process in which an adhesive, placed between the mating surfaces, solidifies to produce an adhesive bond.

Airless Plastic Welding A relatively new method of auto body repair in which the temperature setting of the welder is adjusted to suit the specific plastic.

Alloy Mixture of two or more metals.

Alternating Current (AC) Electricity that reverses its direction of electron flow.

Ampere A measure of electric current used to refer to the input of electrical energy.

Annealing Comprehensive term used to describe the heating and cooling cycle of steel in the solid state, usually implying relatively slow cooling. In annealing, the temperature of the operation, the rate of heating and cooling, and the time the metal is held at heat depend upon the composition, shape, and size of the steel being treated and the purpose of the treatment.

Arc The flow of electricity through an air gap. The arc flowing through the air produces high temperatures.

Arc Brazing A brazing process in which the heat required is obtained from an electric arc.

Arc Burn A temporary but painful eye condition experienced when the eyes are exposed to ultraviolet light for a short period of time: this is also called arc flash.

Arc Cutting A group of cutting processes wherein the severing or removing of metals is brought about by melting with the heat of an arc between an electrode and the base metal.

Arc Gouging An application of arc cutting wherein a bevel or groove is formed.

Arc Voltage The voltage across the welding arc.

Arc Welding A group of welding processes in which coalescence is produced by heating with an arc or arcs, with or without the use of filler metal.

Argon A chemically inert gas that will not combine with the products of the weld zone. It is an excellent shielding gas for the gas metal arc process.

Automatic Welding Welding with equipment that performs the entire welding operation without constant observation and adjustment of the controls by an operator. The equipment may or may not perform the loading and unloading of the work.

Back Goughing The forming of a bevel or groove on the other side of a partially welded joint to assure complete penetration upon subsequent welding from that side.

Backing Material (metal, weld metal, asbestos, carbon, flux, and so on) backing up the joint during welding.

Back Weld A weld deposited at the back of a single groove weld.

Balling Up The formation of globules of molten brazing filler metal or flux due to lack of wetting of the base material.

Base Metal The metal to be welded, soldered, or cut.

Bead The weld; used to describe the neat ripples formed by semiliquid metal.

Bead Weld A type of weld composed of one or more string or weave beads deposited on an unbroken surface.

Bevel An angular type of edge preparation.

Bevel Angle The angle formed between the prepared edge of a member and a plane perpendicular to the surface of the member.

Blind Joint A joint, no portion of which is visible.

Bond The junction of the welding metal and the base metal.

Bonding Strip A piece of fiberglass. Several strips are used as backside support when repairing a large shattered fiberglass panel or when some of the broken pieces are missing.

Braze Welding A method of welding using a filler metal. Unlike brazing, the filler metal is not distributed in the joint by capillary attraction.

Brazing   A group of welding processes that produces coalescence of materials by heating them to a suitable temperature and by using a filler metal. The filler metal is distributed between the closely fitted surfaces of the joint by capillary attraction.

Brazing Filler Metal   The metal that fills the capillary gap and liquefies above 840° F.

Bridging   A welding defect caused by poor penetration. A void at the root of the weld is spanned by weld metal.

Buckling   Distortion caused by the heat of a welding process.

Buildup Sequence   The order in which the weld beads of a multipass weld are deposited with respect to the cross section of a joint.

Burn Test   A means of identifying plastics in which a small piece of the plastic is ignited and its burn characteristics are evaluated.

Butt Joint   A joint between two members aligned approximately in the same plane.

Capillary Attraction   The condition in which adhesion between the molten filler metal and the base metals, together with the surface tension of the molten filler metal, causes distribution of the filler metal between the properly fitted surfaces of the joint to be brazed.

Carburizing Flame   An oxyacetylene flame in which there is an excess of acetylene. Also known as excess acetylene or reducing flame.

Coalescence   The growing together or growth into one body of the materials being welded.

Coating   A relatively thin layer of material applied by surfacing for the purpose of corrosion prevention, resistance to high-temeprature scaling, wear resistance, or lubrication.

Cold Soldered Joint   A joint with incomplete coalescence caused by insufficient application of heat to the base metal during soldering.

Complete Joint Penetration   Joint penetration that extends completely through the joint.

Cone   The conical part of an oxyfuel gas flame next to the orifice of the tip.

Continuous Weld   A weld that extends continuously from one end of a joint to the other.

Corrosive Flux   A flux with a residue that chemically attacks the base metal.

Crack   A fracture-type discontinuity characterized by a sharp tip and high ratio of length and width to opening displacement.

Crater   In arc welding, a depression at the termination of a weld bead or in the weld pool beneath the electrode.

Crown   The surface of the finished bead.

Crush Zones   Buckling points designed into certain structural components for absorbing the energy impact in a collision.

Cutting Tip   That part of an oxygen cutting torch from which the gases are emitted.

Cutting Torch   A device used in oxygen cutting for controlling and directing the gases used for preheating and the oxygen used for cutting the metal.

Cylinder   A portable container used for transportation and storage of a compressed gas.

Cylinder Cart   A portable cart used for moving cylinders.

Defective Weld   A weld containing one or more defects.

Deposited Metal   Filler metal that has been added during a welding operation.

Direct Current (DC)   Electricity that flows in only one direction.

Direct Current Electrode Negative (Straight Polarity)   Electrical current flowing from the electrode to a base metal.

Direct Current Electrode Positive (Reverse Polarity)   The arrangement of direct current arc welding leads wherein the work is the negative pole and the electrode is the positive pole of the welding arc.

Discontinuity   An interruption of the typical structure of a weld. A discontinuity is not necessarily a defect.

Duty Cycle   The number of minutes out of ten that a welding machine can run without overheating.

Edge Joint   A joint between the edges of two or more parallel or nearly parallel members.

Edge Preparation   The contour prepared on the edge of a member for welding.

Electrode   A component of the arc welding circuit through which current is conducted between the electrode holder and the arc.

Electrode Holder   A device used for mechanically holding the electrode and conducting current to it.

Electrode Splitting   The condition that occurs when electrode particles are ejected across the arc.

Face Shield   A protective device to be worn on the head for shielding the face and neck.

Fiberglass Plastic   A very durable and corrosion-resistant material being used increasingly on car and truck bodies.

Filler Metal   The metal to be added in making a welded, brazed, or soldered joint.

Filler Rod   Metal wire that is melted into the puddle of the weld.

Fillet Weld A weld joining two surfaces approximately at right angles to each other in a lap joint, tee joint, or corner joint.

Filter Lens A filter, usually colored glass, used in goggles, helmets, and handshields to exclude harmful light rays.

Flexible Filler A filler that must be applied on top of the weld when repairing a high-stress area of a urethane part.

Flux Material used to prevent, dissolve, or remove oxides and other undesirable substances.

Fuel Gases Gases usually used in addition to oxygen for heating, including acetylene, natural gas, hydrogen, propane, and other synthetic fuels and hydrocarbons.

Fusion The melting together of filler metal and base metal, or of base metal only, which results in coalescence.

Gas Metal Arc Cutting An arc cutting process used to sever metals by melting them with the heat of an arc between a continuous metal (consumable) electrode and the work. Shielding is obtained entirely from an externally supplied gas or gas mixture.

Gas Metal Arc Welding An arc welding process that produces coalescence of metals by heating them with an arc between a continuous filler metal (consumable) electrode and the work. Shielding is obtained entirely from an externally supplied gas or gas mixture.

Globular Metal Transfer A form of metal transfer in MIG welding that occurs in large, irregularly shaped drops.

Goggles A device with colored lenses that protects the eyes from harmful radiation during welding and cutting operations.

Groove The opening provided between two members to be joined by a groove weld.

Ground Connection An electrical connection of the welding machine frame to the earth for safety.

Gun In semiautomatic, machine, and automatic welding, a manipulating device to transfer current and guide the electrode into the arc. It may include provisions for shielding and arc initiation.

Hardening Any process of increasing the hardness of metal by suitable treatment, usually involving heating and cooling.

Heat-Affected Zone The portion of the base metal that has not been melted, but whose mechanical properties or structures have been altered by the heat of welding, brazing, soldering, or cutting.

Helmet A protective device, used in arc welding, for shielding the face and neck. A helmet is equipped with a suitable filter lens and is designed to be worn on the head.

Hold Time The time that pressure is maintained at the electrodes after the welding current has stopped.

Hot-Air Plastic Welding A method of welding automotive plastic in which a ceramic or stainless steel electric heating element produces hot air. The air, in turn, blows through a nozzle and onto the plastic.

HSLA Steel High-strength, low-alloy steel used in the structural components of many domestic vehicles.

HSS Steel High-tensile strength steel whose strength is derived from heat treatment. This steel will tear or fracture if the collision stresses exceed the tensile strength.

Impact Test A test in which one or more blows are suddenly applied to a specimen. The results are usually expressed in terms of energy absorbed or number of blows of a given intensity required to break the specimen.

Inadequate Joint Penetration Joint penetration that is less than what is specified.

Inclusion Nonmetallic material(s) that become trapped in the weld metal, between weld beads, or between the weld and base metal.

Incomplete Fusion Fusion that is less than complete.

Induction Brazing A process in which bonding is produced by the heat obtained from the resistance of the work to the flow of induced electric current and by using a nonferrous filler metal having a melting point above 840° F but below that of the base metals. The filler metal is distributed in the joint by capillary attraction.

Induction Welding A process in which fusion is produced by heat obtained from the resistance of the work to the flow of induced electric current, with or without the application of pressure.

Inert Gas A gas that does not normally combine chemically with the base metal or filler metal.

Infrared Rays Dangerous rays produced by the light of arc welding that are injurious to the eyes and skin.

Intermittent Weld A weld whose continuity is broken by recurring unwelded spaces.

Joint The location where two or more members are to be joined.

Joint Penetration The minimum depth a groove or flange weld extends from its face into a joint, exclusive of reinforcement.

Lamellar Tear Terrace-like separations in the base metal of a weld, usually caused by shrinkage stresses.

Leg of a Fillet Weld   The distance from the root of the joint to the toe of the fillet weld.

Local Preheating   Preheating a specific portion of a structure.

Manual Welding   Welding in which the entire operation is performed and controlled by hand.

MIG (Metallic Inert Gas)   Gas metallic arc welding.

Mixing Chamber   That part of a gas welding or oxygen cutting torch in which the gases are mixed.

Molded Core   A core made from fiberglass mold veil, resin, and hardener that is used to repair a curved portion of a fiberglass panel.

Molten Weld Pool   The liquid state of a weld prior to solidification as weld metal.

Neutral Flame   An oxyfuel gas flame in which the portion used is neither oxidizing nor reducing.

Nondestructive Testing   A method of checking for weld surface defects; can be visual, penetrant, or ultrasonic.

Nonferrous   Metals that contain no iron. Aluminum, brass, bronze, copper, lead, nickel, and titanium are nonferrous.

Nozzle Spray   Used to clean the nozzle of a MIG welding machine; preferred over paste because the spray does not leave a heavy residue inside the nozzle.

One-Sided Weld   In plastic welding, a repair method in which only one side of the damaged part is welded. This technique is used most often for solely cosmetic repairs.

Overhead Position   The position in which the welding is performed from the underside of the joint.

Overlap   Protrusion of weld metal beyond the toe or root of the weld.

Oxidizing Flame   An oxyfuel gas flame having an oxidizing effect (excess oxygen).

Oxyacetylene Cutting   An oxygen cutting process in which the necessary cutting temperature is maintained by flames obtained from the combustion of acetylene with oxygen.

Oxyacetylene Welding   An oxyfuel gas welding process that produces coalescence of metals by heating them with a gas flame or flames obtained from the combustion of acetylene with oxygen. The process may be used with or without the application of pressure and with or without the use of filler metal.

Panel Spotting   The means by which both lap and flange joints can be made with a spliced or full panel installation.

Pass   A single longitudinal progression of a welding operation along a joint or weld deposit. The result of a pass is a weld bead.

Penetration   The depth of fusion into the metal being welded.

Phosgene   A potentially dangerous compound formed when fumes from chlorinated solvents decompose in the welding arc.

Pitch   Center-to-center spacing of welds.

Plasma   A gas that has been heated to an at least partially ionized condition, enabling it to conduct an electric current.

Plasma Arc Cutting   An arc cutting process in which the metal is severed by melting a localized area with a constricted arc and removing the molten material with a high-velocity jet of hot, ionized gas issuing from the orifice.

Plug Weld   A weld made through a hole in one member of a lap or tee joint joining that member to the other.

Polyethylene (PE)   A common thermoplastic whose maximum weld strength is achieved ten hours after the weld is completed.

Polyolefin   A type of automotive plastic used most often for large exterior parts.

Polypropylene (PP)   A thermoplastic that is very similar to polyethylene.

Polyurethane (PUR)   An extremely lightweight and formable plastic available as both a thermoplastic and a thermosetting plastic.

Polyvinyl Chloride (PVC)   A common thermoplastic.

Porosity   Gas pockets or voids in metal.

Positions of Welding   All welding is accomplished in one of four positions: flat, horizontal, overhead, or vertical. The limiting angles of the various positions depend somewhat on whether the weld is a fillet or groove.

Preheat Temperature   The specified temperature that the base metal must attain in the welding, brazing, soldering, or cutting area immediately before these operations are performed.

Pressure Welding   A welding process in which the pieces of metal are heated to a softened state by electrodes. After the pieces are soft, pressure is used to complete the weld.

Preweld Interval   In spot welding, the time between the end of squeeze time and the start of weld time during which the material is preheated.

Puddle   The molten part of the weld where the arc is supplied.

Regulator   A device for controlling the delivery of gas at some substantially constant pressure regardless of variation in the higher pressure at the source.

Residual Stress   Stress remaining in a structure or member as a result of thermal or mechanical treatment or both.

Resistance Welding   A group of welding processes that produces coalescence of metals with the heat obtained from the resistance of the work to electric current in a circuit of which the work is a part, and by the application of pressure.

Reverse Polarity The arrangement of direct current arc welding leads with the work as the negative pole and the electrode as the positive pole of the welding arc.

Root Opening The separation between the members to be joined at the root of the joint.

Root Penetration The depth a weld extends into the root of a joint measured on the centerline of the root cross section.

Seam Weld A continuous weld made between or upon overlapping members in which coalescence may start and occur on the mating surfaces or may have proceeded from the surface of one member. The continuous weld may consist of a single weld bead or a series of overlapping spot welds.

Sectioning A means of replacing parts at factory seams.

Sequence The order in which the beads (passes) are welded on the joint.

Shielded Welding An arc welding process in which protection from the atmosphere is obtained through the use of a flux, decomposition of the electrode covering, or an inert gas.

Short-Circuiting Metal Transfer A form of metal transfer in MIG welding that does not occur across the arc. It occurs instead when the electrode wire contacts the base metals.

Solder A filler metal that liquefies below 840° F.

Spatter In arc and gas welding, the metal particles expelled during welding that do not form a part of the weld.

Speed Welding In plastic welding, a popular method of doing panel work in which a fairly constant rate of speed must be maintained.

Spool Gun A self-contained system consisting of a small drive and a wire supply that allows a welder to move freely around a job.

Spot Weld A weld made between or upon overlapping members in which coalescence may start and occur on the mating surfaces or may proceed from the surface of one member. The weld cross section is approximately circular.

Spray Arc Gas metal arc welding that uses an arc voltage high enough to transfer the electrode metal across the arc in small globules.

Spray Transfer A mode of metal transfer in gas metal arc welding in which the consumable electrode is propelled across the arc in small droplets.

Stitch Welding The use of intermittent welds to join two or more parts.

Straight Polarity The arrangement of direct current arc welding leads in which the work is the positive pole and the electrode is the negative pole of the welding arc.

Stress Cracking Cracking of a weld or base metal containing residual stresses.

Stud Welding A general term for the joining of a metal stud or similar part to a workpiece. Welding can be accomplished by arc, resistance, friction, or other suitable process with or without external gas shielding.

Surface Preparation The operations necessary to produce a desired or specified surface condition.

Tack Weld A weld made to hold parts in proper alignment until the final welds are made.

Thermoplastics Plastics that are capable of being repeatedly softened and hardened by heating or cooling, with no change in their appearance or chemical makeup. They are weldable with a plastic welder.

Thermosetting Plastics Plastics that undergo a chemical change by the action of heating, a catalyst, or ultraviolet light leading to an infusible state. Thermosets are not weldable, although they can be "glued" using an airless welder.

TIG (Tungsten Inert Gas) Gas tungsten arc welding.

Tip-to-Base Metal Distance An important factor in obtaining good welding results, usually 1/4 to 5/8 inch.

Toe to Weld The junction between the face of a weld and the base metal.

Torch Brazing A brazing process in which bonding is produced by heating with a gas flame and by using a nonferrous filler metal having a melting point above 800° F, but below that of the base metal. The filler metal is distributed in the joint by capillary attraction.

Two-Sided Weld In plastic welding, a repair method in which both sides of the damaged part are welded. This is the preferred method whenever the part can be removed from the vehicle.

Ultrasonic Plastic Welding A method of repairing auto plastics in which the welding time is controlled by the power supply. This method is best suited to rigid plastics.

Ultrasonic Stud Welding A variation of the shear joint used to join plastic parts in which the weld is made along the circumference of the stud.

Ultraviolet Rays Harmful energy waves given off by the arc that are dangerous to the eyes and skin.

Undercut A groove melted into the base metal adjacent to the toe or root of a weld and left unfilled by weld metal.

Undercutting An undesirable crater at the edge of the weld caused by poor weaving techniques or excessive welding speed.

Vinyl-Clad Urethane Foam The material used in most padded instrument panels. This foam is repairable using an airless welder.

Voltage Regulator   An automatic electrical control device for maintaining a constant voltage supply to a welding transformer.

Weld   A localized coalescence of metal in which coalescence is produced either by heating to suitable temperatures, with or without the application of pressure, or by the application of pressure alone, and with or without the use of filler metal. The filler metal either has a melting point approximately the same as the base metals or has a melting point below that of the base metals but above 800° F.

Weldability   The capacity of a metal to be welded under the fabrication conditions imposed into a specific, suitable designed structure and to perform satisfactorily in the intended service.

Weld Bead   A weld deposit resulting from a pass.

Welding Current   The current in the welding circuit during the making of a weld.

Welding Machine   Equipment used to perform the welding operation. For example, spot welding machine, arc welding machine, seam welding machine, and so on.

Welding Rod   A form of filler metal used for welding or brazing that does not conduct the electrical current.

Welding Sequence   The order of making the welds in a weldment.

Welding Tip   A welding torch tip designed for welding.

Welding Torch   A device used in oxyfuel gas welding or torch brazing for mixing and controlling the flow of gases.

Weldment   An assembly whose component parts are joined by welding.

Weld Metal   The portion of a weld that has been melted during welding.

Wetting   The bonding or spreading of a liquid filler metal or flux on a solid base metal.

Wire Feed Speed   The rate of speed at which a filler metal is consumed in arc welding.

Work Angle   The angle that the electrode makes with the surface of the base metal in a plane perpendicular to the axis of the weld.

Work Lead   The electric conductor between the source of arc welding current and the work.

REFERENCE TABLES

COMMON TYPES OF WELD DISCONTINUITIES

Discontinuity	Location	Remarks
Porosity (can be uniformly scattered, cluster, or linear type)	Weld	Weld only as described in text.
Nonmetallic slag	Weld	
Metallic tungsten	Weld	
Incomplete fusion (also called lack of fusion)	Weld	At joint boundaries or between passes
Inadequate joint penetration (also called lack of joint preparation)	Weld	Root of weld penetration
Undercut	Base metal	Junction of weld and base metal at surface
Underfill	Weld	Outer surface of joint preparation
Overlap	Weld	Junction of weld and base metal at surface
Laminations	Base metal	Base metal, generally near mid-thickness of section
Delamination	Base metal	Base metal, generally near mid-thickness of section.
Seams and laps	Base metal	Base metal surface almost always longitudinal
Lamellar tears	Base metal	Base metal, near weld, heat-affected zone
Cracks (includes hot cracks and cold cracks described in text)		
Longitudinal	Weld, heat-affected zone	Weld or base metal adjacent to weld fusion boundary
Transverse	Weld, base metal, heat-affected zone	Weld (can spread into heat-affected zone and base metal)
Crater	Weld	Weld, at point where arc is terminated
Throat	Weld	Weld axis
Toe	Heat-affected zone	Junction between face of weld and base metal
Root	Weld	Weld metal, at root
Underbead and heat-affected zone	Heat-affected zone	Base metal in heat-affected zone
Fissures	Weld	Weld metal

ADVANTAGES OF SHIELDING GASES

Metal	Welding Type	Shielding Gas	Advantages
Aluminum and Magnesium	Manual welding	Argon	Better arc starting, cleaning action, and weld quality; lower gas consumption
		Argon-helium	High welding speeds possible
	Machine welding	Argon-helium	Better weld quality; lower gas flow than required with straight helium
		Helium (DCSP)	Deeper penetration and higher weld speeds than can be obtained with argon-helium
	Spot welding	Argon	Generally preferred for longer electrode life; better weld nugget contour; ease of starting; lower gas flows than helium
Carbon steel	Manual welding	Argon	Better pool control, especially for position welding
	Machine welding	Helium	Higher speeds obtained than with argon
Stainless steel	Manual welding	Argon	Permits controlled penetration on thin-guage material (up to 14 gauge)
	Machine welding	Argon	Excellent control of penetration on light-gauge materials
		Argon-helium	Higher heat input; higher welding speeds possible on heavier gauges
		Argon-hydrogen (up to 35% H_2)	Prevents undercutting; produces desirable weld contour at low current levels; requires lower gas flow
		Argon-hydrogen-helium	Excellent for high-speed tube mill operation
		Helium	Provides highest heat input and deepest penetration
Copper, nickel, and Cu-Ni alloys		Argon	Ease of obtaining pool control, penetration, and bead contour on thin-gauge metal
		Argon-helium	Higher heat input to offset high heat conductivity of heavier gauges.
		Helium	Highest heat input for welding speed on heavy metal sections
Silicon-bronze		Argon	Reduces cracking of this "hot short" metal
Aluminum-bronze		Argon	Less penetration of base metal

TROUBLESHOOTING GUIDES

TROUBLESHOOTING GUIDE FOR A TYPICAL ARC WELDING MACHINE

Problem	Cause	Remedy
Welder will not start.	Line switch not turned on	Place line switch in On position
	Supply line fuse blown	Replace (check reason for blown fuse first).
	Power circuit dead	Check input voltage.
	Overload relay tripped	Let cool: remove cause of overloading.
	Loose or broken power, electrode, or ground lead	Replace lead or tighten and repair connection.
	Wrong voltage	Check input voltage against instructions.
	Polarity switch not centered (AC-DC units only)	Center switch handle on +, −, or AC.
	Open circuit to starter button	Repair.
Welder starts but blows fuse after welding begins.	Short circuit in motor or other connection	Check connections and lead insulation.
	Fuse too small	Check instruction manual for correct size.
Welder welds but soon stops.	Proper ventilation hindered	Make sure all case openings are free for proper circulation of air.
	Overloading—welding in excess of rating	Operate welder at rated load and duty cycle.
	Fan inoperative	Check leads and connections.
	Motor generator sets—	
	Wrong direction of rotation	Check connection diagram.
	Brushes worn or missing	Check brushes for pressure on commutator
	Wrong driving speed	Check nameplate for correct motor speed.
	Excessive dust accumulation in welder	Clean thermostat, coils, and other components.
Variable or sluggish welding arc	Current too low	Check recommended currents for rod type and size being used.
	Low line voltage	Check with power company.
	Welding leads too small	Check instruction manual for recommended cable sizes.
	Poor ground, electrode, or control-circuit connections	Check all connections. Clean, repair, or replace as required.
	Motor generator sets—Brushes improper; weak springs; not properly fitted	Check and repair.
	Rough or dirty commutator	Turn down or clean commutator.
Welding arc is loud and spatters excessively.	Current setting is too high	Check setting and output with ammeter
	Polarity wrong	Check polarity: try reversing or an electrode of opposite polarity
Polarity switch will not turn.	Contacts rough and pitted from improper turning under load.	Replace switch.
Welder will not shut off.	Line switch has failed mechanically.	Replace switch.
Arcing at ground clamp.	Loose connection or weak spring.	Tighten connection or replace clamp.
Electrode holder becomes hot.	Loose connection, loose jaw, inadequate duty cycle.	Tighten connection or replace holder.
Touching welder gives shock.	Frame not grounded.	See instruction manual for proper grounding procedure.

329

GUIDE TO EVALUATING POROSITY AND CRACKING IN TIG WELDS

Cause	Contributing Factors	Corrective Measures
Hydrogen	Dirt containing oils or other hydrocarbons; moisture in atmosphere or on metal, or a hydrated oxide film on metal; spatter; moisture in gas or gas lines. Base metal may be source of entrapped hydrogen (the thicker the metal, the greater the possibility for hydrogen).	Degrease and mechanically or chemically remove oxide from weld area. Avoid humidity; use dry metal or wipe dry. Reduce moisture content of gas. Check gas and water lines for leaks. Increase gas flow to compensate for increased hydrogen in thicker sections. To minimize spatter, adjust welding conditions.
Impurities	Cleaning or other compounds, especially those containing calcium.	Use recommended cleaning compounds; keep work free of contaminants.
Incomplete root penetration	Incomplete penetration in heavy sections increases porosity in the weld.	Preheat; use higher welding current, or redesign joint geometry.
Temperature	Running too cool tends to increase porosity due to premature solidification of molten metal.	Maintain proper current, arc length, and torch travel speed relationship
Welding speed	Too great a welding speed may increase porosity.	Decrease welding speed and establish and maintain proper arc length and current relationship.
Solidification time	Quick cooling of weld pool entraps any gases present, causing porosity.	Establish correct welding current and speed. If work is appreciably below room temperature, use supplemental heating.
Chemical composition of weld metal	Pure aluminum weld metal is more susceptible to porosity than an aluminum alloy.	If porosity is excessive, use an alloy filler material.
Cracking	Temperature, welding time, and solidification may be contributing causes of cracking. Other causes include discontinuous welds, welds that intersect, repair welds, cold-working either before or after welding, and weld-metal composition. In general, crack-sensitive alloys include those containing 0.4 to 0.6% Si, 1.5 to 3.0% Mg, or 1.0% Cu.	Lower current and faster speeds often prevent cracking. A change to a filter alloy that brings weld metal composition out of cracking range is recommended where possible.

GUIDE TO GENERAL WELD QUALITY CONTROL TECHNIQUES

Technique	Equipment	Defects Detected	Advantages	Disadvantages	Other Considerations
Visual inspection	Pocket magnifier, welding viewer, flashlight, weld gauge, scale, etc.	Weld preparation, fitup cleaniness, roughness, spatter, undercuts, overlaps, weld contour, and size; welding procedures	Easy to use; fast, inexpensive, usable at all stages of production	For surface conditions only; dependent on subjective opinion of inspector	Most universally used inspection method
Dry penetrant or fluorescent inspection	Fluorescent or visible penetrating liquids and developers; ultraviolet light for the fluorescent type.	Defects open to the surface only; good for leak detection.	Detects very small, tight surface imperfections; easy to apply and to interpret; inexpensive; use on magnetic or nonmagnetic materials.	Time consuming in the various steps of the processes; normally no permanent record	Often used on root pass of highly critical pipe welds. If material is improperly cleaned, some indications can be misleading.
Magnetic particle inspection	Iron particles, wet, dry, or fluorescent; special power source; ultraviolet light for the fluorescent type.	Surface and near-surface discontinuities, cracks, etc.; porosity, slag, etc.	Indicates discontinuities not visible to the naked eye; useful in checking edges prior to welding and also for repairs; no size restriction.	Used on magnetic materials only; surface roughness may distort magnetic field; normally no permanent record	Testing should be from two perpendicular directions to catch discontinuities that might be parallel to one set of magnetic lines of force.
Radiographic inspection	X-ray or gamma ray; source; film processing equipment; film viewing equipment; penetrameters.	Most internal discontinuities and flaws; limited by direction of discontinuity.	Provides permanent record; indicates both surface and internal flaws; applicable on all materials.	Usually not suitable for fillet weld inspection; film exposure and processing are critical; slow and expensive.	Most popular technique for subsurface inspection; required by some codes and specifications.

DECIMAL AND METRIC EQUIVALENTS

DECIMAL AND METRIC EQUIVALENTS

Fractions	Decimal (in.)	Metric (mm)	Fractions	Decimal (in.)	Metric (mm)
1/64	.015625	.397	33/64	.515625	13.097
1/32	.03125	.794	17/32	.53125	13.494
3/64	.046875	1.191	35/64	.546875	13.891
1/16	.0625	1.588	9/16	.5625	14.288
5/64	.078125	1.984	36/64	.578125	14.684
3/32	.09375	2.381	19/32	.59375	15.081
7/64	.109375	2.778	39/64	.609375	15.478
1/8	.125	3.175	5/8	.625	15.875
9/64	.140625	3.572	41/64	.640625	16.272
5/32	.15625	3.969	21/32	.65625	16.669
11/64	.171875	4.366	43/64	.671875	17.066
3/16	.1875	4.763	11/16	.6875	17.463
13/64	.203125	5.159	45/64	.703125	17.859
7/32	.21875	5.556	23/32	.71875	18.256
15/64	.234275	5.953	47/64	.734375	18.653
1/4	.250	6.35	3/4	.750	19.05
17/64	.265625	6.747	49/64	.765625	19.447
9/32	.28125	7.144	25/32	.78125	19.844
19/64	.296875	7.54	51/64	.796875	20.241
5/16	.3125	7.938	13/16	.8125	20.638
21/64	.328125	8.334	53/64	.828125	21.034
11/32	.34375	8.731	27/32	.84375	21.431
23/64	.359375	9.128	55/64	.859375	21.828
3/8	.375	9.525	7/8	.875	22.225
25/64	.390625	9.922	57/64	.890625	22.622
13/32	.40625	10.319	29/32	.90625	23.019
27/64	.421875	10.716	59/64	.921875	23.416
7/16	.4375	11.113	15/16	.9375	23.813
29/64	.453125	11.509	61/64	.953125	24.209
15/32	.46875	11.906	31/32	.96875	24.606
31/64	.484375	12.303	63/64	.984375	25.003
1/2	.500	12.7	1	1.00	25.4

Index